Progress in Nonlinear Differential Equations and Their Applications: Subseries in Control
Volume 88

More information about this series at http://www.springer.com/series/15137

Georges Bastin • Jean-Michel Coron

Stability and Boundary Stabilization of 1-D Hyperbolic Systems

 Birkhäuser

Georges Bastin
Mathematical Engineering, ICTEAM
Université catholique de Louvain
Louvain-la-Neuve, Belgium

Jean-Michel Coron
Laboratoire Jacques-Louis Lions
Université Pierre et Marie Curie
Paris Cedex, France

ISSN 1421-1750   ISSN 2374-0280   (electronic)
Progress in Nonlinear Differential Equations and Their Applications
ISBN 978-3-319-32060-1   ISBN 978-3-319-32062-5   (eBook)
DOI 10.1007/978-3-319-32062-5

Library of Congress Control Number: 2016946174

Mathematics Subject Classification (2010): 35L, 35L-50, 35L-60, 35L-65, 93C, 93C-20, 93D, 93D-05, 93D-15, 93D-20

Printed on acid-free paper

This book is published under the trade name Birkhäuser
The registered company is Springer International Publishing AG, CH

Preface

THE TRANSPORT of electrical energy, the flow of fluids in open channels or in gas pipelines, the light propagation in optical fibres, the motion of chemicals in plug flow reactors, the blood flow in the vessels of mammalians, the road traffic, the propagation of age-dependent epidemics and the chromatography are typical examples of processes that may be represented by hyperbolic partial differential equations (PDEs). In all these applications, described in Chapter 1, the dynamics are usefully represented by *one-dimensional* hyperbolic balance laws although the natural dynamics are three dimensional, because the dominant phenomena evolve in one privileged coordinate dimension, while the phenomena in the other directions are negligible.

From an engineering perspective, for hyperbolic systems as well as for all dynamical systems, the stability of the steady states is a fundamental issue. This book is therefore entirely devoted to the (exponential) stability of the steady states of one-dimensional systems of conservation and balance laws considered over a finite space interval, i.e., where the spatial 'domain' of the PDE is an interval of the real line.

The definition of exponential stability is intuitively simple: starting from an arbitrary initial condition, the system time trajectory has to exponentially converge in spatial norm to the steady state (globally for linear systems and locally for nonlinear systems). Behind the apparent simplicity of this definition, the stability analysis is however quite challenging. First it is because this definition is not so easily translated into practical *tests* of stability. Secondly, it is because the various function norms that can be used to measure the deviation with respect to the steady state are not necessarily equivalent and may therefore give rise to different stability tests.

As a matter of fact, the exponential stability of steady states closely depends on the so-called *dissipativity* of the boundary conditions which, in many instances, is a natural physical property of the system. In this book, one of the main tasks is therefore to derive simple practical tests for checking if the boundary conditions are dissipative.

Linear systems of *conservation* laws are the simplest case. They are considered in Chapters 2 and 3. For those systems, as for systems of linear ordinary differential equations, a (necessary and sufficient) test is to verify that the poles (i.e., the roots of the characteristic equation) have negative real parts. Unfortunately, this test is not very practical and, in addition, not very useful because it is not robust with respect to small variations of the system dynamics. In Chapter 3, we show how a robust (necessary and sufficient) dissipativity test can be derived by using a Lyapunov stability approach, which guarantees the existence of globally exponentially converging solutions for any L^p-norm.

The situation is much more intricate for *nonlinear* systems of *conservation* laws which are considered in Chapter 4. Indeed for those systems, it is well known that the trajectories of the system may become discontinuous in finite time even for smooth initial conditions that are close to the steady state. Fortunately, if the boundary conditions are dissipative and if the smooth initial conditions are sufficiently close to the steady state, it is shown in this chapter that the system trajectories are guaranteed to remain smooth for all time and that they exponentially converge locally to the steady state. Surprisingly enough, due to the nonlinearity of the system, even for smooth solutions, it appears that the exponential stability strongly depends on the considered norm. In particular, using again a Lyapunov approach, it is shown that the dissipativity test of linear systems holds also in the nonlinear case for the H^2-norm, while it is necessary to use a more conservative test for the exponential stability for the C^1-norm.

In Chapters 5 and 6, we move to hyperbolic systems of linear and nonlinear *balance* laws. The presence of the source terms in the equations brings a big additional difficulty for the stability analysis. In fact the tests for dissipative boundary conditions of conservation laws are directly extendable to balance laws only if the source terms themselves have appropriate dissipativity properties. Otherwise, as it is shown in Chapter 5, it is only known (through the special case of systems of two balance laws) that there are intrinsic limitations to the system stabilizability with local controls.

There are also many engineering applications where the dissipativity of the boundary conditions, and consequently the stability, is obtained by using boundary feedback control with actuators and sensors located at the boundaries. The control may be implemented with the goal of stabilization when the system is physically unstable or simply because boundary feedback control is required to achieve an efficient regulation with disturbance attenuation. Obviously, the challenge in that case is to design the boundary control devices in order to have a good control performance with dissipative boundary conditions. This issue is illustrated in Chapters 2 and 5 by investigating in detail the boundary proportional-integral output feedback control of so-called density-flow systems. Moreover Chapter 7 addresses the boundary stabilization of hyperbolic systems of balance laws by *full-state feedback* and by dynamic output feedback in *observer-controller form*, using the backstepping method. Numerous other practical examples of boundary feedback control are also presented in the other chapters.

Finally, in the last chapter (Chapter 8), we present a detailed case study devoted to the control of navigable rivers when the river flow is described by hyperbolic Saint-Venant shallow water equations. The goal is to emphasize the main technological features that may occur in real-life applications of boundary feedback control of hyperbolic systems of balance laws. The issue is presented through the specific application of the control of the Meuse River in Wallonia (south of Belgium).

In our opinion, the book could have a dual audience. In one hand, mathematicians interested in applications of control of 1-D hyperbolic PDEs may find the book a valuable resource to learn about applications and state-of-the-art control design. On the other hand, engineers (including graduate and postgraduate students) who want to learn the theory behind 1-D hyperbolic equations may also find the book an interesting resource. The book requires a certain level of mathematics background which may be slightly intimidating. There is however no need to read the book in a linear fashion from the front cover to the back. For example, people concerned primarily with applications may skip the very first Section 1.1 on first reading and start directly with their favorite examples in Chapter 1, referring to the definitions of Section 1.1 only when necessary. Chapter 2 is basic to an understanding of a large part of the remainder of the book, but many practical or theoretical sections in the subsequent chapters can be omitted on first reading without problem. The book presents many examples that serve to clarify the theory and to emphasize the practical applicability of the results. Many examples are continuation of earlier examples so that a specific problem may be developed over several chapters of the book. Although many references are quoted in the book, our bibliography is certainly not complete. The fact that a particular publication is mentioned simply means that it has been used by us as a source material or that related material can be found in it.

Louvain-la-Neuve, Belgium   Georges Bastin
Paris, France   Jean-Michel Coron
                                                                                        February 2016

Acknowledgements

The material of this book has been developed over the last fifteen years. We want to thank all those who, in one way or another, contributed to this work. We are especially grateful to Fatiha Alabau, Fabio Ancona, Brigitte d'Andrea-Novel, Alexandre Bayen, Gildas Besançon, Michel Dehaen, Michel De Wan, Ababacar Diagne, Philippe Dierickx, Malik Drici, Sylvain Ervedoza, Didier Georges, Olivier Glass, Martin Gugat, Jonathan de Halleux, Laurie Haustenne, Bertrand Haut, Michael Herty, Thierry Horsin, Long Hu, Miroslav Krstic, Pierre-Olivier Lamare, Günter Leugering, Xavier Litrico, Luc Moens, Hoai-Minh Nguyen, Guillaume Olive, Vincent Perrollaz, Benedetto Piccoli, Christophe Prieur, Valérie Dos Santos Martins, Catherine Simon, Paul Suvarov, Simona Oana Tamasoiu, Ying Tang, Alain Vande Wouwer, Paul Van Dooren, Rafael Vazquez, Zhiqiang Wang and Joseph Winkin.

During the preparation of this book, we have benefited from the support of the ERC advanced grant 266907 (CPDENL, European 7th Research Framework Programme (FP7)) and of the Belgian Programme on Inter-university Attraction Poles (IAP VII/19) which are also gratefully acknowledged. The implementation of the Meuse regulation reported in Chapter 8 is carried out by the Walloon region, Siemens and the University of Louvain.

Contents

Preface v

1 Hyperbolic Systems of Balance Laws 1
  1.1   Definitions and Notations 1
         1.1.1   Riemann Coordinates and Characteristic Form 3
         1.1.2   Steady State and Linearization 4
         1.1.3   Riemann Coordinates Around the Steady State 4
         1.1.4   Conservation Laws and Riemann Invariants 5
         1.1.5   Stability, Boundary Stabilization, and the
                Associated Cauchy Problem 6
  1.2   Telegrapher Equations 10
  1.3   Raman Amplifiers 12
  1.4   Saint-Venant Equations for Open Channels 13
         1.4.1   Boundary Conditions 15
         1.4.2   Steady State and Linearization 16
         1.4.3   The General Model 17
  1.5   Saint-Venant-Exner Equations 18
  1.6   Rigid Pipes and Heat Exchangers 19
         1.6.1   The Shower Control Problem 21
         1.6.2   The Water Hammer Problem 22
         1.6.3   Heat Exchangers 23
  1.7   Plug Flow Chemical Reactors 24
  1.8   Euler Equations for Gas Pipes 26
         1.8.1   Isentropic Equations 27
         1.8.2   Steady State and Linearization 28
         1.8.3   Musical Wind Instruments 29
  1.9   Fluid Flow in Elastic Tubes 30
  1.10   Aw-Rascle Equations for Road Traffic 31
         1.10.1   Ramp Metering 33
  1.11   Kac-Goldstein Equations for Chemotaxis 33

1.12 Age-Dependent SIR Epidemiologic Equations 35
     1.12.1 Steady State 36
1.13 Chromatography 38
     1.13.1 SMB Chromatography 39
1.14 Scalar Conservation Laws 43
1.15 Physical Networks of Hyperbolic Systems 45
     1.15.1 Networks of Electrical Lines 46
     1.15.2 Chains of Density-Velocity Systems 47
     1.15.3 Genetic Regulatory Networks 50
1.16 References and Further Reading 52

2 Systems of Two Linear Conservation Laws 55
  2.1 Stability Conditions 55
      2.1.1 Exponential Stability for the L^∞-Norm.................... 57
      2.1.2 Exponential Lyapunov Stability for the L^2-Norm 59
      2.1.3 A Note on the Proofs of Stability in L^2-Norm 64
      2.1.4 Frequency Domain Stability 64
      2.1.5 Example: Stability of a Lossless Electrical Line 65
  2.2 Boundary Control of Density-Flow Systems 67
      2.2.1 Feedback Stabilization with Two Local Controls 68
      2.2.2 Dead-Beat Control 69
      2.2.3 Feedback-Feedforward Stabilization with a
            Single Control 69
      2.2.4 Proportional-Integral Control............................. 70
  2.3 The Nonuniform Case 81
  2.4 Conclusions 83

3 Systems of Linear Conservation Laws................................. 85
  3.1 Exponential Stability for the L^2-Norm............................ 86
      3.1.1 Dissipative Boundary Conditions 88
  3.2 Exponential Stability for the C^0-Norm: Analysis
      in the Frequency Domain... 89
      3.2.1 A Simple Illustrative Example 92
      3.2.2 Robust Stability 94
      3.2.3 Comparison of the Two Stability Conditions 95
  3.3 The Rate of Convergence... 96
      3.3.1 Application to a System of Two Conservation Laws 97
  3.4 Differential Linear Boundary Conditions 97
      3.4.1 Frequency Domain.. 98
      3.4.2 Lyapunov Approach 98
      3.4.3 Example: A Lossless Electrical Line
            Connecting an Inductive Power Supply to a
            Capacitive Load 99
      3.4.4 Example: A Network of Density-Flow Systems
            Under PI Control.. 102
      3.4.5 Example: Stability of Genetic Regulatory Networks....... 106

3.5   The Nonuniform Case 109
3.6   Switching Linear Conservation Laws............................... 110
        3.6.1   The Example of SMB Chromatography 111
3.7   References and Further Reading 115

4   Systems of Nonlinear Conservation Laws 117
4.1   Dissipative Boundary Conditions for the C^1-Norm 119
4.2   Control of Networks of Scalar Conservation Laws 130
        4.2.1   Example: Ramp-Metering Control in Road
                  Traffic Networks 132
4.3   Interlude: Solutions Without Shocks 135
4.4   Dissipative Boundary Conditions for the H^2-Norm................. 136
        4.4.1   Proof of Theorem 4.11...................................... 138
4.5   Stability of General Systems of Nonlinear Conservation
        Laws in Quasi-Linear Form 143
        4.5.1   Stability Condition for the C^1-Norm 145
        4.5.2   Stability Condition for the C^p-Norm
                  for Any $p \in \mathbb{N} \smallsetminus \{0\}$ 153
        4.5.3   Stability Condition for the H^p-Norm
                  for Any $p \in \mathbb{N} \smallsetminus \{0, 1\}$.......................... 156
4.6   References and Further Reading 156

5   Systems of Linear Balance Laws 159
5.1   Lyapunov Exponential Stability..................................... 160
        5.1.1   Example: Feedback Control of an Exothermic
                  Plug Flow Reactor 163
5.2   Linear Systems with Uniform Coefficients.......................... 166
        5.2.1   Application to a Linearized Saint-Venant-Exner Model ... 167
5.3   Existence of a Basic Quadratic Control Lyapunov
        Function for a System of Two Linear Balance Laws 176
        5.3.1   Application to the Control of an Open Channel 181
5.4   Boundary Control of Density-Flow Systems 184
        5.4.1   Transfer Functions 185
        5.4.2   Boundary Feedback Stabilization with Two
                  Local Controls... 187
        5.4.3   Feedback-Feedforward Stabilization with
                  a Single Control 188
        5.4.4   Stabilization with Proportional-Integral Control 190
5.5   Proportional-Integral Control in Navigable Rivers.................. 193
        5.5.1   Dissipative Boundary Condition 195
        5.5.2   Control Error Propagation 195
5.6   Limit of Stabilizability 197
5.7   References and Further Reading 201

6   Quasi-Linear Hyperbolic Systems 203
6.1   Stability of Systems with Uniform Steady States 203

	6.2	Stability of General Quasi-Linear Hyperbolic Systems..............	205
		6.2.1	Stability Condition for the H^2-Norm for	
			Systems with Positive Characteristic Velocities	206
		6.2.2	Stability Condition for the H^p-Norm	
			for Any $p \in \mathbb{N} \setminus \{0, 1\}$.....................................	217
	6.3	References and Further Reading	218

7	**Backstepping Control**	219
	7.1	Motivation and Problem Statement	219
	7.2	Full-State Feedback..	220
	7.3	Observer Design and Output Feedback..............................	223
	7.4	Backstepping Control of Systems of Two Balance Laws...........	226
	7.5	References and Further Reading	227

8	**Case Study: Control of Navigable Rivers**	229
	8.1	Geographic and Technical Data......................................	229
	8.2	Modeling and Simulation	230
	8.3	Control Implementation	233
		8.3.1	Local or Nonlocal Control?.................................	234
		8.3.2	Steady State and Set-Point Selection.......................	235
		8.3.3	Choice of the Time Step for Digital Control...............	236
	8.4	Control Tuning and Performance	238
	8.5	References and Further Reading	240

| **A** | **Well-Posedness of the Cauchy Problem** | |
| | **for Linear Hyperbolic Systems** .... | 243 |

| **B** | **Well-Posedness of the Cauchy Problem for Quasi-Linear** | |
| | **Hyperbolic Systems** .... | 255 |

C	**Properties and Comparisons of the Functions $\overline{\rho}$, ρ_2 and ρ_∞**	261
	C.1	Properties of the Function ρ_2..	261
	C.2	Proof of Theorem 3.12	267
	C.3	Proof of Proposition 4.7	279

| **D** | **Proof of Lemma 4.12 (b) and (c)** .... | 281 |

| **E** | **Proof of Theorem 5.11**.. | 285 |

| **F** | **Notations** .... | 293 |

| **References**... | 295 |

| **Index**... | 305 |

Chapter 1
Hyperbolic Systems of Balance Laws

IN THIS CHAPTER we provide an introduction to the modeling of balance laws by hyperbolic partial differential equations (PDEs). A balance law is the mathematical expression of the physical principle that the variation of the amount of some extensive quantity over a bounded domain is balanced by its flux through the boundaries of the domain and its production/consumption inside the domain. Balance laws are therefore used to represent the fundamental dynamics of many physical open conservative systems.

In the first section, we give the basic definitions and properties that will be used throughout the book. We successively address the characteristic form, the Riemann coordinates, the steady states, the linearization, and the boundary stabilization problem. The remaining of the chapter is then devoted to a presentation of typical examples of hyperbolic systems of balance laws for a wide range of physical engineering applications, with a view to allow the readers to understand the concepts in their most familiar setting. With these examples we also illustrate how the control boundary conditions may be defined for the most commonly used control devices.

1.1 Definitions and Notations

In this section we give the basic definition of one-dimensional systems of **balance laws** as they are used throughout the book. Let $\mathcal{Y}$ be a nonempty connected open subset of $\mathbb{R}^n$. A one-dimensional hyperbolic system of n nonlinear balance laws over a finite space interval is a system of PDEs of the form[1]

$$\partial_t e(\mathbf{Y}(t,x)) + \partial_x f(\mathbf{Y}(t,x)) + g(\mathbf{Y}(t,x)) = \mathbf{0}, \quad t \in [0, +\infty), \quad x \in [0, L], \qquad (1.1)$$

---

[1] The partial derivatives of a function f with respect to the variables x and t are indifferently denoted $\partial_x f$ and $\partial_t f$ or f_x and f_t.

© Springer International Publishing Switzerland 2016
G. Bastin, J.-M. Coron, *Stability and Boundary Stabilization of 1-D Hyperbolic Systems*, Progress in Nonlinear Differential Equations and Their Applications 88, DOI 10.1007/978-3-319-32062-5_1

where

- t and x are the two independent variables: a time variable $t \in [0, +\infty)$ and a space variable $x \in [0, L]$ over a finite interval;
- $\mathbf{Y} : [0, +\infty) \times [0, L] \to \mathcal{Y}$ is the vector of state variables;
- $e \in C^2(\mathcal{Y}; \mathbb{R}^n)$ is the vector of the densities of the balanced quantities; the map e is a diffeomorphism on $\mathcal{Y}$;
- $f \in C^2(\mathcal{Y}; \mathbb{R}^n)$ is the vector of the corresponding flux densities;
- $g \in C^1(\mathcal{Y}; \mathbb{R}^n)$ is the vector of *source terms* representing production or consumption of the balanced quantities inside the system.

Under these conditions, system (1.1) can be written in the form of a quasi-linear system

$$\mathbf{Y}_t + F(\mathbf{Y})\mathbf{Y}_x + G(\mathbf{Y}) = \mathbf{0}, \quad t \in [0, +\infty), \quad x \in [0, L], \tag{1.2}$$

with $F : \mathcal{Y} \to \mathcal{M}_{n,n}(\mathbb{R})$ and $G : \mathcal{Y} \to \mathbb{R}^n$ are of class C^1 and defined as

$$F(\mathbf{Y}) \triangleq (\partial e / \partial \mathbf{Y})^{-1}(\partial f / \partial \mathbf{Y}), \qquad G(\mathbf{Y}) \triangleq (\partial e / \partial \mathbf{Y})^{-1} g(\mathbf{Y}).$$

As usual, $\mathcal{M}_{n,n}(\mathbb{R})$ denotes the set of $n \times n$ real matrices. Also in (1.2), and often in the rest of the book, we drop the argument (t, x) when it does not lead to confusion.

We assume that system (1.2) is **hyperbolic**, i.e., that $F(\mathbf{Y})$ has n real eigenvalues (called *characteristic velocities*) for all $\mathbf{Y} \in \mathcal{Y}$. In this book, it will be also always assumed that these eigenvalues do not vanish in $\mathcal{Y}$. It follows that the number m of positive eigenvalues (counting multiplicity) is independent of $\mathbf{Y}$. Except otherwise stated, we will always use the following notations for the m positive and the $n - m$ negative eigenvalues:

$$\lambda_1(\mathbf{Y}), \ldots, \lambda_m(\mathbf{Y}), -\lambda_{m+1}(\mathbf{Y}), \ldots, -\lambda_n(\mathbf{Y}), \qquad \lambda_i(\mathbf{Y}) > 0 \; \forall \mathbf{Y} \in \mathcal{Y}, \forall i.$$

In the particular case where F is constant (i.e., does not depend on $\mathbf{Y}$), the system (1.2) is called *semi-linear*. Obviously, in that case, the system has constant characteristic velocities denoted:

$$\lambda_1, \ldots, \lambda_m, -\lambda_{m+1}, \ldots, -\lambda_n, \qquad \lambda_i > 0 \; \forall i.$$

Remark that, in contrast with most publications on quasi-linear hyperbolic systems, we use here the notation $\lambda_i(\mathbf{Y})$ to designate the *absolute value* of the characteristic velocities. The reason for using such an heterodox notation is that it simplifies the mathematical writings when the sign of the characteristic velocities matters in the boundary stability analysis which is one of the main concerns of this book.

1.1.1 Riemann Coordinates and Characteristic Form

In this book we shall often focus on the class of hyperbolic systems of balance laws that can be transformed into a *characteristic form* by defining a set of n so-called *Riemann coordinates* (see for instance (Dafermos 2000, Chapter 7, Section 7.3)). The characteristic form is obtained through a change of coordinates $\mathbf{R} = \psi(\mathbf{Y})$ having the following properties:

- The function $\psi : \mathcal{Y} \to \mathcal{R} \subset \mathbb{R}^n$ is a diffeomorphism: $\mathbf{R} = \psi(\mathbf{Y}) \longleftrightarrow \mathbf{Y} = \psi^{-1}(\mathbf{R})$, with Jacobian matrix $\Psi(\mathbf{Y}) \triangleq \partial\psi/\partial Y$.
- The Jacobian matrix $\Psi(\mathbf{Y})$ diagonalizes the matrix $F(\mathbf{Y})$:

$$\Psi(\mathbf{Y})F(\mathbf{Y}) = D(\mathbf{Y})\Psi(\mathbf{Y}), \qquad Y \in \mathcal{Y},$$

with

$$D(\mathbf{Y}) = \mathrm{diag}\big(\lambda_1(\mathbf{Y}), \ldots, \lambda_m(\mathbf{Y}), -\lambda_{m+1}(\mathbf{Y}), \ldots, -\lambda_n(\mathbf{Y})\big).$$

The system (1.2) is then equivalent for C^1-solutions to the following system in characteristic form expressed in the Riemann coordinates:

$$\mathbf{R}_t + \Lambda(\mathbf{R})\mathbf{R}_x + C(\mathbf{R}) = \mathbf{0}, \quad t \in [0, +\infty), \quad x \in [0, L], \tag{1.3}$$

with

$$\Lambda(\mathbf{R}) \triangleq D(\psi^{-1}(\mathbf{R})) \quad \text{and} \quad C(\mathbf{R}) \triangleq \Psi(\psi^{-1}(\mathbf{R}))G(\psi^{-1}(\mathbf{R})).$$

Clearly, this change of coordinates exists for any system of balance laws with linear flux densities (i.e., with $f(\mathbf{Y}) = A\mathbf{Y}$, $A \in \mathcal{M}_{n,n}(\mathbb{R})$ constant) when the matrix A is diagonalizable, in particular when the characteristic velocities are distinct. For systems with nonlinear flux densities, finding the change of coordinates $\mathbf{R} = \psi(\mathbf{Y})$ requires to find a solution of the first order partial differential equation $\Psi(\mathbf{Y})F(\mathbf{Y}) = D(\mathbf{Y})\Psi(\mathbf{Y})$. As it is shown in (Lax 1973, pages 34–35), this partial differential equation can always be solved, at least locally, for systems of size $n = 2$ with distinct characteristic velocities (see also (Li 1994, p. 30)). By contrast, for systems of size $n \geqslant 3$, the change of coordinates exists only in non-generic specific cases. However we shall see in this chapter that there is a multitude of interesting physical models for engineering which have size $n \geqslant 3$ and can nevertheless be written in characteristic form.

1.1.2   Steady State and Linearization

A *steady state* (or equilibrium) is a time-invariant space-varying solution $\mathbf{Y}(t, x) = \mathbf{Y}^*(x)$ $\forall t \in [0, +\infty)$ of the system (1.2). It satisfies the ordinary differential equation

$$F(\mathbf{Y}^*)\mathbf{Y}_x^* + G(\mathbf{Y}^*) = \mathbf{0}, \qquad x \in [0, L]. \tag{1.4}$$

The linearization of the system about the steady state is then

$$\mathbf{Y}_t + A(x)\mathbf{Y}_x + B(x)\mathbf{Y} = \mathbf{0}, \qquad t \in [0, +\infty), \qquad x \in [0, L], \tag{1.5}$$

where

$$A(x) \triangleq F(\mathbf{Y}^*(x)) \quad \text{and} \quad B(x) \triangleq \left[\frac{\partial}{\partial Y} \left(F(\mathbf{Y})\mathbf{Y}_x^* + G(\mathbf{Y}) \right) \right]_{\mathbf{Y}=\mathbf{Y}^*(x)}.$$

In the special case where there is a solution to the algebraic equation $G(\mathbf{Y}^*) = \mathbf{0}$, the system has a *constant steady state* (independent of both t and x) and the linearization is

$$\mathbf{Y}_t + A\mathbf{Y}_x + B\mathbf{Y} = \mathbf{0}, \qquad t \in [0, +\infty), \qquad x \in [0, L], \tag{1.6}$$

where A and B are constant matrices. In this special case where $\mathbf{Y}^*$ is constant, the nonlinear system (1.2) is said to have a *uniform* steady state. In the general case where the steady state $\mathbf{Y}^*(x)$ is space varying, the nonlinear system (1.2) is said to have a *nonuniform* steady state.

1.1.3   Riemann Coordinates Around the Steady State

By definition, the steady state of system (1.3) is

$$\mathbf{R}^*(x) = \psi(\mathbf{Y}^*(x)) \text{ such that } \Lambda(\mathbf{R}^*)\mathbf{R}_x^* + C(\mathbf{R}^*) = \mathbf{0}.$$

Then, alternatively, Riemann coordinates may also be defined *around this steady state* as

$$\mathbf{R} \triangleq \psi(\mathbf{Y}) - \psi(\mathbf{Y}^*).$$

With these coordinates the system is now written in characteristic form as

$$\mathbf{R}_t + \Lambda(\mathbf{R}, x)\mathbf{R}_x + C(\mathbf{R}, x) = \mathbf{0}, \quad t \in [0, +\infty), \quad x \in [0, L], \tag{1.7}$$

with

$$\Lambda(\mathbf{R}(t,x),x) \triangleq D(\psi^{-1}(\mathbf{R}(t,x) + \psi(\mathbf{Y}^*(x))))$$

and

$$C(\mathbf{R}(t,x),x) \triangleq D\big(\psi^{-1}(\mathbf{R}(t,x) + \psi(\mathbf{Y}^*(x)))\big)\psi_x(\mathbf{Y}^*(x))$$
$$+ \Psi\big(\psi^{-1}(\mathbf{R}(t,x) + \psi(\mathbf{Y}^*(x)))\big)G\big(\psi^{-1}(\mathbf{R}(t,x) + \psi(\mathbf{Y}^*(x)))\big).$$

The linearization of the system (1.7) gives:

$$\mathbf{R}_t + \Lambda(x)\mathbf{R}_x + M(x)\mathbf{R} = \mathbf{0}, \quad t \in [0,+\infty), \quad x \in [0,L],$$

with

$$\Lambda(x) \triangleq D(\mathbf{Y}^*(x)) \quad \text{and} \quad M(x) \triangleq \left[\frac{\partial C(\mathbf{R},x)}{\partial \mathbf{R}}\right]_{\mathbf{R}=0}.$$

Remark that this linear model is also the linearization of system (1.3) around the steady state and that it could be obtained as well by transforming directly the linear system (1.5) into Riemann coordinates. In other words the operations of linearization and Riemann coordinate transformation can be inverted.

1.1.4 Conservation Laws and Riemann Invariants

In the special case where there are no source terms (i.e., $G(\mathbf{Y}) = \mathbf{0}, \forall \mathbf{Y} \in \mathcal{Y}$), system (1.1) or (1.2) reduces to

$$\partial_t e(\mathbf{Y}) + \partial_x f(\mathbf{Y}) = \mathbf{0} \quad \text{or} \quad \mathbf{Y}_t + F(\mathbf{Y})\mathbf{Y}_x = \mathbf{0}, \quad t \in [0,+\infty), \quad x \in [0,L], \quad (1.8)$$

A system of this form is a hyperbolic system of *conservation laws*, representing a process where the balanced quantity is conserved since it can change only by the flux through the boundaries. In that case, it is clear that any constant value $\mathbf{Y}^*$ can be a steady state, independently of the value of the coefficient matrix $F(\mathbf{Y})$. Thus such systems have uniform steady states by definition. After transformation in Riemann coordinates (if possible), a system of conservation laws is written in the form

$$\partial_t R_i + \lambda_i(\mathbf{R})\partial_x R_i = 0, \quad i = 1, \ldots, m,$$
$$\partial_t R_i - \lambda_i(\mathbf{R})\partial_x R_i = 0, \quad i = m+1, \ldots, n.$$

The left-hand sides of these equations are the total time derivatives

$$\frac{dR_i}{dt} \triangleq \partial_t R_i + \frac{dx}{dt}\partial_x R_i$$

Fig. 1.1 Characteristic curves

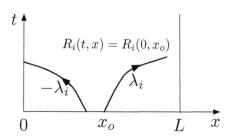

of the Riemann coordinates along the *characteristic curves* which are the integral curves of the ordinary differential equations

$$\frac{dx}{dt} = \lambda_i(\mathbf{R}(t, x)), \qquad i = 1, \ldots, m,$$

$$\frac{dx}{dt} = -\lambda_i(\mathbf{R}(t, x)), \qquad i = m + 1, \ldots, n,$$

in the plane (t, x) as illustrated in Fig. 1.1.

Since $dR_i/dt = 0$, it follows that the Riemann coordinates $R_i(t, x)$ are constant along the characteristic curves and are therefore called *Riemann invariants* for systems of conservation laws.

1.1.5 Stability, Boundary Stabilization, and the Associated Cauchy Problem

In order to have a unique well-defined solution to a quasi-linear hyperbolic system (1.2) over the interval $[0, L]$, initial and boundary conditions must obviously be specified.

In this book, we address the specific issue of identifying and characterizing dissipative boundary conditions which guarantee bounded smooth solutions converging to an equilibrium.

Of special interest is the feedback control problem when the manipulated control input, the controlled outputs and the measured outputs are physically located at the boundaries. Formally, this means that we consider the system (1.2) under n boundary conditions having the general form

$$\mathcal{B}\big(\mathbf{Y}(t, 0), \mathbf{Y}(t, L), \mathbf{U}(t)\big) = \mathbf{0} \qquad (1.9)$$

with the map $\mathcal{B} \in C^1(\mathcal{Y} \times \mathcal{Y} \times \mathbb{R}^q, \mathbb{R}^n)$. The dependence of the map $\mathcal{B}$ on $(\mathbf{Y}(t, 0), \mathbf{Y}(t, L))$ refers to natural physical constraints on the system. The function $\mathbf{U}(t) \in \mathbb{R}^q$ represents a set of q exogenous control inputs that can be used for stabilization, output tracking, or disturbance rejection.

In the case of static feedback control laws $\mathbf{U}(\mathbf{Y}(t,0),\mathbf{Y}(t,L))$, one of our main concerns is to analyze the asymptotic convergence of the solutions of the Cauchy problem:

$$\text{System } \mathbf{Y}_t + F(\mathbf{Y})\mathbf{Y}_x + G(\mathbf{Y}) = 0, \quad t \in [0,+\infty), \quad x \in [0,L],$$

$$\text{B. C. } \mathcal{B}(\mathbf{Y}(t,0),\mathbf{Y}(t,L),\mathbf{U}(\mathbf{Y}(t,0),\mathbf{Y}(t,L))) = 0, \quad t \in [0,+\infty),$$

$$\text{I. C. } \mathbf{Y}(0,x) = \mathbf{Y}_\circ(x), \quad x \in [0,L].$$

Additional constraints on the initial condition (I.C.) and the boundary conditions (B.C.) are needed to have a well-posed Cauchy problem. We examine this issue first in the case when the system can be transformed into characteristic form and then in the general case.

1.1.5.1 The Cauchy Problem in Riemann Coordinates

As we shall see later in this chapter, for many physical systems described by hyperbolic equations written in characteristic form (1.3)

$$\mathbf{R}_t + \Lambda(\mathbf{R})\mathbf{R}_x + C(\mathbf{R}) = 0, \quad t \in [0,+\infty), \quad x \in [0,L],$$

it is a basic property that "at each boundary point the incoming information $\mathbf{R}_{in}$ is determined by the outgoing information $\mathbf{R}_{out}$" (Russell 1978, Section 3), with the definitions

$$\mathbf{R}_{in}(t) \triangleq \begin{pmatrix} \mathbf{R}^+(t,0) \\ \mathbf{R}^-(t,L) \end{pmatrix} \quad \text{and} \quad \mathbf{R}_{out}(t) \triangleq \begin{pmatrix} \mathbf{R}^+(t,L) \\ \mathbf{R}^-(t,0) \end{pmatrix}, \tag{1.10}$$

where $\mathbf{R}^+$ and $\mathbf{R}^-$ are defined as follows[2]:

$$\mathbf{R}^+ = (R_1,\ldots,R_m)^\mathsf{T}, \qquad \mathbf{R}^- = (R_{m+1},\ldots,R_n)^\mathsf{T}.$$

This means that the system (1.3) is subject to boundary conditions having the 'nominal' form

$$\mathbf{R}_{in}(t) = \mathcal{H}(\mathbf{R}_{out}(t)), \tag{1.11}$$

where the map $\mathcal{H} \in C^1(\mathbb{R}^n;\mathbb{R}^n)$.

Moreover, the initial condition

$$\mathbf{R}(0,x) = \mathbf{R}_\circ(x), \quad x \in [0,L], \tag{1.12}$$

must be specified.

---

[2]In this section and everywhere in the book the notation M^T denotes the transpose of the matrix M.

Fig. 1.2 A quasi-linear
hyperbolic systems with
boundary conditions in
nominal form is a closed loop
interconnection of two causal
input-output systems

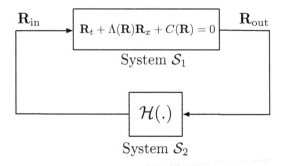

Hence, in Riemann coordinates, the Cauchy problem is formulated as follows:

System $\mathbf{R}_t + \Lambda(\mathbf{R})\mathbf{R}_x + C(\mathbf{R}) = \mathbf{0}, \quad t \in [0, +\infty), \quad x \in [0, L],$

B. C. $\mathbf{R}_{\text{in}}(t) = \mathcal{H}\big(\mathbf{R}_{\text{out}}(t)\big), \quad t \in [0, +\infty),$

I. C. $\mathbf{R}(0, x) = \mathbf{R}_{\text{o}}(x), \quad x \in [0, L].$

The well-posedness of this Cauchy problem may require that the initial condition (1.12) be compatible with the boundary condition (1.11). The compatibility conditions which are necessary for the well-posedness of the Cauchy problem depend on the functional space to which the solutions belong. In this book, we will be mainly concerned with solutions $\mathbf{R}(t, .)$ that may be of class C^0 or L^2 for linear systems and of class C^1 or H^2 for quasi-linear systems. For each case, the required compatibility conditions will be presented at the most suitable place (see also Appendices A and B).

It is also interesting to remark that the hyperbolic system (1.3) under the boundary condition (1.11) can be regarded as the closed loop interconnection of two *causal* input-output systems as represented in Fig. 1.2.

1.1.5.2 The Well-Posedness of the General Cauchy Problem for Strictly Hyperbolic Systems

Let us now consider the case of a general quasi-linear hyperbolic system

$$\mathbf{Y}_t + F(\mathbf{Y})\mathbf{Y}_x + G(\mathbf{Y}) = \mathbf{0}, \quad t \in [0, +\infty), \quad x \in [0, L],$$

$$\mathcal{B}(\mathbf{Y}(t, 0), \mathbf{Y}(t, L)) = \mathbf{0}, \quad t \in [0, +\infty),$$

which cannot be transformed into characteristic form. We assume that the system is *strictly* hyperbolic which means that for each $\mathbf{Y} \in \mathcal{Y}$, the matrix $F(\mathbf{Y})$ has nonzero *distinct* eigenvalues. Therefore, for all $x \in [0, L]$, the matrix $F(\mathbf{Y}^*(x))$, where $\mathbf{Y}^*(x)$ is the steady state as in (1.4), can be diagonalized, i.e., there exists a map $N : x \in [0, L] \to N(x) \in \mathcal{M}_{n,n}(\mathbb{R})$ of class C^1 such that

$$N(x) \text{ is invertible for all } x \in [0, L],$$

$$N(x)F(\mathbf{Y}^*(x)) = \Lambda(x)N(x),$$

with $\Lambda(x) \triangleq \operatorname{diag}\{\lambda_1(x), \ldots, \lambda_m(x), -\lambda_{m+1}(x), \ldots, -\lambda_n(x)\}$.

We define the following change of coordinates:

$$\mathbf{Z}(t, x) \triangleq N(x)\big(\mathbf{Y}(t, x) - \mathbf{Y}^*(x)\big), \qquad \mathbf{Z} = (Z_1, \ldots, Z_n)^{\mathsf{T}}.$$

In the coordinates $\mathbf{Z}$, the system is rewritten

$$\mathbf{Z}_t + A(\mathbf{Z}, x)\mathbf{Z}_x + B(\mathbf{Z}, x) = \mathbf{0},$$

$$\mathcal{B}\big(N(0)^{-1}\mathbf{Z}(t, 0) + \mathbf{Y}^*(0), N(L)^{-1}\mathbf{Z}(t, L) + \mathbf{Y}^*(L)\big) = \mathbf{0},$$

with

$$A(\mathbf{Z}, x) \triangleq N(x)F(N(x)^{-1}\mathbf{Z} + \mathbf{Y}^*(x))N(x)^{-1} \text{ with } A(\mathbf{0}, x) = \Lambda(x),$$

$$B(\mathbf{Z}, x) \triangleq N(x)\big[F(N(x)^{-1}\mathbf{Z} + \mathbf{Y}^*(x))(\mathbf{Y}_x^*(x) - N(x)^{-1}N'(x)N(x)^{-1}\mathbf{Z})$$
$$+ G(N(x)^{-1}\mathbf{Z} + \mathbf{Y}^*(x))\big].$$

Let us now define the incoming and outgoing boundary signals:

$$\mathbf{Z}_{\text{in}}(t) \triangleq \begin{pmatrix} \mathbf{Z}^+(t, 0) \\ \mathbf{Z}^-(t, L) \end{pmatrix} \quad \text{and} \quad \mathbf{Z}_{\text{out}}(t) \triangleq \begin{pmatrix} \mathbf{Z}^+(t, L) \\ \mathbf{Z}^-(t, 0) \end{pmatrix},$$

where $\mathbf{Z}^+$ and $\mathbf{Z}^-$ are as follows:

$$\mathbf{Z}^+ = (Z_1, \ldots, Z_m)^{\mathsf{T}}, \qquad \mathbf{Z}^- = (Z_{m+1}, \ldots, Z_n)^{\mathsf{T}}.$$

Obviously there exists a map $\widehat{\mathcal{B}} \in C^1(\mathbb{R}^n \times \mathbb{R}^n; \mathbb{R}^n)$ such that

$$\mathcal{B}(N(0)^{-1}\mathbf{Z}(t, 0), N(L)^{-1}\mathbf{Z}(t, L)) = \widehat{\mathcal{B}}(\mathbf{Z}_{\text{in}}(t), \mathbf{Z}_{\text{out}}(t)). \tag{1.13}$$

The requirement that, at each boundary point, the incoming information should be determined by the outgoing information imposes that (1.13) can be solved for $\mathbf{Z}_{\text{in}}$:

$$\mathbf{Z}_{\text{in}}(t) = \mathcal{H}(\mathbf{Z}_{\text{out}}(t)).$$

Then, provided the system is strictly hyperbolic and the initial condition is compatible with the boundary condition, the well-posed Cauchy problem is formulated as follows:

System $\mathbf{Z}_t + A(\mathbf{Z}, x)\mathbf{Z}_x + B(\mathbf{Z}, x) = \mathbf{0}$, $t \in [0, +\infty)$, $x \in [0, L]$,

B. C. $\mathbf{Z}_{\text{in}}(t) = \mathcal{H}\big(\mathbf{Z}_{\text{out}}(t)\big)$, $t \in [0, +\infty)$, $\qquad\qquad$ (1.14)

I. C. $\mathbf{Z}(0, x) = \mathbf{Z}_0(x)$, $x \in [0, L]$,

with appropriate compatibility conditions for the initial state $\mathbf{Z}_0$.

* * *

The rest of this chapter is now devoted to presenting typical examples of hyperbolic systems of balance laws for various physical engineering applications. We shall see that in many examples, the system can indeed be transformed into Riemann coordinates. With these examples we also illustrate how the control boundary conditions may be defined for the most commonly used control devices.

1.2   Telegrapher Equations

First published by Heaviside (1892), page 123, the telegrapher equations describe the propagation of current and voltage along electrical transmission lines (see Fig. 1.3). It is a system of two linear hyperbolic balance laws of the following form:

$$\partial_t(L_\ell I) + \partial_x V + R_\ell I = 0,$$
$$\partial_t(C_\ell V) + \partial_x I + G_\ell V = 0,$$
$\qquad\qquad$ (1.15)

where $I(t, x)$ is the current intensity, $V(t, x)$ is the voltage, L_ℓ is the line self-inductance per unit length, C_ℓ is the line capacitance per unit length, R_ℓ is the resistance of the two conductors per unit length, and G_ℓ is the admittance per unit length of the dielectric material separating the conductors.

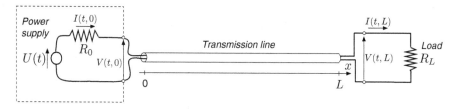

Fig. 1.3 Transmission line connecting a power supply to a resistive load R_L; the power supply is represented by a Thevenin equivalent with electromotive force $U(t)$ and internal resistance R_0

For the circuit represented in Fig. 1.3, the line model (1.15) is to be considered under the following boundary conditions:

$$V(t,0) + R_0 I(t,0) = U(t),$$
$$V(t,L) - R_L I(t,L) = 0,$$

(1.16)

where R_0 is the internal resistance of the power supply and R_L is the load. The telegrapher equations (1.15) coupled with these boundary conditions constitute therefore a boundary control system with the voltage $U(t)$ as control input.

A steady state $I^*(x)$, $V^*(x)$ of system (1.15) is a solution of the differential equation

$$\partial_x \begin{pmatrix} V^* \\ I^* \end{pmatrix} + \begin{pmatrix} R_\ell I^* \\ G_\ell V^* \end{pmatrix} = 0.$$

(1.17)

From equations (1.15) and (1.17), we can write the model, around a steady state, in the general linear form

$$\mathbf{Y}_t + A\mathbf{Y}_x + B\mathbf{Y} = \mathbf{0}$$

(1.18)

with

$$\mathbf{Y}(t,x) \triangleq \begin{pmatrix} I(t,x) - I^*(x) \\ V(t,x) - V^*(x) \end{pmatrix}, \quad A \triangleq \begin{pmatrix} 0 & L_\ell^{-1} \\ C_\ell^{-1} & 0 \end{pmatrix}, \quad B \triangleq \begin{pmatrix} R_\ell L_\ell^{-1} & 0 \\ 0 & G_\ell C_\ell^{-1} \end{pmatrix}.$$

Here, because the physical system (1.15) is linear, we observe that the linear system (1.18) has uniform coefficients although the steady state may be nonuniform.

The system has two characteristic velocities (which are the eigenvalues of the matrix A), one positive and one negative:

$$\lambda_1 = \frac{1}{\sqrt{L_\ell C_\ell}}, \quad -\lambda_2 = -\frac{1}{\sqrt{L_\ell C_\ell}}.$$

Riemann coordinates can be defined as

$$R_1(t,x) = \left(V(t,x) - V^*(x)\right) + \left(I(t,x) - I^*(x)\right)\sqrt{\frac{L_\ell}{C_\ell}},$$

$$R_2(t,x) = \left(V(t,x) - V^*(x)\right) - \left(I(t,x) - I^*(x)\right)\sqrt{\frac{L_\ell}{C_\ell}},$$

with the inverse coordinate transformation

$$I(t,x) - I^*(x) = \frac{R_1(t,x) - R_2(t,x)}{2}\sqrt{\frac{C_\ell}{L_\ell}},$$

$$V(t,x) - V^*(x) = \frac{R_1(t,x) + R_2(t,x)}{2}.$$

With these coordinates, the system (1.15), (1.16) is written as follows in character-istic form:

$$\partial_t R_1 + \lambda \partial_x R_1 + \gamma R_1 + \delta R_2 = 0,$$
$$\partial_t R_2 - \lambda \partial_x R_2 + \delta R_1 + \gamma R_2 = 0,$$

(1.19)

$$R_1(t,0) = \left[(-1 + \lambda R_0 C_\ell) R_2(t,0) + U(t) - U^*\right](1 + \lambda R_0 C_\ell)^{-1},$$
$$R_2(t,L) = \left[(-1 + \lambda R_0 C_\ell) R_1(t,L)\right](1 + \lambda R_0 C_\ell)^{-1},$$

(1.20)

with

$$\lambda \triangleq \frac{1}{\sqrt{L_\ell C_\ell}}, \qquad \gamma \triangleq \frac{1}{2}\left[\frac{G_\ell}{C_\ell} + \frac{R_\ell}{L_\ell}\right], \qquad \delta \triangleq \frac{1}{2}\left[\frac{G_\ell}{C_\ell} - \frac{R_\ell}{L_\ell}\right].$$

1.3 Raman Amplifiers

Raman amplifiers are electro-optical devices that are used for compensating the natural power attenuation of laser signals transmitted along optical fibers in long distance communications. Their operation is based on the *Raman effect* which was discovered by Raman and Krishnan (1928). The simplest implementation of Raman amplification in optical telecommunications is depicted in Fig. 1.4. The transmitted information is encoded by amplitude modulation of a laser signal with wavelength ω_s. The signal is provided by an optical source at the channel input and received by a photo-detector at the output. A pump laser beam with wavelength ω_p is injected backward in the optical fiber. If the wavelengths are appropriately selected, the energy of the pump is transferred to the signal and produces an amplification that counteracts the natural attenuation. The dynamics of the signal and pump powers along the fiber are represented by the following system of two balance laws (Dower and Farrel (2006)):

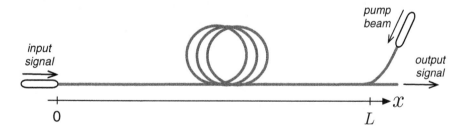

Fig. 1.4 Optical communication with Raman amplification

$$\partial_t S + \lambda_s\left(\partial_x S + \alpha_s S - \beta_s SP\right) = 0,$$

$$\partial_t P - \lambda_p\left(\partial_x P - \alpha_p P - \beta_p PS\right) = 0, \tag{1.21}$$

where $S(t,x)$ is the power of the transmitted signal, $P(t,x)$ is the power of the pump laser beam, λ_s and λ_p are the propagation group velocities of the signal and pump waves respectively, α_s and α_p are the attenuation coefficients per unit length, β_s and β_p are the amplification gains per unit length. All these positive constant parameters α_s and α_p, β_s and β_p, λ_s and λ_p are characteristic of the fiber material and dependent of the wavelengths ω_s and ω_p.

Here, the physical model (1.21) is directly given in characteristic form (1.3). The Riemann coordinates are the powers $R_1 = S(t,x)$ and $R_2 = P(t,x)$. The system is hyperbolic with characteristic velocities $\lambda_s > 0 > -\lambda_p$. As the input signal power and the launch pump power can be exogenously imposed, the boundary conditions are

$$S(t,0) = U_0(t), \qquad P(t,L) = U_L(t). \tag{1.22}$$

Then the system (1.21) coupled to the boundary conditions (1.22) is a *boundary control system* with the boundary control inputs U_0 and U_L.

1.4 Saint-Venant Equations for Open Channels

First proposed by Barré de Saint-Venant (1871), the Saint-Venant equations (also called *shallow water equations*) describe the propagation of water in open channels (see Fig. 1.5). In the simple standard case of a channel with a constant slope, a rectangular cross section and a unit width, the Saint-Venant model is a system of two nonlinear balance laws of the form

$$\partial_t H + \partial_x(HV) = 0,$$

$$\partial_t V + \partial_x\left(\frac{V^2}{2} + gH\right) + \left(C\frac{V^2}{H} - gS_b\right) = 0, \tag{1.23}$$

where $H(t,x)$ is the water depth and $V(t,x)$ is the horizontal water velocity. More precisely, $V(t,x)$ denotes the horizontal velocity averaged across a vertical column of water. S_b is the constant bottom slope, g is the constant gravity acceleration, and C is a constant friction coefficient. The first equation is a mass balance and the second equation is a momentum balance.

This model is written in the general quasi-linear form $\mathbf{Y}_t + F(\mathbf{Y})\mathbf{Y}_x + G(\mathbf{Y}) = \mathbf{0}$ with

$$\mathbf{Y} \triangleq \begin{pmatrix} H \\ V \end{pmatrix}, \quad F(\mathbf{Y}) \triangleq \begin{pmatrix} V & H \\ g & V \end{pmatrix}, \quad G(\mathbf{Y}) \triangleq \begin{pmatrix} 0 \\ CV^2H^{-1} - gS_b \end{pmatrix}.$$

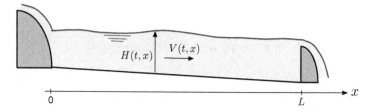

Fig. 1.5 Lateral view of a pool of an open channel with constant bottom slope and rectangular cross section

The eigenvalues of the matrix $F(\mathbf{Y})$ are

$$V + \sqrt{gH} \quad \text{and} \quad V - \sqrt{gH}.$$

The flow is said to be *subcritical (or fluvial)* if the so-called Froude's number

$$Fr = \frac{V(t, x)}{\sqrt{gH(t, x)}} < 1.$$

Under this condition, the system is hyperbolic with characteristic velocities

$$\lambda_1(\mathbf{Y}) = V + \sqrt{gH} > 0 > -\lambda_2(\mathbf{Y}) = V - \sqrt{gH}.$$

Riemann coordinates may be defined as

$$R_1 = V + 2\sqrt{gH}, \qquad R_2 = V - 2\sqrt{gH}$$

and are inverted as

$$H = (R_1 - R_2)^2/16g, \qquad V = (R_1 + R_2)/2.$$

With these coordinates, the system is written in characteristic form $\mathbf{R}_t + \Lambda(\mathbf{R})\mathbf{R}_x + C(\mathbf{R}) = \mathbf{0}$ with

$$\Lambda(\mathbf{R}) \triangleq \begin{pmatrix} \lambda_1(\mathbf{R}) & 0 \\ 0 & -\lambda_2(\mathbf{R}) \end{pmatrix} = \begin{pmatrix} \dfrac{3R_1 + R_2}{4} & 0 \\ 0 & \dfrac{R_1 + 3R_2}{4} \end{pmatrix}$$

and

$$C(\mathbf{R}) \triangleq \left(4gC \left(\frac{R_1 + R_2}{R_1 - R_2} \right)^2 - gS_b \right) \begin{pmatrix} 1 \\ 1 \end{pmatrix}.$$

1.4.1 Boundary Conditions

When the flow is subcritical, two boundary conditions at both ends of the interval $[0, L]$ are needed to close the Saint-Venant equations. These conditions are imposed by physical devices located at the ends of the pool, as for instance the two spillways of the channel in Fig. 1.5.

A very simple situation is when the pool is closed but endowed with pumps that impose the discharges at $x = 0$ and $x = L$. In that case, the boundary conditions are

$$H(t, 0)V(t, 0) = U_0(t), \quad H(t, L)V(t, L) = U_L(t). \tag{1.24}$$

Then the system of the Saint-Venant equations (1.23) coupled to the boundary conditions (1.24) is a boundary control system with the two boundary flow rates U_0 and U_L as command signals.

Another interesting case is when the boundary conditions are assigned by tunable hydraulic gates as in irrigation canals and navigable rivers, see Fig. 1.6.

Standard hydraulic models give the boundary conditions for overflow gates (or mobile spillways):

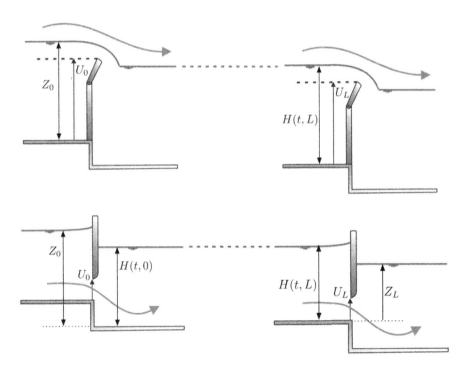

Fig. 1.6 Hydraulic gates at the input and the output of a pool: (above) overflow gates, (below) underflow gates

$$H(t,0)V(t,0) = \left(k_G\sqrt{2g}\right)\sqrt{\left[Z_0(t) - U_0(t)\right]^3},$$

$$H(t,L)V(t,L) = \left(k_G\sqrt{2g}\right)\sqrt{\left[H(t,L) - U_L(t)\right]^3}, \tag{1.25}$$

and for underflow (or sluice) gates:

$$H(t,0)V(t,0) = \left(k_G\sqrt{2g}\right)U_0(t)\sqrt{Z_0(t) - H(t,0)},$$

$$H(t,L)V(t,L) = \left(k_G\sqrt{2g}\right)U_L(t)\sqrt{H(t,L) - Z_L(t)}. \tag{1.26}$$

In these expressions, $H(t,0)$ and $H(t,L)$ denote the water depth at the boundaries inside the pool, $Z_0(t)$ and $Z_L(t)$ are the water levels on the other side of the gates, k_G is a constant adimensional discharge coefficient, $U_0(t)$ and $U_L(t)$ represent either the weir elevation for overflow gates or the height of the aperture for underflow gates. Again the Saint-Venant equations (1.23) coupled to these boundary conditions constitute a boundary control system with U_0 and U_L as command signals, and Z_0 and Z_L as disturbance inputs.

1.4.2   Steady State and Linearization

A steady state $H^*(x)$, $V^*(x)$ is a solution of the differential equations

$$\partial_x(H^*V^*) = 0,$$

$$\partial_x\left(\frac{V^{*2}}{2} + gH^*\right) + \left(C\frac{V^{*2}}{H^*} - gS_b\right) = 0.$$

These equations may also be written as

$$V^*\partial_x H^* = -H^*\partial_x V^* = \frac{H^*V^*(gS_b - CV^{*2}/H^*)}{gH^* - V^{*2}}.$$

In order to linearize the model, we define the deviations of the states $H(t,x)$ and $V(t,x)$ with respect to the steady states $H^*(x)$ and $V^*(x)$:

$$h(t,x) \triangleq H(t,x) - H^*(x), \qquad v(t,x) \triangleq V(t,x) - V^*(x).$$

Then the linearized Saint-Venant equations around the steady state are:

$$\partial_t h + V^*\partial_x h + H^*\partial_x v + (\partial_x V^*)h + (\partial_x H^*)v = 0,$$

$$\partial_t v + g\partial_x h + V^*\partial_x v - (CV^{*2}/H^{*2})h + \left[\partial_x V^* + (2CV^*/H^*)\right]v = 0. \tag{1.27}$$

The Riemann coordinates for the linearized system (1.27) are defined as follows:

$$R_1(t,x) = v(t,x) + h(t,x)\sqrt{\frac{g}{H^*(x)}},$$

$$R_2(t,x) = v(t,x) - h(t,x)\sqrt{\frac{g}{H^*(x)}},$$

with the inverse coordinate transformation

$$h(t,x) = \frac{R_1(t,x) - R_2(t,x)}{2}\sqrt{\frac{H^*(x)}{g}},$$

$$v(t,x) = \frac{R_1(t,x) + R_2(t,x)}{2}.$$

With these definitions and notations, the linearized Saint-Venant equations are written in characteristic form:

$$\partial_t R_1(t,x) + \lambda_1(x)\partial_x R_1(t,x) + \gamma_1(x)R_1(t,x) + \delta_1(x)R_2(t,x) = 0$$

$$\partial_t R_2(t,x) - \lambda_2(x)\partial_x R_2(t,x) + \gamma_2(x)R_1(t,x) + \delta_2(x)R_2(t,x) = 0 \tag{1.28}$$

with the characteristic velocities

$$\lambda_1(x) = V^*(x) + \sqrt{gH^*(x)}, \qquad -\lambda_2(x) = V^*(x) - \sqrt{gH^*(x)}$$

and, using the relation $H^*(\partial_x V^*) = -V^*(\partial_x H^*)$, the coefficients

$$\gamma_1 = \frac{CV^{*2}}{H^*}\left[-\frac{3}{4(\sqrt{gH^*}+V^*)} + \frac{1}{V^*} - \frac{1}{2\sqrt{gH^*}}\right] + \frac{3gS_b}{4(\sqrt{gH^*}+V^*)},$$

$$\delta_1 = \frac{CV^{*2}}{H^*}\left[-\frac{1}{4(\sqrt{gH^*}+V^*)} + \frac{1}{V^*} + \frac{1}{2\sqrt{gH^*}}\right] + \frac{gS_b}{4(\sqrt{gH^*}+V^*)},$$

$$\gamma_2 = \frac{CV^{*2}}{H^*}\left[\frac{1}{4(\sqrt{gH^*}-V^*)} + \frac{1}{V^*} - \frac{1}{2\sqrt{gH^*}}\right] - \frac{gS_b}{4(\sqrt{gH^*}-V^*)},$$

$$\delta_2 = \frac{CV^{*2}}{H^*}\left[\frac{3}{4(\sqrt{gH^*}-V^*)} + \frac{1}{V^*} + \frac{1}{2\sqrt{gH^*}}\right] - \frac{3gS_b}{4(\sqrt{gH^*}-V^*)}.$$

1.4.3 The General Model

To conclude this section, we give a more general version of the Saint-Venant equations which holds for channels with nonconstant slopes and cross-sections. The equations are as follows:

$$\partial_t A + \partial_x Q = 0,$$

$$\partial_t Q + \partial_x \frac{Q^2}{A} + gA\big[\partial_x H - S_b + S_f\big] = 0 \tag{1.29}$$

where $A(t,x)$ is the cross-sectional area of the water in the channel, $Q(t,x)$ is the flow rate (or discharge), $H(t,x)$ is the water depth, S_f is the friction term, $S_b(x)$ is the bottom slope, and g is the constant gravity acceleration. The friction term S_f is usually assumed to be proportional to $V^2 = Q^2/A^2$ and to the perimeter P of the cross-sectional area. Clearly it is natural to assume that both the water depth $H(A)$ and the perimeter $P(A)$ are monotonic increasing functions of A.

1.5 Saint-Venant-Exner Equations

First proposed by Exner (1920) (see also Exner (1925)), the Exner equation is a conservation law that represents the transport of sediments in a water flow in the case where the sediment moves predominantly as bedload. A common modeling of the dynamics of open channels with fluctuating bathymetry is therefore achieved by the coupling of the Exner equation to the Saint-Venant equations.

The state variables of the model (see Fig. 1.7) are the water depth $H(t,x)$ and the average horizontal water velocity $V(t,x)$ as for Saint-Venant equations, and the bathymetry $B(t,x)$ which is the elevation of the sediment bed above a fixed reference datum. For an horizontal channel with a rectangular cross-section and a unit width, the equations are written as follows (see, e.g., Hudson and Sweby (2003)):

$$\partial_t H + \partial_x(HV) = 0,$$

$$\partial_t V + \partial_x \left(\frac{V^2}{2} + g(H+B)\right) + C\frac{V^2}{H} = 0, \tag{1.30}$$

$$\partial_t B + \partial_x \left(a\frac{V^3}{3}\right) = 0.$$

Fig. 1.7 Lateral view of an open channel with a sediment bed

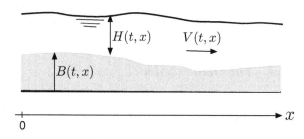

$H(t,x)$ $V(t,x)$

$B(t,x)$

0 x

In these equations, g is the gravity acceleration constant, C is a friction coefficient, and a is a constant parameter that encompasses porosity and viscosity effects on the sediment dynamics. The first two equations are the Saint-Venant equations and the third one is the Exner equation.

This model is in the general quasi-linear form $\mathbf{Y}_t + F(\mathbf{Y})\mathbf{Y}_x + G(\mathbf{Y}) = \mathbf{0}$ with

$$
\mathbf{Y} \triangleq \begin{pmatrix} H \\ V \\ B \end{pmatrix}, \quad F(\mathbf{Y}) \triangleq \begin{pmatrix} V & H & 0 \\ g & V & g \\ 0 & aV^2 & 0 \end{pmatrix}, \quad G(\mathbf{Y}) \triangleq \begin{pmatrix} 0 \\ C\dfrac{V^2}{H} \\ 0 \end{pmatrix}.
$$

The characteristic polynomial of the matrix $F(\mathbf{Y})$ is

$$
\lambda^3 - 2V\lambda^2 + (V^2 - g(aV^2 + H))\lambda + agV^3.
$$

From this polynomial, analytic expressions of the eigenvalues of $F(\mathbf{Y})$ are not easily derived. However, as shown by Hudson and Sweby (2003), good approximations can be obtained for small values of the parameter a under the subcritical flow condition $V^2 < gH$. As $a \to 0$, the eigenvalues of $F(\mathbf{Y})$ tend to

$$
\lambda_1 \to V + \sqrt{gH}, \quad \lambda_2 \to 0, \quad -\lambda_3 \to V - \sqrt{gH}.
$$

The determinant of $F(\mathbf{Y})$ is

$$
\det(F(\mathbf{Y})) = -\lambda_1\lambda_2\lambda_3 = -agV^3. \tag{1.31}
$$

Then, for small values of a, we have the following realistic approximations

$$
\left[\lambda_1 \approx V + \sqrt{gH}\right] \gg \left[\lambda_2 \approx \frac{agV^3}{gH - V^2}\right] > 0 > \left[-\lambda_3 \approx V - \sqrt{gH}\right] \tag{1.32}
$$

where the value of λ_2 is obtained by substituting the values of λ_1 and λ_3 in (1.31). Here λ_1 and λ_3 are the characteristic velocities of the water flow and λ_2 the characteristic velocity of the sediment motion. Obviously the sediment motion is much slower than the water flow.

Thus, the Saint-Venant-Exner model (1.30) is a hyperbolic system of three balance laws with characteristic velocities approximately given by (1.32).

1.6 Rigid Pipes and Heat Exchangers

The management of hydro-electric plants, the design of water supply networks with water hammer prevention, or the temperature control in heat exchangers are typical engineering issues that require dynamic models of water flow in pipes. Under the assumptions of axisymmetric flow and negligible radial fluid velocity, a standard model for the motion of water in a rigid cylindrical pipe is given by the following system of three balance laws:

$$\partial_t \left(\exp\left(\frac{gH}{c^2} \right) \right) + \partial_x \left(V \exp\left(\frac{gH}{c^2} \right) \right) = 0,$$

$$\partial_t V + \partial_x \left(gH + \frac{V^2}{2} \right) + \frac{C}{2d} V|V| = 0, \tag{1.33}$$

$$\partial_t T + \partial_x(VT) + k_o(T_e - T) = 0,$$

where $H(t,x)$ is the piezometric head, $V(t,x)$ is the water velocity, $T(t,x)$ is the water temperature, c is the sound velocity in water, C is a constant friction coefficient, g is the gravity acceleration, d is the pipe diameter, and T_e is the external atmospheric temperature. The first equation is a mass conservation law, the second equation is a momentum balance, and the third equation is a heat balance.

The piezometric head H is defined as

$$H(t,x) = Z(x) + \frac{P(t,x)}{\rho g},$$

where $Z(x)$ is the elevation of the pipe, $P(t,x)$ is the pressure, and ρ is the density. For an horizontal pipe, the piezometric head is just proportional to the pressure.

The constant parameter k_o is defined as

$$k_o \triangleq \frac{\alpha}{c_p \rho A},$$

where α is the thermal conductance of the pipe wall, c_p is the water specific heat, and $A = \pi d^2/4$ is the cross-sectional area of the pipe.

This kind of model based on one-dimensional mass, momentum, or heat balances was already present in the engineering scientific literature by the late nineteenth century (see, e.g., the paper by Allievi (1903) and also other references quoted in the survey paper by Ghidaoui et al. (2005)).

The model (1.33) is written in the general quasi-linear form $\mathbf{Y}_t + F(\mathbf{Y})\mathbf{Y}_x + G(\mathbf{Y}) = \mathbf{0}$ with

$$\mathbf{Y} \triangleq \begin{pmatrix} H \\ V \\ T \end{pmatrix}, \quad F(\mathbf{Y}) \triangleq \begin{pmatrix} V & c^2/g & 0 \\ g & V & 0 \\ 0 & T & V \end{pmatrix}, \quad G(\mathbf{Y}) \triangleq \begin{pmatrix} 0 \\ CV|V|/2d \\ k_o(T_e - T) \end{pmatrix}.$$

The characteristic polynomial of the matrix $F(\mathbf{Y})$ is

$$(V - \lambda)((V - \lambda)^2 - c^2).$$

The roots of this polynomial are V, $V + c$, and $V - c$.

In practice, the sound velocity is about 1400 m/s and the flow velocity is much lower. In that case, the system is hyperbolic with characteristic velocities (which are the eigenvalues of the matrix $F(\mathbf{Y})$):

$$\lambda_1 = V + c > \lambda_2 = V > 0 > -\lambda_3 = V - c.$$

Riemann coordinates may then be defined as

$$R_1 = V + \frac{g}{c}H, \qquad R_2 = \frac{g}{c}H - c\ln T, \qquad R_3 = V - \frac{g}{c}H,$$

and are inverted as

$$H = \frac{c}{g}\frac{R_1 - R_3}{2}, \qquad V = \frac{R_1 + R_3}{2} \qquad \ln T = \frac{R_1 - 2R_2 - R_3}{2c}.$$

With these coordinates, the system is written in characteristic form $\mathbf{R}_t + \Lambda(\mathbf{R})\mathbf{R}_x + C(\mathbf{R}) = \mathbf{0}$ with

$$\Lambda(\mathbf{R}) \triangleq \begin{pmatrix} \lambda_1(\mathbf{R}) & 0 & 0 \\ 0 & \lambda_2(\mathbf{R}) & 0 \\ 0 & 0 & -\lambda_3(\mathbf{R}) \end{pmatrix} = \frac{R_1 + R_2}{2}\begin{pmatrix} 1 & 0 & 0 \\ 0 & 1 & 0 \\ 0 & 0 & 1 \end{pmatrix} + \begin{pmatrix} c & 0 & 0 \\ 0 & 0 & 0 \\ 0 & 0 & -c \end{pmatrix}.$$

and

$$C(\mathbf{R}) = \begin{pmatrix} \dfrac{C(R_1 + R_3)|R_1 + R_3|}{8d} \\[2ex] k_o c\left(1 - T_e \exp\left(\dfrac{-R_1 + 2R_2 + R_3}{2c}\right)\right) \\[2ex] \dfrac{C(R_1 + R_3)|R_1 + R_3|}{8d} \end{pmatrix}.$$

1.6.1 The Shower Control Problem

Everybody knows the shower control problem which is the problem of simultaneously regulating the temperature and the flow rate of a shower by manipulating the two valves of hot and cold water as illustrated in Fig. 1.8. The system is described by the model (1.33) with L being the length of the pipe between the valves and the shower outlet. This control problem may be analyzed under the following boundary conditions:

$$AV(t, 0) = Q_c(t) + Q_h(t), \qquad P(t, L) = \frac{P_a}{\rho g},$$

$$T(t, 0) = \frac{Q_c(t)T_c(t) + Q_h(t)T_h(t)}{Q_c(t) + Q_h(t)}. \tag{1.34}$$

The first condition represents the flow conservation at the junction of the valves, with $Q_c(t)$ and $Q_h(t)$ the cold and hot flow rates assigned by the two valves respectively. The second condition is that the atmospheric pressure P_a is imposed at the outlet. The third condition expresses that the inlet temperature is an average of the cold T_c and hot T_h temperatures.

Then the system of the shower equations (1.33) with the boundary conditions (1.34) is a boundary control system with the flow rates Q_c and Q_h as command signals.

1.6.2 The Water Hammer Problem

The device of Fig. 1.9 is a typical example of a system that may have a water hammer problem if the valve is closed too quickly or the pump is started up too quickly, see, e.g., Van Pham et al. (2014). Such a problem can be analyzed with the first two equations of (1.33) and appropriate boundary conditions imposed by the pump and the valve respectively, see, e.g., Luskin and Temple (1982). For instance, the pump

Fig. 1.8 The shower control problem

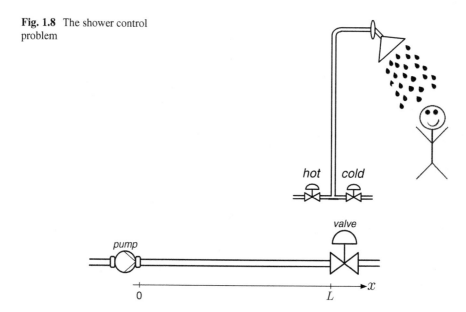

Fig. 1.9 A pipe connecting a pump and a valve

may be regarded as a device which is able to deliver a desired pressure drop no matter the flow rate:

$$H_{in}(t) - H(t,0) = U(t). \tag{1.35}$$

Moreover, the valve is typically modeled by a quadratic relationship between the pressure drop and the velocity:

$$H(t,L) - H_{out}(t) = k_v V(t,L)|V(t,L)|, \tag{1.36}$$

where k_v is a constant characteristic parameter. The pipe equation (1.33) coupled with these boundary conditions form a boundary control system with the pump command signal $U(t)$ as control input and the external piezometric heads $H_{in}(t)$ and $H_{out}(t)$ as disturbance inputs.

1.6.3   Heat Exchangers

A simple tubular heat exchanger is depicted in Fig. 1.10. It is made up of two counter current concentric pipes. Clearly a dynamical model is obtained by duplicating the basic balance equations (1.33) supplemented with appropriate interconnection terms as follows:

$$\partial_t H_1 + V_1 \partial_x H_1 + \frac{c^2}{g} \partial_x V_1 = 0,$$

$$\partial_t V_1 + \partial_x \left(g H_1 + \frac{V_1^2}{2} \right) + \frac{C}{2d} V_1 |V_1| = 0,$$

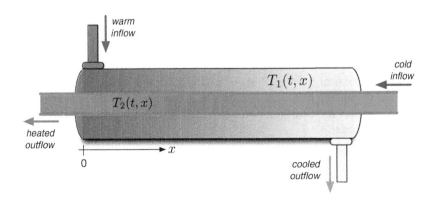

Fig. 1.10   A tubular heat exchanger

$$\partial_t T_1 + \partial_x(V_1 T_1) - k_1(T_1 - T_2) - k_o(T_1 - T_e) = 0,$$

$$\partial_t H_2 + V_2 \partial_x H_2 + \frac{c^2}{g} \partial_x V_2 = 0,$$

$$\partial_t V_2 + \partial_x \left(g H_2 + \frac{V_2^2}{2} \right) + \frac{C}{2d} V_2 |V_2| = 0,$$

$$\partial_t T_2 + \partial_x(V_2 T_2) + k_2(T_1 - T_2) = 0,$$

where $T_1(t,x)$, $T_2(t,x)$ are the water temperatures, $V_1(t,x)$, $V_2(t,x)$ the water velocities and $H_1(t,x)$, $H_2(t,x)$ the piezometric heads in the heating and heated tubes respectively and T_e is the external atmospheric temperature. The constant parameters k_o, k_1, and k_2 are defined as

$$k_o \triangleq \frac{\alpha_1}{c_p \rho A_1}, \qquad k_1 \triangleq \frac{\alpha_2}{c_p \rho A_1}, \qquad k_2 \triangleq \frac{\alpha_2}{c_p \rho A_2},$$

where α_i ($i = 1, 2$) are the thermal conductivities of the tube walls and A_i ($i = 1, 2$) are the effective cross-sections of the tubes.

The system (1.10) is hyperbolic with the characteristic velocities

$$\lambda_1 = V_1 + c, \quad \lambda_2 = V_1, \quad -\lambda_3 = V_1 - c, \quad \lambda_4 = V_2 + c, \quad \lambda_5 = V_2 \quad -\lambda_6 = V_2 - c,$$

and the corresponding Riemann coordinates

$$R_1 = V_1 + \frac{g}{c} H_1, \qquad R_2 = \frac{g}{c} H_1 - c \ln T_1, \qquad R_3 = V_1 - \frac{g}{c} H_1,$$

$$R_4 = V_2 + \frac{g}{c} H_2, \qquad R_5 = \frac{g}{c} H_2 - c \ln T_2, \qquad R_6 = V_2 - \frac{g}{c} H_2.$$

1.7 Plug Flow Chemical Reactors

A plug flow chemical reactor (PFR) is a tubular reactor where a liquid reaction mixture circulates. The reaction proceeds as the reactants travel through the reactor. Here, we consider the case of a horizontal PFR where a simple mono-molecular reaction takes place:

$$A \rightleftarrows B.$$

A is the reactant species and B is the desired product. The reaction is supposed to be exothermic and a jacket is used to cool the reactor. The cooling fluid flows around the wall of the tubular reactor. Therefore, the dynamics of the system are naturally described by the model (1.33) of the flow in a heat exchanger supplemented with the mass balance equations for the concerned chemical species. However it is usual to assume, for simplicity, that the dynamics of velocity and pressure in the reactor and the jacket are negligible. The dynamics of the PFR are then described by the following semi-linear system of balance laws:

$$\partial_t T_c - V_c \partial_x T_c - k_0 (T_c - T_r) = 0,$$

$$\partial_t T_r + V_r \partial_x T_r + k_0 (T_c - T_r) - k_1 r(T_r, C_A, C_B) = 0,$$

$$\partial_t C_A + V_r \partial_x C_A + r(T_r, C_A, C_B) = 0,$$

$$\partial_t C_B + V_r \partial_x C_B - r(T_r, C_A, C_B) = 0,$$

(1.37)

where $V_c(t)$ is the coolant velocity in the jacket, $V_r(t)$ is the reactive fluid velocity in the reactor, $T_c(t, x)$ is the coolant temperature, $T_r(t, x)$ is the reactor temperature. The variables $C_A(t, x)$ and $C_B(t, x)$ denote the concentrations of the chemicals in the reaction medium. The function $r(T_r, C_A, C_B)$ represents the reaction rate. A typical form of this function is:

$$r(T_r, C_A, C_B) = (a C_A - b C_B) \exp\left(-\frac{E}{RT_r}\right),$$

where a and b are rate constants, E is the activation energy, and R is the Boltzmann constant.

This model is in the general quasi-linear form $\mathbf{Y}_t + F(\mathbf{Y})\mathbf{Y}_x + G(\mathbf{Y}) = 0$ with

$$\mathbf{Y} \triangleq \begin{pmatrix} T_c \\ T_r \\ C_A \\ C_B \end{pmatrix}, \quad F(\mathbf{Y}) \triangleq \begin{pmatrix} -V_c & 0 & 0 & 0 \\ 0 & V_r & 0 & 0 \\ 0 & 0 & V_r & 0 \\ 0 & 0 & 0 & V_r \end{pmatrix},$$

$$G(\mathbf{Y}) \triangleq \begin{pmatrix} -k_0 (T_c - T_r) \\ k_0 (T_c - T_r) - k_1 r(T_r, C_A, C_B) \\ r(T_r, C_A, C_B) \\ -r(T_r, C_A, C_B) \end{pmatrix}.$$

It is a hyperbolic system of four balance laws with characteristic velocities V_c and V_r. This system is not strictly hyperbolic because it has three identical characteristic velocities. It is nevertheless endowed with Riemann coordinates because $F(\mathbf{Y})$ is diagonal.

1.8 Euler Equations for Gas Pipes

The motion of an inviscid ideal gas in a rigid cylindrical pipe with a unit cross-section is most often described by the classical Euler (1755) equations which have the form of a system of three balance laws:

$$\partial_t \varrho + \partial_x(\varrho V) = 0, \tag{1.38a}$$

$$\partial_t(\varrho V) + \partial_x(\varrho V^2 + P) + C\varrho V|V| = 0, \tag{1.38b}$$

$$\partial_t \left[\varrho \frac{V^2}{2} + \frac{P}{\gamma - 1} \right] + \partial_x \left[V \left(\varrho \frac{V^2}{2} + \frac{\gamma P}{\gamma - 1} \right) \right] = 0, \tag{1.38c}$$

where $\varrho(t, x)$ is the gas density, $V(t, x)$ is the gas velocity, $P(t, x)$ is the gas pressure, C is a constant friction coefficient, and $\gamma > 1$ is the constant heat capacity ratio. The first equation is a mass balance, the second equation is a momentum balance, and the third equation is an energy balance, with the total energy defined as

$$E \triangleq \varrho \frac{V^2}{2} + \frac{P}{\gamma - 1}.$$

Using ϱ, V and P as state variables, the model (1.38) is equivalent to

$$\partial_t \varrho + \partial_x(\varrho V) = 0,$$

$$\partial_t V + \partial_x \frac{V^2}{2} + \frac{1}{\varrho} \partial_x P + CV|V| = 0, \tag{1.39}$$

$$\partial_t P + V \partial_x P + \gamma P \partial_x V - (\gamma - 1)C\varrho|V|^3 = 0.$$

This model is in the general quasi-linear form $\mathbf{Y}_t + F(\mathbf{Y})\mathbf{Y}_x + G(\mathbf{Y}) = \mathbf{0}$ with

$$\mathbf{Y} \triangleq \begin{pmatrix} \varrho \\ V \\ P \end{pmatrix}, \quad F(\mathbf{Y}) \triangleq \begin{pmatrix} V & \varrho & 0 \\ 0 & V & 1/\varrho \\ 0 & \gamma P & V \end{pmatrix}, \quad G(\mathbf{Y}) \triangleq \begin{pmatrix} 0 \\ CV|V| \\ -(\gamma - 1)C\varrho|V|^3 \end{pmatrix}.$$

The characteristic polynomial of the matrix $F(\mathbf{Y})$ is

$$(V - \lambda)((V - \lambda)^2 - c^2) \quad \text{with} \quad c \triangleq \sqrt{\frac{\gamma P}{\varrho}}.$$

The quantity c is the sound velocity in the concerned medium. In subsonic conditions (i.e., $|V| < c$) the system is hyperbolic with characteristic velocities (which are the roots of the characteristic polynomial):

$$\lambda_1 = V + \sqrt{\frac{\gamma P}{\varrho}}, \quad \lambda_2 = V, \quad -\lambda_3 = V - \sqrt{\frac{\gamma P}{\varrho}}.$$

The Euler equations (1.38) are a typical example of a hyperbolic system of size > 2 which cannot be transformed into an equivalent system expressed in Riemann coordinates.

1.8.1   Isentropic Equations

For the modeling and the analysis of gas pipeline networks, a common model is made of the isentropic equations which correspond to the special case where the dynamics of the energy balance are neglected. The model reduces to the first two equations of (1.39):

$$\partial_t \varrho + \partial_x(\varrho V) = 0,$$

$$\partial_t V + \partial_x \frac{V^2}{2} + \frac{1}{\varrho} \partial_x P(\varrho) + CV|V| = 0, \tag{1.40}$$

where ϱ, V, C are defined as above. The gas pressure $P(\varrho)$ is a monotonically increasing function of the gas density ($P'(\varrho) > 0$). This model is equivalent to the general quasi-linear form $\mathbf{Y}_t + F(\mathbf{Y})\mathbf{Y}_x + G(\mathbf{Y}) = \mathbf{0}$ with

$$\mathbf{Y} \triangleq \begin{pmatrix} \varrho \\ V \end{pmatrix}, \quad F(\mathbf{Y}) \triangleq \begin{pmatrix} V & \varrho \\ c^2(\varrho)/\varrho & V \end{pmatrix}, \quad G(\mathbf{Y}) \triangleq \begin{pmatrix} 0 \\ CV|V| \end{pmatrix},$$

where $c(\varrho) \triangleq \sqrt{P'(\varrho)}$ is the sound velocity. Under subsonic conditions (i.e., $V^2 < c^2(\varrho)$), the system is hyperbolic with characteristic velocities

$$\lambda_1(\mathbf{Y}) = V + c(\varrho) > 0 > -\lambda_2(\mathbf{Y}) = V - c(\varrho).$$

Riemann coordinates may then be defined as

$$R_1 = V + \phi(\varrho), \quad R_2 = V - \phi(\varrho),$$

and inverted as

$$\varrho = \phi^{-1}(R_1 - R_2)/2, \quad V = (R_1 + R_2)/2,$$

where $\phi(\varrho)$ is a primitive of $c(\varrho)/\varrho$, i.e., a function such that

$$\phi'(\varrho) \triangleq \frac{c(\varrho)}{\varrho}.$$

1.8.2   Steady State and Linearization

A steady state $\varrho^*(x)$, $V^*(x)$ is a solution of the differential equations

$$\partial_x(\varrho^* V^*) = 0,$$

$$\partial_x \frac{V^{*2}}{2} + \frac{1}{\varrho^*}\partial_x P(\varrho^*) + CV^*|V^*| = 0.$$

In order to linearize the model, we define the deviations of the states $\varrho(t,x)$ and $V(t,x)$ with respect to the steady states $\varrho^*(x)$ and $V^*(x)$:

$$p(t,x) \triangleq \varrho(t,x) - \varrho^*(x), \qquad v(t,x) \triangleq V(t,x) - V^*(x).$$

Then the linearized isentropic equations around the steady state are

$$\partial_t p + V^* \partial_x p + \varrho^* \partial_x v + (\partial_x V^*)p + (\partial_x \varrho^*)v = 0$$

$$\partial_t v + \phi'(\varrho^*)c(\varrho^*)\partial_x p + V^*\partial_x v,$$

$$+ \left(\left[2\phi'(\varrho^*)c'(\varrho^*) - (\phi'(\varrho^*))^2 \right]\partial_x\varrho^* \right)p + (\partial_x V^* + 2C|V^*|)v = 0.$$
$$(1.41)$$

The Riemann coordinates for the linearized system (1.41) are defined as follows:

$$R_1(t,x) = v(t,x) + p(t,x)\phi'(\varrho^*(x)),$$

$$R_2(t,x) = v(t,x) - p(t,x)\phi'(\varrho^*(x)),$$

with the inverse coordinate transformation

$$p(t,x) = \frac{R_1(t,x) - R_2(t,x)}{2\phi'(\varrho^*(x))},$$

$$v(t,x) = \frac{R_1(t,x) + R_2(t,x)}{2}.$$

With these definitions and notations, the linearized isentropic Euler equations are written in characteristic form:

$$\partial_t R_1(t,x) + \lambda_1(x)\partial_x R_1(t,x) + \gamma_1(x)R_1(t,x) + \delta_1(x)R_2(t,x)] = 0$$

$$\partial_t R_2(t,x) - \lambda_2(x)\partial_x R_2(t,x) + \gamma_2(x)R_1(t,x) + \delta_2(x)R_2(t,x)] = 0$$

with the characteristic velocities

$$\lambda_1(x) = V^*(x) + c(\varrho^*(x)), \qquad -\lambda_2(x) = V^*(x) - c(\varrho^*(x))$$

and the coefficients

$$\gamma_1(x) \triangleq c'(\varrho^*(x))\partial_x\varrho^*(x) + \partial_x V^*(x) + C|V^*(x)|,$$

$$\delta_1(x) \triangleq \Big(-c'(\varrho^*(x)) + \phi'(\varrho^*(x))\Big)\partial_x\varrho^*(x) + C|V^*(x)|,$$

$$\gamma_2(x) \triangleq \Big(+c'(\varrho^*(x)) - \phi'(\varrho^*(x))\Big)\partial_x\varrho^*(x) + C|V^*(x)|,$$

$$\delta_2(x) \triangleq -c'(\varrho^*(x))\partial_x\varrho^*(x) + \partial_x V^*(x) + C|V^*(x)|.$$

1.8.3 Musical Wind Instruments

Musical wind instruments (flute, trumpet, organ, etc.) are a nice example of devices that use air motion in pipes to produce a pleasant entertainment. Here, as a matter of illustration, we consider the case of a simple 'slide flute' as described by d'Andréa-Novel et al. (2010). A slide flute (Fig. 1.11) is a recorder without finger holes which is ended by a piston mechanism (D) to modify the length of the resonator chamber and consequently the sound pitch. The air motion dynamics in the tube are assumed to be described by the isentropic equations (1.40).

The player's breath is transformed into a linear airstream in the channel (B) called the 'windway', through the mouthpiece (A) of the instrument. Exiting from the windway, the airflow is directed against a sharp edge (C), called the 'labium'. This structure is a nonlinear oscillator which, amplified by the acoustic resonator, produces a stationary wave at the desired frequency determined by the pipe length L. The pitch of a note may therefore be adjusted by moving the piston at the back of the instrument.

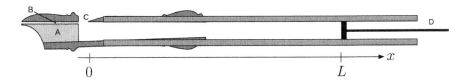

Fig. 1.11 Cross-section of a slide flute (adapted from en.wikipedia.org/Recorder)

The boundary condition at $x = 0$ is given by a second order nonlinear differential relation

$$\frac{d^2 V(t,0)}{dt^2} + c_1 \frac{dV(t,0)}{dt} + c_2 V(t,0)|V(t,0)| + c_3 P(\varrho(t,0)) + c_4 P_f = 0, \qquad (1.42)$$

where c_i $(i = 1,\ldots,4)$ are constant parameters. This equation is motivated in (d'Andréa-Novel et al. 2010, Section 5.3.3) and describes the resonance phenomenon that occurs in the labium under the external pressure P_f imposed by the player's breath. At $x = L$, a moving boundary condition is determined by the piston motion represented by the expression

$$m\frac{d^2 L}{dt^2} = F - s_p P(\varrho(t,L)), \qquad (1.43)$$

where F is the force exerted on the piston, m is the piston mass, and s_p is the pipe section. Then the isentropic equations (1.40) coupled to the boundary conditions (1.42) and (1.43) constitute a boundary control system with P_f and F as command signals.

1.9 Fluid Flow in Elastic Tubes

The laminar flow of an incompressible fluid in an elastic tube is of special interest because of its relevance to the dynamics of blood flow in arteries. Most often, one-dimensional hyperbolic balance law models are adopted under the assumption of cylindrical tubes with axisymmetric flow and negligible radial fluid velocity. The general form of such models (e.g., Barnard et al. (1966) and Li and Canic (2009)) is as follows:

$$\partial_t A + \partial_x (AV) = 0,$$

$$\partial_t (AV) + \partial_x \left(\alpha AV^2\right) + \frac{A}{\delta}\partial_x P(A) + CV = 0, \qquad (1.44)$$

where $A(t,x)$ is the cross-sectional area, $V(t,x)$ is the average fluid velocity, $P(A)$ is the pressure, C is a constant friction coefficient, δ is the fluid density, and $\alpha > 1$ is a constant depending on the shape of the axial velocity profile. The first equation is a mass balance and the second equation is a momentum balance. The pressure function $P(A)$ is a monotonically increasing function of the area ($P'(A) > 0$) such that $P(A_o) = 0$ with A_o denoting the cross-sectional area at rest.

Using A and AV as state variables, the model (1.44) is equivalent to the general quasi-linear form $\mathbf{Y}_t + F(\mathbf{Y})\mathbf{Y}_x + G(\mathbf{Y}) = \mathbf{0}$ with

$$\mathbf{Y} \triangleq \begin{pmatrix} A \\ AV \end{pmatrix}, \quad F(\mathbf{Y}) \triangleq \begin{pmatrix} 0 & 1 \\ c^2(A) - \alpha V^2 & 2\alpha V \end{pmatrix}, \quad G(\mathbf{Y}) \triangleq \begin{pmatrix} 0 \\ CV \end{pmatrix},$$

where the function $c(A) \triangleq \sqrt{AP'(A)\delta^{-1}}$ has the dimension of a velocity. The system is hyperbolic with characteristic velocities (which are the eigenvalues of $F(\mathbf{Y})$)

$$\lambda_1(\mathbf{Y}) = \alpha V + \sqrt{(\alpha^2 - \alpha)V^2 + c^2(A)}, \quad -\lambda_2(\mathbf{Y}) = \alpha V - \sqrt{(\alpha^2 - \alpha)V^2 + c^2(A)}$$

that have *opposite signs* for numerical values corresponding to human arteries. For $(\alpha - 1)$ sufficiently small, the Riemann coordinates may be approximated as follows:

$$R_{1,2} \approx V \pm \left[\int_{A_0}^{A} \frac{c(a)}{a} da + (\alpha - 1)V\left(\frac{V}{2} - c(A)\right) \right].$$

1.10 Aw-Rascle Equations for Road Traffic

In the fluid paradigm for road traffic modeling, the traffic is described in terms of two basic macroscopic state variables: the density $\varrho(t, x)$ and the speed $V(t, x)$ of the vehicles at position x along the road at time t. The following dynamical model for road traffic was proposed by Aw and Rascle (2000):

$$\partial_t \varrho + \partial_x(\varrho V) = 0,$$

$$(\partial_t + V\partial_x)V + (\partial_t + V\partial_x)P(\varrho) + \sigma(V - V_0(\varrho)) = 0. \tag{1.45}$$

In this model the first equation is a continuity equation representing the conservation of the number of vehicles on the road. The second equation is a phenomenological model describing the speed variations induced by the driver's behavior. The function $V_0(\varrho)$ represents the monotonically decreasing relation between the average speed of the vehicles and the density: the larger the density, the smaller the average speed. A typical experimental example of this function is shown in Fig. 1.12. The parameter σ is a relaxation constant. The function $P(\varrho)$ is a monotonically increasing function of the density $(P'(\varrho) > 0)$, called *traffic pressure*, which is selected such that the term $(\partial_t + Vp_x)V + (\partial_t + V\partial_x)P(\varrho)$ represents the dynamics of the transient speed variations around the average $V_0(\varrho)$ when the density is changing. The use of the Lagrangian derivative $\partial_t + V\partial_x$ allows to account for the density variations that are 'really' perceived by the drivers in front of them.

Now, multiplying the first equation of (1.45) by $V + P(\varrho)$ and the second equation by ϱ, and adding the two, we obtain a system of two nonlinear balance laws of the form

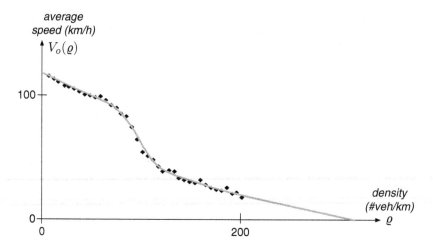

Fig. 1.12 Relation between average speed and density on a three lane highway. The experimental data have been recorded on the E411 highway (Belgium) at ten measurement stations between Namur and Brussels from October 1 to October 31, 2003. (Data provided by the Service Public de Wallonie)

$$\partial_t \varrho + \partial_x(\varrho V) = 0,$$

$$\partial_t\big(\varrho V + \varrho P(\varrho)\big) + \partial_x\big(\varrho V^2 + \rho VP(\varrho)\big) + \sigma\varrho(V - V_0(\varrho)) = 0. \tag{1.46}$$

The model can also be written in the general quasi-linear form $\mathbf{Y}_t + F(\mathbf{Y})\mathbf{Y}_x + G(\mathbf{Y}) = \mathbf{0}$ with

$$\mathbf{Y} \triangleq \begin{pmatrix} \varrho \\ V \end{pmatrix}, \quad F(\mathbf{Y}) \triangleq \begin{pmatrix} V & \varrho \\ 0 & V - \varrho P'(\varrho) \end{pmatrix}, \quad G(\mathbf{Y}) \triangleq \begin{pmatrix} 0 \\ \sigma(V - V_0(\varrho)) \end{pmatrix}.$$

Therefore, for a positive density $\varrho > 0$, the system is hyperbolic with characteristic velocities

$$\lambda_1(\mathbf{Y}) = V > \lambda_2(\mathbf{Y}) = V - \varrho P'(\varrho),$$

which are the eigenvalues of $F(\mathbf{Y})$. Moreover Riemann coordinates may be defined as

$$R_1 = V + P(\varrho), \quad R_2 = V,$$

with the inverse coordinate transformation:

$$V = R_2, \quad \varrho = P^{-1}(R_1 - R_2).$$

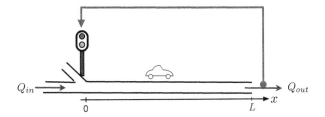

Fig. 1.13 Ramp metering on a stretch of a motorway

1.10.1   Ramp Metering

Ramp metering is a strategy that uses traffic lights to regulate the flow of traffic entering freeways according to measured traffic conditions as illustrated in Fig. 1.13. For the stretch of motorway represented in this figure, the boundary conditions are

$$\varrho(t,0)V(t,0) = Q_{in}(t) + U(t), \quad \varrho(t,L)V(t,L) = Q_{out}(t),$$

where $U(t)$ is the inflow rate controlled by the traffic lights. The Aw-Rascle equations (1.46) coupled to these boundary conditions form a boundary control system with $U(t)$ as the command signal. In a feedback implementation of the ramp metering strategy, $U(t)$ may be a function of the measured disturbances $Q_{int}(t)$ or $Q_{out}(t)$ that are imposed by the traffic conditions.

1.11   Kac-Goldstein Equations for Chemotaxis

Chemotaxis refers to the motion of certain living microorganisms (bacteria, slime molds, leukocytes, etc.) in response to the concentrations of chemicals. A simple model for one-dimensional chemotaxis, known as the Kac-Goldstein model, has been proposed by Goldstein (1951) in order to explain the spatial pattern formations in chemosensitive populations. Revisited by Kac (1956), this model, in its simplest form, is a semi-linear hyperbolic system of two balance laws of the form:

$$\partial_t \varrho^+ + \gamma \partial_x \varrho^+ + \mu(\varrho^+, \varrho^-)(\varrho^- - \varrho^+) = 0,$$
$$\partial_t \varrho^- - \gamma \partial_x \varrho^- + \mu(\varrho^+, \varrho^-)(\varrho^+ - \varrho^-) = 0,$$

$$(1.47)$$

where ϱ^+ denotes the density of right-moving cells and ϱ^- the density of left-moving cells. The function $\mu(\varrho^+, \varrho^-)$ is called the 'turning function'. The constant parameter γ is the velocity of the cell motion. With the change of coordinates $\varrho \triangleq \varrho^+ + \varrho^-$, $q \triangleq \gamma(\varrho^+ - \varrho^-)$, we have the following alternative equivalent model:

$$\partial_t \varrho + \partial_x q = 0, \tag{1.48a}$$

$$\partial_t q + \gamma^2 \partial_x \varrho - 2\mu\left(\frac{\varrho}{2} + \frac{q}{2\gamma}, \frac{\varrho}{2} - \frac{q}{2\gamma}\right)q = 0, \tag{1.48b}$$

where ϱ is the total density and q is a flux proportional to the difference of densities of right and left moving cells. Remark that we have $q = \varrho V$ where

$$V \triangleq \gamma \frac{\varrho^+ - \varrho^-}{\varrho^+ + \varrho^-}$$

can be interpreted as the average group velocity of the moving cells.

Various possible turning functions are reviewed in Lutscher and Stevens (2002). A typical example is

$$\mu(\varrho^+, \varrho^-) = \alpha \varrho^+ \varrho^- - \mu_o,$$

where α and μ_o are positive constants. It is an evidence that the system (1.47) is directly written in characteristic form (1.3) with the coordinates $R_1 \triangleq \varrho^+$ and $R_2 \triangleq \varrho^-$ as Riemann coordinates. The system is hyperbolic with characteristic velocities $\lambda_1 = \gamma$ and $-\lambda_2 = -\gamma$.

A special case of interest (see, e.g., Lutscher (2002)) is when the cells are confined in the domain $[0, L]$. This situation may be represented by 'no-flow boundary conditions' of the form:

$$q(t, 0) = \gamma\left(\varrho^+(t, 0) - \varrho^-(t, 0)\right) = 0,$$
$$q(t, L) = \gamma\left(\varrho^+(t, L) - \varrho^-(t, L)\right) = 0. \tag{1.49}$$

These boundary conditions can be written in the nominal form

$$\begin{pmatrix} \varrho^+(t, 0) \\ \varrho^-(t, L) \end{pmatrix} = \begin{pmatrix} 0 & 1 \\ 1 & 0 \end{pmatrix} \begin{pmatrix} \varrho^+(t, L) \\ \varrho^-(t, 0) \end{pmatrix}, \tag{1.50}$$

which is required for the well-posedness of the associated Cauchy problem (see page 7). We remark that, under these boundary conditions, the total amount of cells in the domain $[0, L]$ is constant since, from (1.48a) and (1.49), we have

$$\partial_t \int_0^L \varrho(t, x)dx = q(t, 0) - q(t, L) = 0. \tag{1.51}$$

Let us now consider the chemotaxis system (1.47) and (1.50) under an initial condition

$$\varrho^+(0, x) \triangleq \varrho_0^+(x), \qquad \varrho^-(0, x) \triangleq \varrho_0^-(x). \tag{1.52}$$

Then, using (1.51), it can be verified that the Cauchy problem (1.47), (1.50), and (1.52) has a single equilibrium state

$$\varrho^{+*} = \varrho^{-*} \triangleq \frac{1}{2} \int_0^L \left(\varrho_o^+(x) + \varrho_o^-(x) \right) dx.$$

In this example, we observe that the steady state depends on the initial condition. This is in contrast with most other examples of this book where the values of the steady states are generally independent of the initial conditions.

1.12 Age-Dependent SIR Epidemiologic Equations

In the field of epidemiology, mathematical models are currently used to explain epidemic phenomena and to assess vaccination strategies. For infectious diseases (where individuals are infected by pathogen microorganisms like viruses or bacteria), a first fundamental model was formulated by Kermack and McKendrick (1927). In this model, the population is classified into three groups : (i) the group of individuals who are uninfected and susceptible (S) of catching the disease, (ii) the group of individuals who are infected (I) by the concerned pathogen, (iii) the group of recovered (R) individuals who have acquired a permanent immunity to the disease.

The propagation of the disease is represented by a compartmental diagram shown in Fig. 1.14.

The model is derived under three main assumptions : (i) a closed population without immigration or emigration, (ii) spatial homogeneity, (iii) disease transmission by contact between susceptible and infected individuals. In the case where the age of patients is an important factor to be taken into account, $S(t, a)$ represents the age distribution of the population of susceptible individuals at time t. This means that

$$\int_{a_1}^{a_2} S(t, a) da \tag{1.53}$$

is the number of susceptible individuals with ages between a_1 and a_2. Similar definitions are introduced for the age distributions $I(t, a)$ of infected individuals and $R(t, a)$ of recovered individuals. The dynamics of the disease propagation in the population are then described by the following set of hyperbolic partial integro-differential equations, e.g., (Hethcote 2000, Section 5), (Thieme 2003, Chapter 22), (Perthame 2007, Chapter 1):

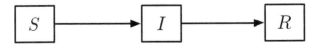

Fig. 1.14 Compartmental diagram of disease propagation

$$\partial_t S(t,a) + \partial_a S(t,a) + \mu(a)S(t,a) + \beta(a)S(t,a) \int_0^L I(t,b)db = 0,$$

$$\partial_t I(t,a) + \partial_a I(t,a) + \gamma(a)I(t,a) + \mu(a)I(t,a) - \beta(a)S(t,a) \int_0^L I(t,b)db = 0,$$

$$\partial_t R(t,a) + \partial_a R(t,a) - \gamma(a)I(t,a) + \mu(a)I(t,a) = 0,$$

(1.54)

under the boundary conditions

$$S(t,0) = B(t), \qquad I(t,0) = 0, \qquad R(t,0) = 0.$$

In these equations, $\mu(a) > 0$ denotes the natural age-dependent per capita death rate in the population. In the first and second equations, the term

$$\beta(a)S(t,a) \int_0^L I(t,b)db$$

represents the disease transmission rate by contact between susceptible and infected individuals which is assumed to be proportional to the sizes of both groups with $\beta(a) > 0$ the age-dependent transmission coefficient between all infected individuals and susceptibles having age a. In the second and third equations, the parameter $\gamma(a) > 0$ is the age-dependent rate at which infected individuals recover from the disease. Obviously, L denotes here the maximal life duration in the considered population. In the boundary conditions, $B(t) > 0$ is the inflow of newborn individuals in the susceptible part of the population at time t.

The system (1.54) is a semi-linear system of balance laws in Riemann coordinates $\mathbf{R}_t + \mathbf{R}_x + C(\mathbf{R}) = \mathbf{0}$ with $\mathbf{R}^\mathsf{T} \triangleq (S,I,R)^\mathsf{T}$. In contrast with all other models presented in this chapter, $C(\mathbf{R})$ is here a so-called 'nonlocal' source term because it depends on the spatial integral of the state variable $I(t,x)$ over the interval $[0, L]$. Moreover, it can be also observed that, in accordance with the physical reality, the system (1.54) is *positive*, which means that, if the initial state is nonnegative, i.e., $S(0,a) \geqslant 0$, $I(0,a) \geqslant 0$, $R(0,a) \geqslant 0$, then the solution is nonnegative, i.e., $S(t,a) \geqslant 0$, $I(t,a) \geqslant 0$, $R(t,a) \geqslant 0$ for all $t \in [0, +\infty)$.

1.12.1 Steady State

For a constant birth rate $B(t) = B^*$ for all t, a steady state $S^*(a), I^*(a), R^*(a)$ is a solution of the system of integro-differential equations

$$\partial_a S^*(a) + \mu(a)S^*(a) + \beta(a)S^*(a) \int_0^L I^*(b)db = 0,$$

(1.55a)

$$\partial_a I^*(a) + \gamma(a)I^*(a) + \mu(a)I^*(a) - \beta(a)S^*(a) \int_0^L I^*(b)db = 0, \qquad (1.55b)$$

$$\partial_a R^*(a) - \gamma(a)I^*(a) + \mu(a)R^*(a) = 0, \qquad (1.55c)$$

with the initial conditions $S^*(0) = B^*, I^*(0) = 0, R^*(0) = 0$.

There is one trivial *disease free* steady state:

$$S^*(a) = B \exp\left(-\int_0^a \mu(b)db\right), \qquad I^*(a) = 0, \qquad R^*(a) = 0.$$

In order to determine a nontrivial *endemic* steady state, let us integrate equations (1.55a) and (1.55b):

$$S^*(a) = B \exp\left(-\int_0^a (\mu(b) + \beta(b)\psi^*)db\right)$$

$$I^*(a) = \psi(a)\psi^* \int_0^a \frac{\beta(b)S^*(b)}{\psi(b)} db,$$

with

$$\psi(a) \triangleq \exp\left(-\int_0^a (\mu(c) + \gamma(c))dc\right), \qquad \psi^* \triangleq \int_0^L I^*(a)da.$$

Substituting for the expression of I^* into ψ^*, we get

$$\psi^* = \int_0^L \psi(a)\psi^* \left(\int_0^a \frac{\beta(b)S^*(b)}{\psi(b)} db\right) da.$$

Clearly, an endemic steady state may exists only if $\psi^* > 0$ which implies that

$$1 = \int_0^L \psi(a) \left(\int_0^a \frac{\beta(b)S^*(b)}{\psi(b)} db\right) da. \qquad (1.56)$$

Let us now substitute for the expression of S^* into (1.56):

$$1 = \int_0^L \psi(a) \left(\int_0^a \frac{\beta(b)}{\psi(b)} B \exp\left(-\int_0^b (\mu(c) + \beta(c)\psi^*)dc\right) db\right) da \triangleq \mathcal{R}(\psi^*). \qquad (1.57)$$

Clearly $\mathcal{R}(\psi^*)$ is a positive exponentially decreasing function of ψ^* such that $\mathcal{R}(+\infty) = 0$. It follows that the equality (1.57) may be satisfied if and only if $\mathcal{R}(0) \geq 1$ and therefore that there exists an endemic steady state if and only if $\mathcal{R}(0) \geq 1$. Furthermore, it can be shown that this endemic equilibrium, when it exists, is uniquely determined (see, e.g., Thieme (2003, Theorem 22.1)).

1.13 Chromatography

In chromatography, a fluid carrying dissolved chemical species flows through a porous solid fixed bed. The carried species are partially adsorbed on the bed. The fluid flow is supposed to have a constant velocity V. Let us first consider the case of a single carried species with concentration C_f in the fluid and concentration C_s deposited on the solid. Then we have the conservation equation

$$\partial_t(C_f + C_s) + V\partial_x C_f = 0. \tag{1.58}$$

A standard model for the net exchange rate between the fluid and the solid is

$$k_1 C_f\left(1 - \frac{C_s}{C_{\max}}\right) - k_2 C_s$$

where k_1, k_2 are positive kinetic constants. The first term represents the deposition from the fluid to the solid at a rate proportional to the amount in the fluid, but limited by the amount already on the solid up to capacity $C_{\max}$. The second term is the reverse transfer from the solid to the fluid. Here we assume quasi steady state conditions such that the net exchange rate vanishes and, consequently, such that C_s is a function of C_f:

$$C_s = L(C_f) = \frac{hC_f}{1 + bC_f}$$

where $h \triangleq k_1/k_2$ is the so-called Henry coefficient and $b \triangleq k_1/(k_2 C_{\max})$ is the adsorption equilibrium coefficient. This function was proposed by Langmuir (1916) and is known under the name of *Langmuir isotherm*. Using this expression in (1.58), we obtain the following scalar hyperbolic conservation law:

$$\partial_t C_f + \frac{V}{1 + L'(C_f)}\partial_x C_f = 0, \quad \text{with } L'(C_f) = \frac{h}{\left(1 + bC_f\right)^2}. \tag{1.59}$$

The characteristic velocity λ such that

$$0 < \lambda = \frac{V}{1 + L'(C_f)} < V$$

is the propagation speed of the carried species. It is slower for species with larger affinity for the solid material which is measured by $L'(C_f)$.

Let us now suppose that a pulse of a mixture of species with different affinities is injected in the carrying fluid at the entrance of the process as illustrated in Fig. 1.15. Clearly, the various substances will travel at different propagation speeds and will ultimately be separated in different bands. Obviously, the scalar

Fig. 1.15 Principle of chromatography

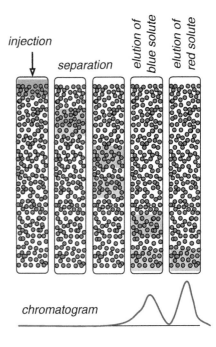

conservation law (1.59) for a single component applies only after separation. Before the separation, the dynamics of the mixture are described by a system of coupled conservation laws that generalizes the scalar case in the following way:

$$\partial_t \left(C_i + \frac{h_i C_i}{1 + \sum_{j=1}^{n} b_j C_j} \right) + V \partial_x C_i = 0 \quad i = 1, \ldots, n, \tag{1.60}$$

where C_i $(i = 1, \ldots, n)$ denote the densities of the n carried species.

1.13.1 SMB Chromatography

Simulated moving bed (SMB) chromatography is a technology where several interconnected chromatographic columns are switched periodically against the fluid flow. This allows for a continuous separation with a better performance than the discontinuous single-column chromatography. A standard SMB chromatography process (Suvarov et al. (2012)) is represented in Fig. 1.16. The input flows (feed mixture and solvent) and output flows (extract and raffinate) divide the system in four zones each containing one chromatographic column. The four (mobile) columns are labeled $1, 2, 3, 4$. The four operating zones are labeled I, II, III, IV. Pumps connected at each port determine the liquid phase flow rates in all the zones. The feed mixture composed of species A and B is injected between zone I and IV.

Fig. 1.16 SMB chromatography (adapted from www.mpi-magdeburg. mpg.de/research/projects/ 1119/1127/DSMBC)

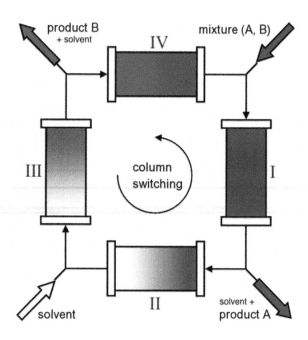

The adsorbent is chosen in such a way that the two components are adsorbed at different rates, allowing them to travel with different velocities. The less adsorbed component (A) is collected at the raffinate port and the more adsorbed one (B) at the extract port. The separation of the two components is performed in zone I and IV, whereas zones II and III are dedicated to adsorbent regeneration and solvent recycling. A liquid-solid counter-current movement is obtained by a periodic circular switching of the columns containing in the opposite direction to the liquid phase flow as shown in Fig. 1.16.

We introduce the following notations:

- The switching time period is T and the column length is L;
- $C_i^\ell(t, x)$, $0 \leqslant x \leqslant L$, $t \geqslant 0$, is the density of species $\ell \in \{A, B\}$ in the column $i \in \{1, 2, 3, 4\}$;
- V_I is the fluid velocity in the columns located in zones I and III; V_{II} is the fluid velocity in the columns located in zones II and IV with $V_I > V_{II} > 0$;
- V_F is the fluid velocity and C_F^ℓ is the density of species $\ell \in \{A, B\}$ in the feeding stream;
- h_A and h_B are the Henry coefficients, b_A and b_B are the adsorption equilibrium coefficients.

We consider the operation of the SMB process in the standard conditions where V_I, V_{II}, V_F and C_F^ℓ are all constant. We first state the dynamical model during the first time period, assuming that column 1 is in zone I, column 2 in zone II, column 3 in zone III, and column 4 in zone IV. The process dynamics are represented by the following set of conservation laws of the form (1.60):

For $0 \leq t < T$,

$$\partial_t \left[C_1^A + \frac{h_A C_1^A}{1 + b_A C_1^A + b_B C_1^B} \right] + V_{\mathrm{I}} \partial_x C_1^A = 0,$$

$$\partial_t \left[C_2^A + \frac{h_A C_2^A}{1 + b_A C_2^A + b_B C_2^B} \right] + V_{\mathrm{II}} \partial_x C_2^A = 0,$$

$$\partial_t \left[C_3^A + \frac{h_A C_3^A}{1 + b_A C_3^A + b_B C_3^B} \right] + V_{\mathrm{I}} \partial_x C_3^A = 0,$$

$$\partial_t \left[C_4^A + \frac{h_A C_4^A}{1 + b_A C_4^A + b_B C_4^B} \right] + V_{\mathrm{II}} \partial_x C_4^A = 0,$$

$$\partial_t \left[C_1^B + \frac{h_B C_1^B}{1 + b_A C_1^A + b_B C_1^B} \right] + V_{\mathrm{I}} \partial_x C_1^B = 0,$$

$$\partial_t \left[C_2^B + \frac{h_B C_2^B}{1 + b_A C_2^A + b_B C_2^B} \right] + V_{\mathrm{II}} \partial_x C_2^B = 0,$$

$$\partial_t \left[C_3^B + \frac{h_B C_3^B}{1 + b_A C_3^A + b_B C_3^B} \right] + V_{\mathrm{I}} \partial_x C_3^B = 0,$$

$$\partial_t \left[C_4^B + \frac{h_B C_4^B}{1 + b_A C_4^A + b_B C_4^B} \right] + V_{\mathrm{II}} \partial_x C_4^B = 0.$$

with the boundary conditions:

$$V_{\mathrm{I}} C_1^A(t, 0) = V_{\mathrm{II}} C_4^A(t, L) + V_F C_F^A,$$

$$C_2^A(t, 0) = C_1^A(t, L),$$

$$V_{\mathrm{I}} C_3^A(t, 0) = V_{\mathrm{II}} C_2^A(t, L),$$

$$C_4^A(t, 0) = C_3^A(t, L).$$

$$V_{\mathrm{I}} C_1^B(t, 0) = V_{\mathrm{II}} C_4^B(t, L) + V_F C_F^B,$$

$$C_2^B(t, 0) = C_1^B(t, L),$$

$$V_{\mathrm{I}} C_3^B(t, 0) = V_{\mathrm{II}} C_2^B(t, L),$$

$$C_4^B(t, 0) = C_3^B(t, L).$$

We introduce the following vector and matrix notations:

$$\mathbf{C}^\ell(t,x) \triangleq (C_1^\ell(t,x), C_2^\ell(t,x), C_3^\ell(t,x), C_4^\ell(t,x))^\mathsf{T}, \quad \ell \in \{A, B\},$$

$$U^\ell \triangleq (V_F C_F^\ell, 0, 0, 0)^\mathsf{T},$$

$$\mathbf{C}(t,x) = \begin{pmatrix} \mathbf{C}^A(t,x) \\ \mathbf{C}^B(t,x) \end{pmatrix}, \quad D(\mathbf{C}) \triangleq \mathrm{diag}\{(1 + b_A C_i^A + b_B C_i^B)^{-1}; i = 1, 2, 3, 4\},$$

$$\Upsilon \triangleq \mathrm{diag}\{V_\mathrm{I}, V_\mathrm{II}, V_\mathrm{I}, V_\mathrm{II}\}, \quad K \triangleq \begin{pmatrix} 0 & 0 & 0 & V_\mathrm{II}/V_\mathrm{I} \\ 1 & 0 & 0 & 0 \\ 0 & V_\mathrm{II}/V_\mathrm{I} & 0 & 0 \\ 0 & 0 & 1 & 0 \end{pmatrix}.$$

With these notations, the model equations are written in compact form as follows:

$$0 \leqslant t < T, \quad \ell \in \{A, B\},$$

$$\partial_t \Big((\mathbf{I} + h_\ell D(\mathbf{C}))\mathbf{C}^\ell \Big) + \Upsilon \partial_x \mathbf{C}^\ell = \mathbf{0},$$

$$\mathbf{C}^\ell(t,0) = K\mathbf{C}^\ell(t,L) + U^\ell.$$

We now consider the second time period when the columns have been shifted by one position such that column 1 is now located in zone IV, column 2 in zone I, etc. To take the shifting process into account in a systematic way, we introduce the following permutation matrix:

$$P = \begin{pmatrix} 0 & 0 & 0 & 1 \\ 1 & 0 & 0 & 0 \\ 0 & 1 & 0 & 0 \\ 0 & 0 & 1 & 0 \end{pmatrix}.$$

Then, noticing also that $P^{-1} = P^\mathsf{T}$, it can be checked that the model equations during the second period become

$$T \leqslant t < 2T, \quad \ell \in \{A, B\},$$

$$\partial_t \Big((I + h_\ell D(\mathbf{C}))\mathbf{C}^\ell \Big) + P\Upsilon P^\mathsf{T} \partial_x \mathbf{C}^\ell = \mathbf{0},$$

$$\mathbf{C}^\ell(t,0) = PKP^\mathsf{T}\mathbf{C}^\ell(t,L) + PU^\ell.$$

It is then clear that, by iteration, we have the following general form for the hyperbolic system of conservation laws describing the periodic SMB chromatography process:

$$mT \leqslant t < (m+1)T, \quad m = 0, 1, 2, 3, 4, 5, \ldots, \infty, \quad \ell \in \{A, B\},$$

$$\partial_t \left((\mathbf{I} + h_\ell D(\mathbf{C})) \mathbf{C}^\ell \right) + (P^m) \Upsilon (P^m)^{\mathsf{T}} \partial_x \mathbf{C}^\ell = \mathbf{0}, \tag{1.61}$$

$$\mathbf{C}^\ell(t, 0) = (P^m) \mathbf{K} (P^m)^{\mathsf{T}} \mathbf{C}^\ell(t, L) + (P^m) U^\ell.$$

This system (1.61) has a periodic solution $\mathbf{C}^* \triangleq (C^{*A}, C^{*B})$ such that $(\mathbf{C}^*(t, x) = \mathbf{C}^*(t + 4T, x))$, $x \in [0, L]$, $t \geqslant 0$.

1.14   Scalar Conservation Laws

In several examples of hyperbolic systems presented above, the first equation represents a mass balance with the form

$$\partial_t \varrho + \partial_x(\varrho V) = 0. \tag{1.62}$$

The variable ϱ is a density and the variable V is a velocity such that $q = \varrho V$ is a flux density. For instance, it is the case for the Saint Venant equations (1.23) (with $\varrho = H$), for the Euler equations (1.38), for the elastic tube model (1.44) (with $\varrho = A$), for the Aw-Rascle traffic model (1.45), or for the chemotaxis model (1.48). Equation (1.62) is the basis for many physical models of interest in engineering where momentum and energy balances are assumed to be quasi balanced. The assumption that the mass balance is the dominant effect then relies on the quasi steady state simplification of considering the velocity V as a static function of the density ϱ:

$$V = V_0(\varrho) \quad \Longrightarrow \quad \partial_t \varrho + \partial_x q(\varrho) = \partial_t \varrho + \partial_x(\varrho V_0(\varrho)) = 0. \tag{1.63}$$

Obviously, this is just a scalar conservation law since it involves only one dependent variable ϱ. This scalar conservation law (1.63) can also be written in a quasi-linear form:

$$\partial_t \varrho + \lambda(\varrho) \partial_x \varrho = 0, \quad \lambda(\varrho) = q'(\varrho) = V_0(\varrho) + \varrho V_0'(\varrho),$$

where $\lambda(\varrho)$ is the characteristic velocity of the system.

As a first example, in road traffic modeling, we all have experienced that the larger is the density and the smaller is the speed. The simplest model is to assume that the drivers instantaneously adapt their speed to the local traffic density such that the relation $V_0(\varrho)$ between the vehicle speed and the traffic density is linearly decreasing (see, e.g., Garavello and Piccoli (2006, Chapter 3)):

$$V_0(\varrho) = V_{max}\left(1 - \frac{\varrho}{\varrho_{max}}\right). \tag{1.64}$$

Here $V_0(0) = V_{max}$ is the maximal velocity of the vehicles when the road is (almost) empty while $V_0(\varrho_{max}) = 0$ means that the velocity drops to zero when the density is maximal and the traffic is totally congested. This model was first proposed by Greenshields (1935) and gives rise to the well-known scalar LWR model (Lighthill and Whitham (1955)) for traffic flow:

$$\partial_t\varrho + \partial_x\left(V_{max}\left(\varrho - \frac{\varrho^2}{\varrho_{max}}\right)\right) = 0.$$

A direct improvement of this model is to adopt the experimental nonlinear function of Fig. 1.12 for $V_0(\varrho)$ instead of the affine function (1.64). Non-monotonic variants of the function $V_0(\varrho)$ are also suggested for pedestrian flow modeling (e.g., Chalons et al. (2013)).

Another classical example is the scalar model for open channels. Let us consider again the Saint-Venant equations (1.23). Let us assume that the momentum dynamics are negligible so that the momentum equation reduces to the static source term which links the water velocity to the slope S_b and the viscous friction C:

$$g\left[C\frac{V^2}{H} - S_b\right] = 0.$$

From this equation, we recover the Torricelli's formula between the velocity V and the water depth H:

$$V = V_0(H) = k\sqrt{gH} \quad \text{with} \quad k = \sqrt{\frac{S_b}{C}}. \tag{1.65}$$

By substituting this expression into (1.63), we get

$$\partial_t H + \partial_x(kH\sqrt{gH}) = 0.$$

which is the basic scalar equation for open channels.

A further example of the model (1.63) is the Buckley and Leverett (1942) equation:

$$V = V_0(\varrho) = \left(\frac{\varrho}{\varrho^2 + (1 - \varrho)^2}\right) \implies \partial_t\varrho + \partial_x\left(\frac{\varrho^2}{\varrho^2 + (1 - \varrho)^2}\right) = 0 \tag{1.66}$$

for modeling a two phase (e.g., oil and water) fluid flow in a porous medium. Here $0 \leq \varrho \leq 1$ represents the saturation of water: $\varrho = 0$ is a flow of pure oil while $\varrho = 1$ is a flow of pure water. This kind of model has applications in oil reservoir studies.

Let us now observe that, in all these examples, the function V_0 is monotone: decreasing for road models (1.64) and increasing for open channels (1.65) and two-phase fluid flow (1.66). Hence the function $V_0(\varrho)$ may be supposed to be invertible as

$$\varrho = \varrho_0(V)$$

in such a way that the system may also be written in the inverse form

$$\partial_t V + c(V)\partial_x V = 0, \tag{1.67}$$

with

$$c(V) = V_0'(\varrho_0(V))\lambda(\varrho_0(V))\varrho_0'(V).$$

We see that, depending on the application, ϱ or V can be equally taken as the state variable. In both cases the system is a scalar conservation law with either $\lambda(\varrho)$ or $c(V)$ as the characteristic velocity.

The simplest physical model which is usually given in the form (1.67) is the well-known equation of Burgers (1939) with $c(V) = V$ such that

$$\partial_t V + V\partial_x V = 0. \tag{1.68}$$

This equation may be considered as a simplified model of isentropic gas motion because it can be derived from the momentum balance Euler equation (1.38b) by assuming that the friction is negligible ($C \approx 0$) and that the pressure is almost constant with respect to x ($\partial_x P(\varrho) \approx 0$).

1.15   Physical Networks of Hyperbolic Systems

The operation of many physical networks having an engineering relevance can be represented by hyperbolic systems of balance laws in one space dimension. Typical examples are hydraulic networks (for water supply, irrigation, or navigation), road traffic networks, electrical line networks, gas transportation networks, networks of heat exchangers, communication networks, blood flow networks, etc.

Such physical networks can generally be schematized by using a graph representation. The edges of the network represent the physical links (for instance the pipes, the canals, the roads, the electrical lines, etc.) that are governed by hyperbolic systems of balance laws. Without loss of generality and for simplicity, it can always be assumed that, by an appropriate scaling, all the links have exactly the same length L. Typically, the links carry some kind of flow and the network has only a few nodes where flows enter or leave the network. The other nodes of the network represent the physical junctions between the links. The mechanisms that occur at the junctions are described by 'junction models' under the form of algebraic or differential relations that determine the boundary conditions of the PDEs.

Fig. 1.17 An electrical line junction

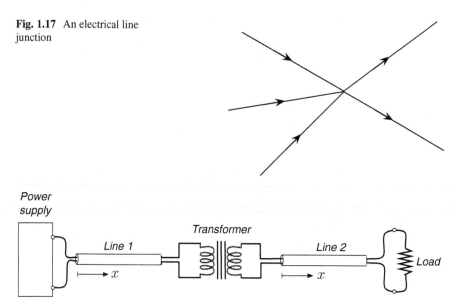

Fig. 1.18 Two electrical lines interconnected with a transformer

1.15.1 Networks of Electrical Lines

Let us consider an electrical line network in which several lines meet at a given junction as shown in Fig. 1.17. In this figure, the directions of the lines represented by arrows are arbitrary. For convenience, each arrow denotes both the direction of the increasing space coordinate x and the direction of a positive current $I(t, x)$ on the concerned line.

According to the Kirchhoff's law, we know that the following conditions hold at the junction:

$$\sum_j I_j(t, L) = \sum_k I_k(t, 0) \quad \text{and} \quad V_j(t, L) = V_k(t, 0) \quad \forall (j, k)$$

where j and k index the incoming and outgoing lines respectively. In words, these conditions mean that the voltage is identical on all lines at the junction and that the currents balance. These expressions are the boundary conditions for the hyperbolic system of balance laws that describes the network. We observe that the number of independent boundary conditions at each junction is equal to the number of incident lines at the junction. Then it is clear that, if the whole network has the structure of a simply connected graph, the hyperbolic system is well-posed.

Now, it is evident that we may also have more complex devices at the junctions, such as transformers or amplifiers. In such cases, we have to use the appropriate equations of lumped electrical circuits as boundary conditions. For example, let us consider two electrical lines connected by a transformer as shown in Fig. 1.18.

This system is described by two electrical line equations:

$$\partial_t(L_1 I_1) + \partial_x V_1 + R_1 I_1 = 0, \qquad \partial_t(L_2 I_2) + \partial_x V_2 + R_2 I_2 = 0,$$

$$\partial_t(C_1 V_1) + \partial_x I_1 + G_1 V_1 = 0, \qquad \partial_t(C_2 V_2) + \partial_x I_2 + G_2 V_2 = 0.$$

coupled by the boundary conditions

$$V_1(t,0) + R_g I_1(t,0) = U(t), \quad V_1(t,L) = L_{1s}(dI_1(t,L)/dt) + M(dI_2(t,0)/dt),$$

$$V_2(t,L) - R_\ell I_2(t,L) = 0, \qquad V_2(t,0) = L_{2s}(dI_2(t,0)/dt) + M(dI_1(t,L)/dt),$$

where L_{1s} and L_{2s} denote the self-inductance of the transformer coils and M is the mutual inductance (see Section 1.2 for the meaning of the other notations). Remark that this is a typical example of a situation where some of the boundary conditions have the form of ordinary differential equations.

1.15.2 Chains of Density-Velocity Systems

We consider a chain of two-by-two hyperbolic density-velocity systems of the form

$$\partial_t \varrho_j + \partial_x(\varrho_j V_j) = 0,$$
$$\partial_t V_j + \partial_x f(\varrho_j, V_j) + g(\varrho_j, V_j) = 0, \qquad j = 1,\dots,n. \qquad (1.69)$$

The system is subject to $n - 1$ physical boundary conditions

$$\varrho_j(t,L)V_j(t,L) = \varrho_{j+1}(t,0)V_{j+1}(t,0) \quad j = 1,\dots,n-1 \qquad (1.70)$$

which are the expression of the natural constraint of flow conservation between two successive elements of the chain. The form of the remaining $n + 1$ boundary conditions then depends on the particular application. We illustrate the issue with two typical examples: gas pipe lines and navigable rivers.

1.15.2.1 Gas Pipe Lines

In gas pipes, the pressure decreases along the pipes due to friction. Compressors are therefore used from place to place in order to amplify the pressure. A gas pipe line is thus a chain of n pipes separated by compressor stations (Gugat and Herty (2011), Dick et al. (2010)) as illustrated in Fig. 1.19. The gas flow in the pipes is described by isentropic Euler equations:

$$\partial_t \varrho_j + \partial_x(\varrho_j V_j) = 0,$$
$$\partial_t V_j + \partial_x \frac{V_j^2}{2} + \frac{1}{\varrho_j}\partial_x P(\varrho_j) + C V_j |V_j| = 0, \qquad j = 1,\dots,n. \qquad (1.71)$$

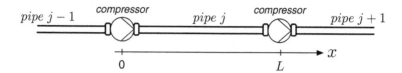

Fig. 1.19 The structure of a gas pipe line

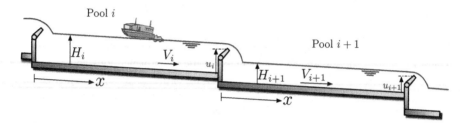

Fig. 1.20 Lateral view of successive pools of a navigable river with overflow gates

In addition to the flow conservation conditions (1.70), the compressors impose the following $n - 1$ boundary conditions:

$$k\varrho_{j+1}(t,0)V_{j+1}(t,0)\left(\left(\frac{P(\varrho_{j+1}(t,0))}{P(\varrho_j(t,L))}\right)^m - 1\right) = U_j(t), \quad j = 1,\ldots,n-1,$$

where k and m are positive constant parameters. This is a common static model of the pressure amplification which is achieved with a centrifugal adiabatic compressor under a power supply $U_j(t)$ which can be considered as a control input.

Two additional boundary conditions are needed to close the system. For instance:

$$P(\varrho_1(t,0)) = P_o(t) \quad \text{and} \quad \varrho_n(t,L)V_n(t,L) = Q_L(t),$$

where $P_o(t)$ is the input pressure imposed by the producer and $Q_L(t)$ is an outflow rate constraint at the consumer side.

1.15.2.2 Navigable Rivers and Irrigation Channels

In navigable rivers or irrigation canals (see, e.g., Litrico et al. (2005), Cantoni et al. (2007), Bastin et al. (2009)) the water is transported along the channel under the power of gravity through successive pools separated by automated gates that are used to regulate the water flow, as illustrated in Figs. 1.20 and 1.21. Here, we consider a channel with n pools the dynamics of which are described by Saint-Venant equations (1.23)

$$\partial_t \begin{pmatrix} H_i \\ V_i \end{pmatrix} + \partial_x \begin{pmatrix} H_iV_i \\ \frac{1}{2}V_i^2 + gH_i \end{pmatrix} + \begin{pmatrix} 0 \\ g[C_iV_i^2H_i^{-1} - S_i] \end{pmatrix} = \mathbf{0}, \quad i = 1\ldots,n. \quad (1.72)$$

Fig. 1.21 Automated control gates in the Sambre river (Belgium). The left gate is in operation. The right gate is lifted for maintenance (©L. Moens)

In this model, for simplicity, we assume that all the pools have a rectangular section with the same width W. The system (1.72) is subject to a set of $2n$ boundary conditions that are distributed into three subsets:

1) A first subset of $n - 1$ conditions expresses the flow-rate conservation at the junction of two successive pools (the flow that exits pool i is equal to the flow that enters pool $i + 1$):

$$H_i(t,L)V_i(t,L) = H_{i+1}(t,0)V_{i+1}(t,0), \quad i = 1,\ldots,n-1. \tag{1.73}$$

2) A second subset of n boundary conditions is made up of the equations that describe the gate operations. A standard gate model is given by the algebraic relation

$$H_i(t,L)V_i(t,L) = \left(k_G\sqrt{2g}\right)\sqrt{\left[H_i(t,L) - u_i(t)\right]^3}, \quad i = 1,\ldots,n. \tag{1.74}$$

where k_G is a positive constant coefficient and $u_i(t)$ denotes the weir elevation which is a control input (see Fig. 1.20).

3) The last boundary condition imposes the value of the canal inflow rate $Q_0(t)$:

$$Q_0(t) = WH_1(t,0)V_1(t,0). \tag{1.75}$$

Depending on the application, $Q_0(t)$ may be viewed as a control input (in irrigation channels) or as a disturbance input (in navigable rivers).

A steady state (or equilibrium) is a constant state H_i^*, V_i^* $(i = 1, \ldots, n)$ that satisfies the relations

$$S_i H_i^* = C(V_i^*)^2, \qquad i = 1, \ldots, n.$$

The subcritical flow condition is

$$gH_i^* - (V_i^*)^2 > 0, \qquad i = 1, \ldots, n. \tag{1.76}$$

The characteristic velocities are

$$\lambda_i = V_i + \sqrt{gH_i} > 0, \qquad -\lambda_{n+i} = V_i - \sqrt{gH_i} < 0, \qquad i = 1, \ldots, n.$$

1.15.3 Genetic Regulatory Networks

Systems of nonlinear delay-differential equations constitute an obvious special case of quasi-linear hyperbolic systems which are represented by a set of scalar conservation laws (transport equations) coupled by nonlinear ordinary differential equations. An early pioneering reference on this topic is the fundamental book Bellman and Cooke (1963). A typical and important example of this class of systems is given by the continuous time models of genetic regulatory networks when "time delays are included to allow for the time required for transcription, translation, and transport" (Smolen et al. (2000)).

For a genetic regulatory network which involves n genes interconnected through activator or repressor proteins, the expression of the i-th gene in the network ($i = 1, \ldots, n$) is represented by the following standard delay-differential system (see, e.g., Bernot et al. (2013)):

$$\frac{dM_i(t)}{dt} = b_i + h_i(P_k(t - \tau_k)) - \delta_i M_i(t),$$

$$\frac{dP_i(t)}{dt} = \alpha_i M_i(t - \tau_{n+i}) - \beta_i P_i(t), \tag{1.77}$$

where, at time t, $M_i(t)$ is the density of mRNA molecules, P_i and P_k are the densities of proteins expressed by the i-th and k-th genes respectively. Here, for simplicity, the model is restricted to the case where the i-th gene is controlled by only one protein

Fig. 1.22 Scheme of genetic transcription and translation with activation or repression

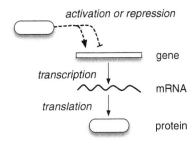

expressed by the k-th gene (k depends on i), see Fig. 1.22 for an illustration. The constants b_i and α_i denote respectively the basal transcription rate and the specific translation rate. The constants β_i and δ_i are the natural degradation rate coefficients. The constant delays τ_i and τ_{n+i} are the times needed for transcription and translation respectively. The function $h_i(P_k)$ describes how the transcription of the i-th gene is activated or repressed by the density P_k of the protein expressed by the k-th. It may therefore be either an activation Hill function of the form

$$h_i(P_k) = \frac{\upsilon_i P_k^{c_i}}{k_i^{c_i} + P_k^{c_i}},$$

or an inhibition Hill function of the form

$$h_i(P_k) = \frac{\upsilon_i k_i^{c_i}}{k_i^{c_i} + P_k^{c_i}},$$

where υ_i, k_i, and c_i are positive constant parameters with υ_i the maximal transcription rate, k_i the half-saturation coefficient and c_i the so-called Hill coefficient.

A steady state of the system is a constant solution M_i^*, P_i^*, $i = 1, \ldots, n$, of the dynamical system (1.77), i.e., a solution of the algebraic system

$$b_i + h_i(P_k^*) - \delta_i M_i^* = 0,$$

$$\alpha_i M_i^* - \beta_i P_i^* = 0.$$

It is well known that these systems may have multiple steady states or equilibria. A critical issue is to determine the stability of these equilibria. This issue will be addressed in Section 3.4.

Let us define the deviations of P_i and M_i with respect to a steady state M_i^*, P_i^*:

$$m_i(t) = M_i(t) - M_i^*, \qquad p_i(t) = P_i(t) - P_i^*.$$

With these coordinates, the model (1.77) is alternatively written under the form

$$\frac{dm_i(t)}{dt} = g_i(p_k(t - \tau_k))p_k(t - \tau_k) - \delta_i m_i(t),$$

$$\frac{dp_i(t)}{dt} = \alpha_i m_i(t - \tau_{n+i}) - \beta_i p_i(t),$$

where the function g_i is defined such that

$$g_i(p)p = h_i(P^* + p) - h_i(P^*).$$

This system is then a special case of a general hyperbolic system of linear conservation laws in Riemann coordinates

$$\mathbf{R}_t + \Lambda_{\mathbf{R}}\mathbf{R}_x = \mathbf{0}, \quad \mathbf{S}_t + \Lambda_{\mathbf{S}}\mathbf{S}_x = \mathbf{0}, \quad t \in [0, +\infty), \quad x \in [0, 1], \quad (1.78)$$

with the maps $\mathbf{R} : [0, +\infty) \times [0, 1] \to \mathbb{R}^n$ and $\mathbf{S} : [0, +\infty) \times [0, 1] \to \mathbb{R}^n$ such that

$$\mathbf{R}(t, 0) \triangleq \big(p_1(t), \ldots, p_n(t)\big)^\mathsf{T} \quad \text{and} \quad \mathbf{S}(t, 0) \triangleq \big(m_1(t), \ldots, m_n(t)\big)^\mathsf{T},$$

and the maps $\Lambda_{\mathbf{R}} : \mathbb{R}^n \to \mathcal{D}^+$ and $\Lambda_{\mathbf{S}} : \mathbb{R}^n \to \mathcal{D}^+$ such that

$$\Lambda_{\mathbf{R}} \triangleq \mathrm{diag}\left\{\frac{1}{\tau_1}, \ldots, \frac{1}{\tau_n}\right\}, \quad \Lambda_{\mathbf{S}} \triangleq \mathrm{diag}\left\{\frac{1}{\tau_{n+1}}, \ldots, \frac{1}{\tau_n}\right\}.$$

The system (1.78) is subject to nonlinear differential boundary conditions of the form

$$\frac{d\mathbf{R}(t, 0)}{dt} = A\mathbf{S}(t, 1) - B\mathbf{R}(t, 0),$$

$$\frac{d\mathbf{S}(t, 0)}{dt} = G(\mathbf{R}(t, 1))\mathbf{R}(t, 1) - D\mathbf{S}(t, 0) \quad (1.79)$$

with the notations $A \triangleq \mathrm{diag}\{\alpha_1, \ldots, \alpha_n\}$, $B \triangleq \mathrm{diag}\{\beta_1, \ldots, \beta_n\}$, $D \triangleq \mathrm{diag}\{\delta_1, \ldots, \delta_n\}$ and $G(\mathbf{R}) \in \mathcal{M}_{n,n}(\mathbb{R})$ the matrix with entry $[G(\mathbf{R})]_{ik} \triangleq g_i(R_k)$ if $k \sim i$ and zero otherwise. The notation $k \sim i$ means that the protein expressed by the k-th gene is an activator or a repressor of the i-th gene transcription.

Here, for simplicity, the model has been restricted to the case of networks where each gene can be controlled by one protein at most. The model (1.78), (1.79) is obviously also valid for situations where a gene can be controlled by several proteins simultaneously provided the definition of the matrix $G(\mathbf{R})$ is adequately extended.

1.16   References and Further Reading

In this chapter, we have presented various examples of physical systems or engineering problems that are usefully represented by one-dimensional hyperbolic systems of conservation and balance laws. Among many other examples which can be found in the literature, let us mention the following interesting additional references.

- Multi-layer Saint-Venant equations with interfacial exchanges of mass for the modeling of suspended matters transported by shallow-water streams, see, e.g., Audusse et al. (2011). This kind of model is applied for example in the design of the so-called raceway processes used for the cultivation of micro-algae where the water flows around a circular channel, driven by a paddlewheel, see, e.g., Bernard et al. (2013).
- Alternative models of conservation laws for bedload sediment transport in shallow waters are presented and compared in Castro Diaz et al. (2008).
- Isentropic gas flow models generalized to the case of flow through porous media where the velocity obeys the Darcy's law, see, e.g., Dafermos and Pan (2009).
- Relativistic isentropic Euler equations for the description of one-dimensional gas flow at speeds where the relativistic effects become significant as in high energy particle beams for instance, see, e.g., Smoller and Temple (1993), Chen and Li (2004) and LeFloch and Yamazaki (2007).
- For general gas transportation networks, the determination of steady states may be a complicated task. Gugat et al. (2015) present a method for the computation of steady states in certain pipeline networks that involve cycles and are represented by isentropic Euler equations.
- Balance law models with nonlocal source terms for the description of crystal growth or multilane road traffic with nonlocal flux density, see, e.g., Colombo et al. (2007), Lee and Liu (2015).
- Aw-Rascle traffic flow models with phase transitions, see, e.g., Goatin (2006).
- Conservation laws similar to the models of car traffic for the description of packets flow on telecommunication networks with routing algorithms at the nodes, e.g., D'Apice et al. (2006).
- Conservation law models for supply chains and other highly re-entrant manufacturing systems as encountered, for instance, in semi-conductor production. These models may have nonlocal characteristic velocities, see, e.g., Armbruster et al. (2003), Armbruster et al. (2006), Shang and Wang (2011), Coron et al. (2010), Armbruster et al. (2011), Coron and Wang (2013).
- Semi-linear hyperbolic systems with nonlinear reaction interactions of Lotka-Volterra type, e.g., Pavel (2013).
- In addition to age-dependent epidemiologic models, Perthame (2007) describes also other interesting examples of biological and ecological systems represented by hyperbolic equations, such as intra-host virus or sea phytoplankton dynamics (see also Calvez et al. (2010), Calvez et al. (2012)).
- Balance law models for extrusion processes, e.g., Dos Santos Martins (2013), Diagne et al. (2016a), Diagne et al. (2016b).
- Balance law models for oil well drilling processes, e.g., Aamo (2013), Hasan and Imsland (2014).
- Balance law models for incompressible two-phase flow with an interfacial pressure and a drag force as the coupling terms between the two phases, e.g., Djordjevic et al. (2010) and Djordjevic et al. (2011).

- Quasi-linear hyperbolic models for the onset of bubbling in fluidized bed chemical reactors, e.g., Hsiao and Marcati (1988).

Finally, readers interested in numerical simulations of one-dimensional hyperbolic equations may refer to the classical textbooks of LeVeque (1992) and Godlewski and Raviart (1996).

Chapter 2
Systems of Two Linear Conservation Laws

IN THIS CHAPTER, we start the analysis of the stability and the boundary stabilization design with the simple case of systems of two linear conservation laws. There are two good reasons for beginning in this way.

The first reason is that a system of two linear conservation laws is simple enough to allow an explicit and complete mathematical analysis of many aspects of the stability issue. It is therefore an excellent pedagogical starting point for more complex studies on general systems of conservation and balance laws. The first section of this chapter is devoted to a thorough presentation of the necessary and sufficient stability conditions for systems of two linear conservation laws under local boundary conditions, both in the time and the frequency domain.

A second reason is that, in many instances as we have illustrated with numerous examples in Chapter 1, a system of two linear conservation laws may be a valid approximation of a physical system having a direct engineering interest. In particular, in the second section of this chapter, we shall develop in great details the issue of the boundary feedback stabilization of the so-called density-flow systems.

Finally, in the third section, we examine how the stability conditions are extended to the case of nonuniform characteristic velocities.

2.1 Stability Conditions

We consider a hyperbolic system of two linear conservation laws in Riemann coordinates:

$$\partial_t R_1 + \lambda_1 \partial_x R_1 = 0,$$
$$\qquad\qquad t \in [0, +\infty), \quad x \in [0, L], \quad \lambda_1 > 0 > -\lambda_2, \qquad (2.1)$$
$$\partial_t R_2 - \lambda_2 \partial_x R_2 = 0,$$

© Springer International Publishing Switzerland 2016
G. Bastin, J.-M. Coron, *Stability and Boundary Stabilization of 1-D Hyperbolic Systems*, Progress in Nonlinear Differential Equations and Their Applications 88,
DOI 10.1007/978-3-319-32062-5_2

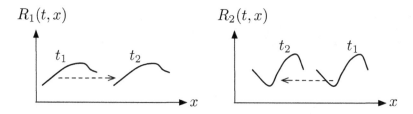

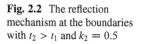

Fig. 2.1 The motion of the solutions of linear conservation laws in Riemann coordinates ($t_2 > t_1$)

Fig. 2.2 The reflection
mechanism at the boundaries
with $t_2 > t_1$ and $k_2 = 0.5$

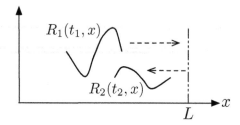

with constant characteristic velocities $\lambda_1 > 0$ and $-\lambda_2 < 0$. If the solutions R_1
and R_2 are differentiable with respect to t and x, it is immediate to check that the
system (2.1) is equivalent to a pair of scalar delay equations such that

$$R_1\left(t + \frac{x' - x}{\lambda_1}, x'\right) = R_1(t, x),$$

$$\forall t \geqslant 0, \quad \forall\, (x, x') \text{ such that } 0 \leqslant x < x' \leqslant L.$$

$$R_2\left(t + \frac{x' - x}{\lambda_2}, x\right) = R_2(t, x'), \tag{2.2}$$

The behavior of the solutions is illustrated in Fig. 2.1 where they are seen as waves
moving to the right or to the left without change of shape at the constant velocities
λ_1 and λ_2.

It is however evident that the delay equations (2.2) make sense also if the
solutions are not differentiable and even not continuous with respect to t and x.
In such case, the dynamics of the system may still be represented by the hyperbolic
PDEs (2.1) albeit with appropriate definitions of 'weak' partial derivatives for ∂_t
and ∂_x.

Let us define linear boundary conditions of the form

$$R_1(t, 0) = k_1 R_2(t, 0), \qquad R_2(t, L) = k_2 R_1(t, L), \tag{2.3}$$

where k_1 and k_2 are constant real coefficients. Observe that this is a special
case of the general nominal boundary conditions (1.11) that we have introduced
in Chapter 1. The mechanism described by these boundary conditions can be
interpreted as a 'reflection' of the waves at the boundaries with amplifications
(or attenuations) of size k_1 and k_2, as illustrated in Fig. 2.2.

When $k_1 k_2 \neq 1$, the unique equilibrium solution of the system (2.1), (2.3) is $R_1^* = 0, R_2^* = 0$ which can be stable or unstable depending on the values of k_1 and k_2. When $k_1 k_2 = 1$, the system has infinitely many non-isolated equilibria: $R_1^* = k_1 R_2^*$ for every $R_2^* \in \mathbb{R}$ which are therefore not asymptotically stable.

Our concern, in this section, is to show that the equilibrium solution is exponentially stable if and only if $|k_1 k_2| < 1$.

2.1.1 Exponential Stability for the L^∞-Norm

In this subsection, we use the method of characteristics to analyze the exponential convergence of the system solutions in the L^∞ and C^0 spaces.

We consider the system (2.1) under an initial condition for $t = 0$,

$$R_1(0, x) = R_{1o}(x), \quad R_2(0, x) = R_{2o}(x), \quad x \in [0, L], \tag{2.4}$$

and, for $t > 0$, under the boundary conditions

$$R_1(t, 0) = k_1 R_2(t, 0), \quad R_2(t, L) = k_2 R_1(t, L), \quad t \in (0, +\infty). \tag{2.5}$$

For a function $\varphi = (\varphi_1, \varphi_2)^\mathsf{T} \in L^\infty((0, L); \mathbb{R}^2)$, we define the L^∞-norm:

$$\|\varphi\|_{L^\infty((0,L);\mathbb{R}^2)} \triangleq \max \left(\|\varphi_1\|_{L^\infty((0,L);\mathbb{R})}, \|\varphi_2\|_{L^\infty((0,L);\mathbb{R})} \right) < +\infty.$$

We assume that the functions[1] $R_{1o} : [0, L] \to \mathbb{R}$ and $R_{2o} : [0, L] \to \mathbb{R}$ are bounded, and therefore that $(R_{1o}, R_{2o})^\mathsf{T} \in L^\infty((0, L); \mathbb{R}^2)$.

Theorem 2.1. There exist positive constants C, ν such that, for any bounded functions R_{1o} and R_{2o}, the solution of the Cauchy problem (2.1), (2.4), (2.5) defined by the delay equations (2.2) satisfies

$$\|(R_1(t, .), R_2(t, .))^\mathsf{T}\|_{L^\infty((0,L);\mathbb{R}^2)} \leq C e^{-\nu t} \|(R_{1o}, R_{2o})^\mathsf{T}\|_{L^\infty((0,L);\mathbb{R}^2)}, \quad t \in [0, +\infty), \tag{2.6}$$

if and only if $|k_1 k_2| < 1$.

Proof. The mechanism of the proof is illustrated in Fig. 2.3. As shown in this figure, since the system (2.1) is linear, the two characteristic curves C_1 and C_2 are straight lines with slopes λ_1 and $-\lambda_2$ respectively.

By using twice alternatively the formula (2.2) and the boundary condition (2.5), for a given $x \in [0, L]$, we have

---

[1]Here, and in the rest of the book except otherwise stated, we always assume that these functions are *measurable*.

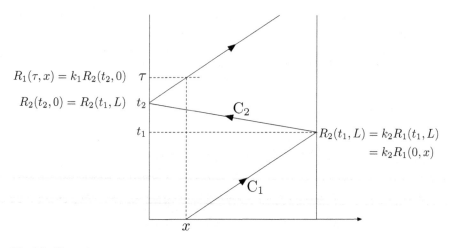

Fig. 2.3 Illustration of the proof of Theorem 2.1

$$R_1(\tau, x) = R_1(t_2, 0) = k_1 R_2(t_2, 0) \quad \text{with } \tau \triangleq \frac{L}{\lambda_1} + \frac{L}{\lambda_2} \text{ and } t_2 \triangleq \tau - \frac{x}{\lambda_1}$$

$$= k_1 R_2(t_1, L) = k_1 k_2 R_1(t_1, L) = k_1 k_2 R_1(0, x) \quad \text{with } t_1 = \frac{L - x}{\lambda_1}$$

$$= k_1 k_2 R_{1o}(x).$$

By continuing this recurrence for increasing time delays $n\tau$, we get

$$R_1(n\tau, x) = (k_1 k_2)^n R_{1o}(x) \quad n = 2, 3, 4, \ldots. \tag{2.7}$$

From this expression, we conclude readily that $R_1(t, x)$ cannot converge to zero as $t \to \infty$ if $|k_1 k_2| \geq 1$. A similar argument can obviously be developed for $R_2(t, x)$.

On the other hand, using again (2.2) and (2.5) appropriately, the following upper bounds can be established:

$$\forall x \in [0, L] \ \forall t \text{ s.t. } n\tau \leq t - \frac{x}{\lambda_1} < (n+1)\tau \ : \ |R_1(t, x)| \leq (1 + |k_1|)(|k_1 k_2|)^n R_o^{\max},$$

$$\forall x \in [0, L] \ \forall t \text{ s.t. } n\tau \leq t - \frac{L - x}{\lambda_2} < (n+1)\tau \ : \ |R_2(t, x)| \leq (1 + |k_2|)(|k_1 k_2|)^n R_o^{\max},$$

with $R_o^{\max} \triangleq \|(R_{1o}, R_{2o})^\top\|_{L^\infty((0,L);\mathbb{R}^2)}$. From these expressions, if $0 < |k_1 k_2| < 1$, it follows readily that inequality (2.6) is satisfied with

$$C \triangleq (2 + |k_1| + |k_2|) R_o^{\max} \quad \text{and} \quad \nu \triangleq \frac{1}{\tau} \ln\left(\frac{1}{|k_1 k_2|}\right)$$

and therefore that the solutions exponentially converge to zero for the L^∞-norm.

In the special case where $k_1 k_2 = 0$, $R_1(t, x) = 0$, and $R_2(t, x) = 0$ for all $(t, x) \in [\tau, +\infty) \times [0, L]$ (see also page 69).

This concludes the proof of Theorem 2.1. □

Let us now consider the special case when the initial condition is not only bounded but also continuous, i.e., when $(R_{1o}, R_{2o})^\mathsf{T} \in C^0([0, L]; \mathbb{R}^2)$, and when it is desired to have a so-called C^0-solution, i.e., a solution $(R_1(t, .), R_2(t, .))^\mathsf{T} \in C^0([0, L]; \mathbb{R}^2)$ which is continuous with respect to x for all $t \in [0, +\infty)$. In that case, to avoid discontinuities with respect to x at the initial time instant and to get a C^0-solution, it is clearly necessary that the initial condition (2.4) be compatible with the boundary condition (2.5) as follows:

$$R_{1o}(0) = k_1 R_{2o}(0), \qquad R_{2o}(L) = k_2 R_{1o}(L). \tag{2.8}$$

Under this condition, it follows readily not only that the solution is unique, but also that it inherits the regularity of the initial condition, i.e., that it is a C^0-solution. Then, since by definition $R_o^{\max}$ is also the C^0-norm of $(R_{1o}, R_{2o})^\mathsf{T}$, we have the following corollary of Theorem 2.1.

Corollary 2.2. For every $(R_{1o}, R_{2o})^\mathsf{T} \in C^0([0, L]; \mathbb{R}^2)$ satisfying the compatibility condition (2.8), the Cauchy problem (2.1), (2.4), (2.5) has a unique C^0-solution. Furthermore, the solutions exponentially converge to zero for the C^0-norm if and only if $|k_1 k_2| < 1$.

2.1.2 Exponential Lyapunov Stability for the L^2-Norm

In this subsection, using a Lyapunov stability approach, we shall now examine how the convergence to zero of the solutions of the system (2.1), (2.5) can be analyzed in the L^2-space. We consider again the system (2.1), (2.5) under a nonzero initial condition

$$R_1(0, x) = R_{1o}(x), \qquad R_2(0, x) = R_{2o}(x), \tag{2.9}$$

but we assume now that the function $(R_{1o}, R_{2o})^\mathsf{T} \in L^2((0, L); \mathbb{R}^2)$ with a L^2-norm

$$\|(R_{1o}, R_{2o})^\mathsf{T}\|_{L^2((0,L);\mathbb{R}^2)} \triangleq \left(\int_0^L \left(R_{1o}^2(x) + R_{2o}^2(x) \right) dx \right)^{1/2} < +\infty. \tag{2.10}$$

Thus here, provided it is in L^2, the initial condition may be unbounded (in contrast with the assumption of Theorem 2.1) and does not need to satisfy any compatibility condition (in contrast with the assumption of Corollary 2.2).

We will now give the definition of a solution to the Cauchy problem (2.1), (2.5), (2.9) in $L^2((0, L); \mathbb{R}^2)$. In order to motivate this definition let us multiply (2.1) on the left by $(\varphi_1, \varphi_2) \in C^1([0, T] \times [0, L]; \mathbb{R}^2)$ where T is given.

We get the equation

$$\varphi_1(\partial_t R_1 + \lambda_1 \partial_x R_1) + \varphi_2(\partial_t R_2 - \lambda_2 \partial_x R_2) = 0.$$

Let us now integrate this equation on $(0, T) \times (0, L)$. Assuming for a while that the solution R_1, R_2 is of class C^1 with respect to both t and x, we have, using integrations by parts and (2.9):

$$0 = \int_0^L \int_0^T \left(\varphi_1(\partial_t R_1 + \lambda_1 \partial_x R_1) + \varphi_2(\partial_t R_2 - \lambda_2 \partial_x R_2) \right) dt dx$$

$$= \int_0^L \left(\varphi_1(T, x)R_1(T, x) + \varphi_2(T, x)R_2(T, x) \right) dx$$

$$- \int_0^L \left(\varphi_1(0, x)R_{1o}(x) + \varphi_2(0, x)R_{2o}(x) \right) dx$$

$$+ \int_0^T \left(\lambda_1 \varphi_1(t, L)R_1(t, L) - \lambda_2 \varphi_2(t, L)R_2(t, L) \right) dt$$

$$- \int_0^T \left(\lambda_1 \varphi_1(t, 0)R_1(t, 0) - \lambda_2 \varphi_2(t, 0)R_2(t, 0) \right) dt$$

$$- \int_0^L \int_0^T \left((\partial_t \varphi_1 + \lambda_1 \partial_x \varphi_1)R_1 + (\partial_t \varphi_2 - \lambda_2 \partial_x \varphi_2)R_2 \right) dt dx.$$

Then, using the boundary condition (2.5), we get

$$0 = \int_0^L \left(\varphi_1(T, x)R_1(T, x) + \varphi_2(T, x)R_2(T, x) \right) dx$$

$$- \int_0^L \left(\varphi_1(0, x)R_{1o}(x) + \varphi_2(0, x)R_{2o}(x) \right) dx$$

$$+ \int_0^T \left(\lambda_1 \varphi_1(t, L) - k_2 \lambda_2 \varphi_2(t, L) \right) R_1(t, L) dt$$

$$- \int_0^T \left(k_1 \lambda_1 \varphi_1(t, 0) - \lambda_2 \varphi_2(t, 0) \right) R_2(t, 0) dt$$

$$- \int_0^L \int_0^T \left((\partial_t \varphi_1 + \lambda_1 \partial_x \varphi_1)R_1 + (\partial_t \varphi_2 - \lambda_2 \partial_x \varphi_2)R_2 \right) dt dx.$$

Now, if the functions φ_1 and φ_2 are selected such that

$$k_1\lambda_1\varphi_1(t,0) - \lambda_2\varphi_2(t,0) = 0,$$

$$\lambda_1\varphi_1(t,L) - k_2\lambda_2\varphi_2(t,L) = 0,$$

(2.11)

we get

$$0 = \int_0^L \Big(\varphi_1(T,x)R_1(T,x) + \varphi_2(T,x)R_2(T,x)\Big)dx$$

$$- \int_0^L \Big(\varphi_1(0,x)R_{1o}(x) + \varphi_2(0,x)R_{2o}(x)\Big)dx$$

$$- \int_0^L \int_0^T \Big((\partial_t\varphi_1 + \lambda_1\partial_x\varphi_1)R_1 + (\partial_t\varphi_2 - \lambda_2\partial_x\varphi_2)R_2\Big)dtdx.$$

(2.12)

The key point here is that, although this latter equation has been derived under the assumption that the functions R_1 and R_2 are of class C^1 with respect to t and x, it appears that it makes sense also even if the functions R_1 and R_2 are *not* differentiable and can therefore be considered as 'weak' solutions of the system. The L^2-solutions are then defined as the functions (R_1, R_2) which satisfy (2.12) for all (φ_1, φ_2) verifying (2.11), when the initial conditions are in L^2. The technical definition is as follows.

Definition 2.3. Let $(R_{1o}, R_{2o}) \in L^2((0,L); \mathbb{R}^2)$. A map $(R_1, R_2) : [0, +\infty) \times (0,L) \rightarrow \mathbb{R}^2$ is a L^2-solution of the Cauchy problem (2.1), (2.5), (2.9) if

$$(R_1, R_2) \in C^0([0, +\infty); L^2((0,L); \mathbb{R}^2))$$

is such that (2.12) is satisfied for every $T \in [0, +\infty)$ and for every $(\varphi_1, \varphi_2) \in C^1([0, T]) \times [0, L]; \mathbb{R}^2)$ satisfying (2.11). $\qquad\square$

We then have the following stability theorem.

Theorem 2.4. For any function $(R_{1o}, R_{2o}) \in L^2((0,L); \mathbb{R}^2)$, the Cauchy problem (2.1), (2.5), (2.9) has one and only one solution. Furthermore, there exist positive constants C, ν such that

$$\|(R_1(t,.), R_2(t,.))^\top\|_{L^2((0,L);\mathbb{R}^2)} \leq Ce^{-\nu t}\|(R_{1o}, R_{2o})^\top\|_{L^2((0,L);\mathbb{R}^2)}, \quad t \in [0, +\infty),$$

if and only if $|k_1 k_2| < 1$.

Proof. The existence of a unique L^2-solution follows, as a special case, from Theorem A.4 in Appendix A.

For the convergence analysis, the following candidate Lyapunov function is introduced:

$$\mathbf{V}(t) \triangleq \int_0^L \left[\frac{p_1}{\lambda_1}R_1^2(t,x)\exp(-\frac{\mu x}{\lambda_1}) + \frac{p_2}{\lambda_2}R_2^2(t,x)\exp(+\frac{\mu x}{\lambda_2})\right]dx$$

(2.13)

with positive constant coefficients p_1, p_2, and μ.

The time derivative of $\mathbf{V}(t)$ along the C^1 solutions of the Cauchy problem (2.1), (2.5), (2.9) is

$$\frac{d\mathbf{V}(t)}{dt} = \int_0^L \left[2\frac{p_1}{\lambda_1} R_1(\partial_t R_1)\exp(-\frac{\mu x}{\lambda_1}) + 2\frac{p_2}{\lambda_2} R_2(\partial_t R_2)\exp(+\frac{\mu x}{\lambda_2}) \right] dx$$

$$= \int_0^L \left[-2p_1 R_1(\partial_x R_1)\exp(-\frac{\mu x}{\lambda_1}) + 2p_2 R_2(\partial_x R_2)\exp(+\frac{\mu x}{\lambda_2}) \right] dx$$

$$= \int_0^L \left[-p_1(\partial_x R_1^2)\exp(-\frac{\mu x}{\lambda_1}) + p_2(\partial_x R_2^2)\exp(+\frac{\mu x}{\lambda_2}) \right] dx$$

$$= -\mu \int_0^L \left(\frac{p_1}{\lambda_1} R_1^2 \exp(-\frac{\mu x}{\lambda_1}) + \frac{p_2}{\lambda_2} R_2^2 \exp(+\frac{\mu x}{\lambda_2}) \right) dx$$

$$\quad - \left[p_1 R_1^2 \exp(-\frac{\mu x}{\lambda_1}) \right]_0^L + \left[p_2 R_2^2 \exp(+\frac{\mu x}{\lambda_2}) \right]_0^L$$

$$= -\mu \mathbf{V}(t) - p_1 \left[R_1^2(t,L)\exp(-\frac{\mu L}{\lambda_1}) - R_1^2(t,0) \right]$$

$$\quad - p_2 \left[R_2^2(t,0) - R_2^2(t,L)\exp(+\frac{\mu L}{\lambda_2}) \right].$$

Using the boundary condition (2.5), we have:

$$\frac{d\mathbf{V}(t)}{dt} = -\mu \mathbf{V}(t) - \left[p_1 \exp(-\frac{\mu L}{\lambda_1}) - p_2 k_2^2 \exp(+\frac{\mu L}{\lambda_2}) \right] R_1^2(t,L) - \left[p_2 - p_1 k_1^2 \right] R_2^2(t,0).$$

If $|k_1 k_2| < 1$, we can select $\mu > 0$ such that

$$\exp(\frac{\mu L}{\lambda_1} + \frac{\mu L}{\lambda_2}) k_1^2 k_2^2 < 1.$$

Then, we can select p_1 and p_2 such that

$$\exp(\frac{\mu L}{\lambda_1} + \frac{\mu L}{\lambda_2}) k_2^2 < \frac{p_1}{p_2} < \frac{1}{k_1^2}$$

which implies that

$$p_1 \exp(-\frac{\mu L}{\lambda_1}) - p_2 k_2^2 \exp(+\frac{\mu L}{\lambda_2}) > 0 \quad \text{and} \quad p_2 - p_1 k_1^2 > 0.$$

Hence, we see that $d\mathbf{V}/dt \le -\mu\mathbf{V}$ along the trajectories of the system (2.1), (2.5) which are of class C^1. By density (see the explanation given in the next section), this inequality also holds in the distribution sense for every solution of (2.1), (2.5) which is in $C^0([0, +\infty); L^2(0, L))$.

Then, since there exists $\gamma > 0$ such that

$$\frac{1}{\gamma} \int_0^L \left(R_1^2(t, x) + R_2^2(t, x) \right) dx \leqslant \mathbf{V}(t) \leqslant \gamma \int_0^L \left(R_1^2(t, x) + R_2^2(t, x) \right) dx, \qquad (2.14)$$

we get that

$$\left\| \left(R_1(t, .), R_2(t, .) \right)^{\mathsf{T}} \right\|_{L^2((0,L);\mathbb{R}^2)} \leqslant \gamma e^{-\mu t/2} \left\| \left(R_{1o}, R_{2o} \right)^{\mathsf{T}} \right\|_{L^2((0,L);\mathbb{R}^2)}, \quad \forall t \in [0, +\infty).$$

Consequently, if $|k_1 k_2| < 1$, the solutions converge in L^2-norm and the equilibrium is exponentially stable. We thus have proved the sufficient condition.

In order to prove the necessary condition, we assume that $|k_1 k_2| \geqslant 1$ and, as Xu and Sallet (2014), we use the same Lyapunov function as above albeit with $\mu = 0$:

$$\mathbf{W}(t) = \int_0^L \left[\frac{p_1}{\lambda_1} R_1^2(t, x) + \frac{p_2}{\lambda_2} R_2^2(t, x) \right] dx.$$

Along the C^1-solutions of the system (2.1), (2.5), the time derivative of $\mathbf{W}$ is

$$\frac{d\mathbf{W}(t)}{dt} = \left[p_2 k_2^2 - p_1 \right] R_1^2(t, L) + \left[p_1 k_1^2 - p_2 \right] R_2^2(t, 0).$$

Since $|k_1 k_2| \geqslant 1$, we can select p_1 and p_2 such that

$$k_1^2 \geqslant \frac{p_1}{p_2} \geqslant \frac{1}{k_2^2},$$

which implies that $d\mathbf{W}/dt \geqslant 0$ along the system trajectories which, therefore, cannot exponentially converge to zero. This completes the proof of Theorem 2.4. $\qquad \square$

From this proof it follows also that the maximum admissible value of the parameter μ

$$\mu_{max} = \frac{2}{\tau} \ln \left(\frac{1}{|k_1 k_2|} \right) \qquad (2.15)$$

is an estimate of the fastest possible decay rate of $\mathbf{V}$. We observe that it is identical to the convergence rate ν which was obtained in Theorem 2.1 with the method of characteristics.

Remark 2.5. The weights $e^{-\mu x/\lambda_1}$ and $e^{\mu x/\lambda_2}$ in (2.13) are essential to get a strict Lyapunov function in Theorem 2.4. The use of such terms in a quadratic Lyapunov function was originally introduced in Coron (1999) for the stabilization of the Euler equation of incompressible fluids.

2.1.3   A Note on the Proofs of Stability in L^2-Norm

In the course of the proof of Theorem 2.4, we have seen that the Lyapunov stability analysis was technically derived for the C^1-solutions of the system. At the end of the proof, it was then stated that the analysis is actually also valid for L^2-solutions. Roughly speaking, the reason is that the C^1-solutions are dense in the set of L^2-solutions of the system. Indeed, since the set of C^1 functions vanishing at 0 and L is dense in $L^2((0, L); \mathbb{R}^2)$, there exists a sequence of initial conditions $(R_{1o}^k, R_{2o}^k)^\mathsf{T}$ of class C^1 vanishing at 0 and L (and therefore satisfying the compatibility condition (2.8)) which converge to $(R_{1o}, R_{2o})^\mathsf{T}$ in $L^2((0, L); \mathbb{R}^2)$. Let $(R_1^k, R_2^k)^\mathsf{T}$: $[0, +\infty) \times (0, L) \to \mathbb{R}^2$ be the L^2-solution of the Cauchy problem (2.1), (2.5), for the initial condition $(R_{1o}^k, R_{2o}^k)^\mathsf{T}$. By Theorem A.1 in Appendix A, we have $(R_1^k, R_2^k)^\mathsf{T} \in C^1([0, +\infty); L^2(0, L)^2) \cap C^0([0, +\infty); H^1(0, L)^2)$. This regularity is sufficient to get the inequality $d\mathbf{V}^k/dt \leq -\mu \mathbf{V}^k$ with

$$\mathbf{V}^k(t) \triangleq \int_0^L \left[\frac{p_1}{\lambda_1} (R_1^k)^2(t, x) \exp(-\frac{\mu x}{\lambda_1}) + \frac{p_2}{\lambda_2} (R_2^k)^2(t, x) \exp(+\frac{\mu x}{\lambda_2}) \right] dx.$$

Then letting $k \to +\infty$, we get $d\mathbf{V}/dt \leq -\mu \mathbf{V}$ in the sense of distributions. (See also Comment 4.6 in Chapter 4 for an explanation of this assertion in a more general case.)

In the sequel of the book, each time we will establish stability conditions for the L^2-norm, we will follow the same line of deriving the Lyapunov analysis for smooth solutions but, somewhat abusively, we will generally not explicitly mention that the analysis is, obviously, also valid for general weak solutions.

2.1.4   Frequency Domain Stability

Taking the (two-sided) Laplace transform of equations (2.2), we have the following representation in the frequency domain (with $s \triangleq \sigma + j\omega$ the Laplace complex variable):

$$R_1(s, x') = \exp(-\frac{x' - x}{\lambda_1} s) R_1(s, x),$$

$$\forall (x, x') \text{ such that } 0 \leq x < x' \leq L.$$

$$R_2(s, x) = \exp(-\frac{x' - x}{\lambda_2} s) R_2(s, x'), \tag{2.16}$$

For $x = 0$ and $x' = L$, the system (2.1) of linear conservation laws endowed with the linear boundary conditions (2.5) can be represented under the form of a feedback loop of two scalar delay systems as shown in Fig. 2.4 with the notations $\tau_1 \triangleq L/\lambda_1$ and $\tau_2 \triangleq L/\lambda_2$.

Fig. 2.4 Representation of the hyperbolic system of two linear conservation laws (2.1), (2.5) as a feedback delay system

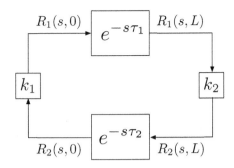

The poles $\sigma_n + j\omega_n$ of the system of Fig. 2.4 are the roots of the characteristic equation:

$$e^{s\tau} - k_1 k_2 = 0, \quad (\text{with } \tau \triangleq \tau_1 + \tau_2). \tag{2.17}$$

There is a countable infinity of poles lying on a vertical line in the complex plane. Their values depend on the sign of $k_1 k_2$ as follows:

$$k_1 k_2 > 0 \;\Rightarrow\; \sigma_n = -\frac{1}{\tau} \ln\left(\frac{1}{k_1 k_2}\right), \quad \omega_n = \pm\frac{2n\pi}{\tau}, \quad n = 0, 1, 2, \dots$$

$$k_1 k_2 < 0 \;\Rightarrow\; \sigma_n = -\frac{1}{\tau} \ln\left(-\frac{1}{k_1 k_2}\right), \quad \omega_n = \pm\frac{(2n+1)\pi}{\tau}.$$

The following stability theorem follows immediately.

Theorem 2.6. The poles of the system (2.1), (2.5) have a strictly negative real part $\sigma_n < 0$ if and only if $|k_1 k_2| < 1$. □

From Theorems 2.1, 2.4, and 2.6, we see that having a pole spectrum strictly located in the complex left-half plane is equivalent to the global exponential stability of the equilibrium for the system (2.1), (2.5), whatever the considered norm L^∞ or L^2. Moreover, the rate of exponential convergence of the solutions given by the absolute real part $|\sigma_n|$ of the poles is again identical to the previous estimates obtained with the method of characteristics and the Lyapunov approach.

2.1.5 Example: Stability of a Lossless Electrical Line

Let us come back to the example of an electrical line that we have already presented in Section 1.2 and which is illustrated in Fig. 2.5. Assuming a lossless line (i.e., with zero resistance R_ℓ and zero conductance G_ℓ), the dynamics are described by the following system of two conservation laws:

$$\partial_t I + \frac{1}{L_\ell} \partial_x V = 0, \qquad \partial_t V + \frac{1}{C_\ell} \partial_x I = 0, \tag{2.18}$$

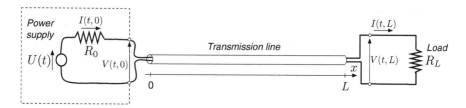

Fig. 2.5 Transmission line connecting a power supply to a resistive load

with the boundary conditions

$$R_0 I(t,0) + V(t,0) = U(t), \qquad V(t,L) = R_L I(t,L). \tag{2.19}$$

For a given constant input voltage $U(t) = U^*$, the system has a unique constant steady state

$$I^* = \frac{U^*}{R_0 + R_L}, \qquad V^* = \frac{R_L U^*}{R_0 + R_L}.$$

For this system, as we have seen in Section 1.2, the Riemann coordinates around the steady state are defined as

$$R_1 \triangleq (V - V^*) + (I - I^*) \sqrt{L_\ell / C_\ell},$$

$$R_2 \triangleq (V - V^*) - (I - I^*) \sqrt{L_\ell / C_\ell},$$

and the characteristic velocities are

$$\lambda_1 = \frac{1}{\sqrt{L_\ell C_\ell}}, \qquad -\lambda_2 = -\frac{1}{\sqrt{L_\ell C_\ell}}.$$

Then, expressing the boundary conditions (2.19) in Riemann coordinates, we have:

$$R_1(t,0) = k_1 R_2(t,0), \qquad R_2(t,L) = k_2 R_1(t,L),$$

with

$$k_1 = \frac{R_0 \sqrt{C_\ell} - \sqrt{L_\ell}}{R_0 \sqrt{C_\ell} + \sqrt{L_\ell}}, \qquad k_2 = \frac{R_L \sqrt{C_\ell} - \sqrt{L_\ell}}{R_L \sqrt{C_\ell} + \sqrt{L_\ell}}.$$

It is easy to see that $|k_1 k_2| < 1$ for any (positive) value of R_0, R_L, C_ℓ and L_ℓ. Consequently, the equilibrium I^*, V^* is exponentially stable in C^0-norm according to Theorem 2.1, in L^2-norm according to Theorem 2.4 and in the frequency domain according to Theorem 2.6. This is obviously a natural property for the device of Fig. 2.5 since it is a passive electrical circuit. However, the analysis provides here the accurate value of the exponential decay-rate.

2.2   Boundary Control of Density-Flow Systems

In this section, we shall now give a first example of the use of boundary feedback control for the stabilization of an hyperbolic system. We consider a system of two linear conservation laws of the general form:

$$\partial_t H + \partial_x Q = 0,$$

$$\partial_t Q + \lambda_1 \lambda_2 \partial_x H + (\lambda_1 - \lambda_2) \partial_x Q = 0, \tag{2.20}$$

where λ_1 and λ_2 are two real positive constants. The first equation can be interpreted as a mass conservation law with H the density and Q the flow density. The second equation can then be interpreted as a momentum conservation law.

The model (2.20) may be used to represent many physical systems. For instance, it may be used as a valid approximate linearized model for the motion of liquid fluids in pipes interconnected by pumps where H is the piezometric head and Q is the flow rate, while $\lambda_1 = \lambda_2 = c$ is the sound velocity. A detailed justification of this model is nicely presented in (Nicolet 2007, Chapter 2). In such fluid distribution networks, it may be relevant to provide the system with feedback controllers that regulate the piezometric head at certain places in order, for instance, to prevent water hammer phenomena.

We are concerned with the solutions of the Cauchy problem for the system (2.20) under an initial condition:

$$H(0, x) = H_o(x), \qquad Q(0, x) = Q_o(x), \qquad x \in [0, L],$$

and two boundary conditions of the form:

$$Q(t, 0) = Q_0(t), \qquad Q(t, L) = Q_L(t), \qquad t \in [0, +\infty). \tag{2.21}$$

Any pair of constant states H^*, Q^* is a potential steady state of the system. We assume that one of them has been selected as the desired steady state or *set point*.

The Riemann coordinates are defined around the set point by the following change of coordinates:

$$R_1 = Q - Q^* + \lambda_2 (H - H^*), \qquad R_2 = Q - Q^* - \lambda_1 (H - H^*),$$

with the inverse change of coordinates:

$$H = H^* + \frac{R_1 - R_2}{\lambda_1 + \lambda_2}, \qquad Q = Q^* + \frac{\lambda_1 R_1 + \lambda_2 R_2}{\lambda_1 + \lambda_2}.$$

With these coordinates, the system (2.20) is written in characteristic form:

$$\partial_t R_1 + \lambda_1 \partial_x R_1 = 0, \qquad \partial_t R_2 - \lambda_2 \partial_x R_2 = 0. \tag{2.22}$$

Then, assuming a constant flow rate $Q_0(t) = Q_L(t) = Q^*$ and expressing the boundary conditions (2.21) in Riemann coordinates, we have:

$$R_1(t,0) = k_1 R_2(t,0), \quad R_2(t,L) = k_2 R_1(t,L) \quad \text{with} \quad k_1 = -\frac{\lambda_2}{\lambda_1}, \quad k_2 = -\frac{\lambda_1}{\lambda_2}.$$

Consequently $|k_1 k_2| = 1$ and the equilibrium (H^*, Q^*) is not asymptotically stable. In fact, all the system poles are located on the imaginary axis as we have seen in the previous section (see also (Litrico and Fromion 2009, Section 3.2)).

It is therefore relevant to study the boundary feedback stabilization of the control system (2.20), (2.21). It will be the main concern of this section. More precisely, we are looking for controls $Q_0(t)$ and $Q_L(t)$ that are functions of the state variables at the boundaries and that guarantee the asymptotic stability of the steady state (H^*, Q^*).

2.2.1 Feedback Stabilization with Two Local Controls

From the stability analysis of Section 2.1, a simple and natural solution is to select control laws that realize boundary conditions of the form

$$R_1(t,0) = k_1 R_2(t,0), \quad R_2(t,L) = k_2 R_1(t,L),$$

in Riemann coordinates. This is easily achieved by defining the following feedback control laws:

$$Q_0(t) \triangleq Q^* + k_0(H^* - H(t,0)), \qquad k_0 \triangleq \frac{\lambda_1 k_1 + \lambda_2}{1 - k_1},$$

$$\tag{2.23}$$

$$Q_L(t) \triangleq Q^* - k_L(H^* - H(t,L)), \qquad k_L \triangleq \frac{\lambda_2 k_2 + \lambda_1}{1 - k_2}.$$

Then, the steady state (H^*, Q^*) of the closed loop system (2.20), (2.21), (2.23) is exponentially stable if and only if the control tuning parameters k_0 and k_L are chosen such that

$$|k_1 k_2| = \left| \left(\frac{k_0 - \lambda_2}{k_0 + \lambda_1} \right) \left(\frac{k_L - \lambda_1}{k_L + \lambda_2} \right) \right| < 1.$$

A block diagram representation of the control system is given in Fig. 2.6. It can be seen that each control Q has the form of a feedback of the density H at the same boundary: the implementation of $Q_0(t)$ requires only the on-line measurement of $H(t,0)$ and the implementation of $Q_L(t)$ requires only the on-line measurement of $H(t,L)$. For this reason, the controls are said to be 'local'.

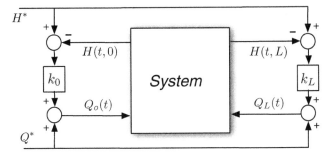

Fig. 2.6 Block diagram of the control system with two local controllers

2.2.2   Dead-Beat Control

An interesting very special choice of the control tuning parameters is $k_0 = \lambda_2$ and $k_L = \lambda_1$. Indeed with this choice we have $k_1 k_2 = 0$ and, therefore, a so-called 'dead-beat' control because, starting from an arbitrary initial condition, the steady state is reached as fast as possible in a finite time τ given by

$$\tau \triangleq \frac{L}{\lambda_1} + \frac{L}{\lambda_2}.$$

It must however be pointed out that dead-beat control may have the severe drawback of producing excessively big transients and feedback control actions that are too strong to be achieved with the available physical actuators.

2.2.3   Feedback-Feedforward Stabilization with a Single Control

We now consider the situation where there is only one boundary control input, say $Q_0(t)$, available for feedback stabilization. The other boundary flow $Q_L(t)$ perturbs the system in an unpredictable manner and cannot be manipulated. But we assume that this disturbance can be measured on-line. Then, from the previous section, a natural candidate control law is:

$$Q_0(t) \triangleq Q_L(t) + k_P(H^* - H(t,0)), \qquad k_P \triangleq \frac{\lambda_1 k_1 + \lambda_2}{1 - k_1}, \qquad (2.24)$$

where k_P is a tuning parameter and H^* is the density set point. This control law involves a 'feedforward' term $Q_L(t)$ (which compensates for the measured disturbance) and a proportional feedback term $k_P(H^* - H(t,0))$ for the density regulation. The control system is illustrated in Fig. 2.7.

Fig. 2.7 Block diagram of
the closed-loop system with a
feedback-feedforward control

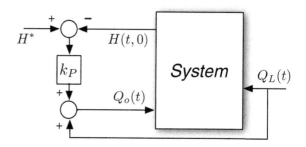

Assuming a constant disturbance $Q_L(t) = Q^*$, if $k_P \neq 0$ the closed-loop system
has a unique steady state (H^*, Q^*) and can be written in Riemann coordinates with
boundary conditions

$$R_1(t, 0) = k_1 R_2(t, 0), \qquad k_1 = \frac{k_P - \lambda_2}{k_P + \lambda_1},$$

$$R_2(t, L) = k_2 R_1(t, L), \qquad k_2 = -\frac{\lambda_1}{\lambda_2}.$$

Then, the steady state (H^*, Q^*) of the closed loop system (2.20), (2.21), (2.24) is
exponentially stable if and only if the control tuning parameter k_P is selected such
that

$$|k_1 k_2| = \left| \left(\frac{k_P - \lambda_2}{k_P + \lambda_1} \right) \frac{\lambda_1}{\lambda_2} \right| < 1.$$

2.2.4 Proportional-Integral Control

In the previous section we have used a controller that involves a feedforward action
when the flow $Q_L(t)$ is measurable. But, obviously, it could arise that this flow
is a so-called 'load disturbance' which cannot be measured and cannot therefore
be directly compensated in the control. In such case it is useful to implement
an 'integral' action in order to eliminate offsets and to attenuate the incidence of
the load disturbance. A so-called *Proportional-Integral* control law may be of the
following form:

$$Q_0(t) \triangleq Q_R + k_P(H^* - H(t, 0)) + k_I \int_0^t (H^* - H(s, 0)) ds. \qquad (2.25)$$

The first term Q_R is a constant reference value for the flow which is arbitrary and
freely chosen by the designer. The second term is the proportional correction action
with the tuning parameter k_P. The last term is the integral action with the tuning

Fig. 2.8 Block diagram of
the closed-loop system with a
Proportional-Integral control

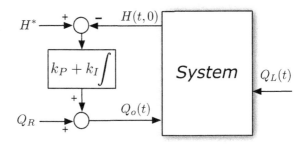

parameter $k_I \neq 0$. In case of a constant (unknown) disturbance $Q_L(t) = Q^*$,
the closed-loop system has a unique steady state (H^*, Q^*). The control system is
illustrated in Fig. 2.8. As it is explained in detail in Chapter 11 of the textbook
Feedback Systems (Åström and Murray (2009)), PI control is by far the most popular
way of using feedback in engineering systems because it is the simplest way to
cancel offset errors and to attenuate load disturbances in a robust way. The integral
gain k_I is a measure of the disturbance attenuation but a too large value of k_I may
lead to instability in some instances. It is therefore of interest to characterize the
range of values of k_I for which the closed-loop system is guaranteed to be stable.

In Riemann coordinates, the control law (2.25) gives a first boundary condition
at $x = 0$:

$$R_1(t, 0) = k_1 R_2(t, 0) + k_3 X(t) \tag{2.26}$$

$$\text{with } k_1 \triangleq \frac{k_P - \lambda_2}{k_P + \lambda_1}, \quad k_3 \triangleq \frac{k_I}{k_P + \lambda_1}$$

$$\text{and } X(t) \triangleq \frac{Q_R - Q^*}{k_I}(\lambda_1 + \lambda_2) + \int_0^t (R_2(\tau, 0) - R_1(\tau, 0)) d\tau.$$

The constant disturbance $Q_L(t) = Q^*$ gives the second boundary condition at $x = L$:

$$R_2(t, L) = k_2 R_1(t, L) \text{ with } k_2 = -\frac{\lambda_1}{\lambda_2}. \tag{2.27}$$

From (2.27), since $R_1(t, x)$ and $R_2(t, x)$ are constant along their respective characteristic lines, we have that

$$R_2(t + \tau, 0) = k_2 R_1(t, 0) \quad \text{with} \quad \tau \triangleq \frac{L}{\lambda_1} + \frac{L}{\lambda_2} \tag{2.28}$$

and therefore that

$$\frac{dR_2(t + \tau, 0)}{dt} = k_2 \frac{dR_1(t, 0)}{dt}. \tag{2.29}$$

Moreover, by differentiating (2.26) with respect to time, the first boundary condition is rewritten as:

$$\frac{dR_1(t,0)}{dt} = k_1 \frac{dR_2(t,0)}{dt} + k_3 \big(R_2(t,0) - R_1(t,0)\big). \tag{2.30}$$

Then, by eliminating $R_1(t,0)$ and $dR_1(t,0)/dt$ between (2.28), (2.29), and (2.30), we get that $R_2(t,0)$ is the solution of the following delay-differential equation of neutral type:

$$\frac{dR_2(t+\tau,0)}{dt} - k_1 k_2 \frac{dR_2(t,0)}{dt} + k_3 \Big(R_2(t+\tau,0) - k_2 R_2(t,0)\Big) = 0. \tag{2.31}$$

The Laplace transform of this equation is:

$$\Big[(e^{s\tau} - k_1 k_2)s + k_3(e^{s\tau} - k_2)\Big]R_2(s,0) = 0. \tag{2.32}$$

This is a so-called *neutral delay-differential* equation. The roots of the characteristic equation

$$(e^{s\tau} - k_1 k_2)s + k_3(e^{s\tau} - k_2) = 0 \tag{2.33}$$

are called *the poles* of the system (2.22), (2.26), (2.27).

2.2.4.1 Stability Analysis in the Frequency Domain

In the next theorem, we give necessary and sufficient conditions to have stable poles, i.e., poles located in the left-half complex plane and bounded away from the imaginary axis. The stability of the poles is equivalent, in the time domain, to the exponential stability of the equilibrium (when the disturbance $Q_L(t) = Q^*$ is constant) and therefore the input-to-state stability (when the disturbance $Q_L(t)$ is bounded time-varying) for the C^0-norm (see, e.g., Hale and Verduyn-Lunel (2002, Section 9.3) and Michiels and Niculescu (2007, Section 1.2)).

In the proof of the theorem we use a variant of the Walton-Marshall procedure (see Walton and Marshall (1987) and (Silva et al. 2005, Section 5.6)).

Theorem 2.7. There exist $\delta > 0$ such that the poles of the system (2.32) are in the half plane $(-\infty, -\delta] \times \mathbb{R}$ if and only if

- when $\lambda_1 \leqslant \lambda_2$ (i.e., $-1 \leqslant k_2 < 0$),

$$|k_1 k_2| < 1 \quad \text{and} \quad 0 < k_3;$$

- when $\lambda_1 > \lambda_2$ (i.e., $k_2 < -1$),

$$|k_1 k_2| < 1 \quad \text{and} \quad 0 < k_3 < \omega_0 \sin \omega_0 \tau \frac{k_2(1+k_1)}{1-k_2^2}$$

where ω_0 is the smallest positive ω such that $\cos(\omega\tau) = \dfrac{1+k_1 k_2^2}{k_2(1+k_1)}$.

Proof. For $|s| \longrightarrow +\infty$, the equation (2.33) is approximated by $e^{s\tau} - k_1 k_2 = 0$, from which it follows that $|k_1 k_2| < 1$ is a necessary condition to have stable poles (i.e., $\Re(s) < -\delta$, see, e.g., Hale and Verduyn-Lunel (2002) and Michiels and Vyhlidal (2005)). From now on, we assume that $|k_1 k_2| < 1$. It is also easily checked that, for every k_1 and k_2, for every $\eta > \ln(|k_1 k_2|)$ and for every $C_0 > 0$, there exists $C_1 > 0$ such that

$$\left\{ |k_3| \leqslant C_0, \ |s| \geqslant C_1 \text{ and } (2.33) \right\} \Rightarrow \left\{ \Re(s) \leqslant \eta \right\}. \tag{2.34}$$

Indeed the existence of C_1 results from rewriting (2.33) under the form

$$e^{s\tau} = \frac{k_1 k_2 s + k_3 k_2}{s + k_3} \tag{2.35}$$

which implies

$$\tau \Re(s) = \ln \left| \frac{k_1 k_2 s + k_3 k_2}{s + k_3} \right| \xrightarrow{|s| \to \infty} \ln |k_1 k_2|$$

where the convergence is uniform for $|k_0| \leqslant C_0$.

With the notation $s \triangleq \sigma + i\omega$, the poles satisfy the following equation:

$$\begin{aligned}
k_3 &= -\frac{(e^{s\tau} - k_1 k_2)s}{e^{s\tau} - k_2} \\
&= \frac{[\omega a(\sigma, \omega) - \sigma b(\sigma, \omega)] - i[\sigma a(\sigma, \omega) + \omega b(\sigma, \omega)]}{e^{2\sigma\tau} + k_2^2 - 2k_2 e^{\sigma\tau} \cos(\omega\tau)}
\end{aligned} \tag{2.36}$$

with

$$a(\sigma, \omega) \triangleq k_2 e^{\sigma\tau}(k_1 - 1)\sin(\omega\tau) \quad \text{and} \tag{2.37}$$

$$b(\sigma, \omega) \triangleq e^{2\sigma\tau} - k_2(1 + k_1)e^{\sigma\tau}\cos(\omega\tau) + k_1 k_2^2. \tag{2.38}$$

Since the left-hand side of equation (2.36) is real, it follows that the imaginary part of the right-hand side must be zero. Therefore we are looking for the values of σ and ω such that

$$\sigma a(\sigma, \omega) + \omega b(\sigma, \omega) = 0. \tag{2.39}$$

Let us now consider the poles with non-positive real parts, i.e., $\sigma \leqslant 0$. If $k_3 = 0$, we see that the poles are roots of $(e^{s\tau} - k_1 k_2)s = 0$. This means that there is a pole $s = 0$ at the origin and the other poles are stable if and only if $|k_1 k_2| < 1$. Now for small nonzero k_3, we have:

$$(1 - k_1 k_2)s + k_3(1 - k_2) \approx 0,$$

that is

$$s \approx -k_3 \frac{1 - k_2}{1 - k_1 k_2}.$$

This approximation can be justified by using the implicit function theorem applied to the map

$$(s, k_3) \in \mathbb{C} \times \mathbb{R} \longrightarrow F(s, k_3) = (e^{s\tau} - k_1 k_2)s + k_3(e^{s\tau} - k_2)$$

since $\partial F / \partial s(0, 0) = 1 - k_1 k_2 \neq 0$.

Then, since $|k_1 k_2| < 1$ and $k_2 = -\lambda_1 / \lambda_2 < 0$, it follows that for small $k_3 > 0$ the pole at zero moves inside the negative half-plane while the other poles stay inside the negative half-plane. Moreover, for small $k_3 < 0$, the pole at zero moves inside the right half plane. As k_3 decreases, this simple pole cannot come back on the imaginary axis (since $k_2 \neq 0$) and therefore it remains in the right half plane for all $k_3 < 0$.

Now, in order to analyze what happens when $k_3 > 0$ becomes larger, we consider the conditions for having poles on the imaginary axis, i.e., $\sigma = 0$. Since $k_3 \neq 0$, the case $\sigma = 0, \omega = 0$ is excluded. Therefore $\sigma = 0$ implies $b = 0$ from (2.39), which together with (2.38) gives:

$$\cos(\omega \tau) = \frac{1 + k_1 k_2^2}{k_2(1 + k_1)}.$$

In this case, it can be readily verified that, since $|k_1 k_2| < 1$,

$$\lambda_1 < \lambda_2 \quad \Leftrightarrow \quad |k_2| < 1 \quad \Leftrightarrow \quad 1 - k_2^2 > 0$$

$$\Leftrightarrow \quad (1 - k_2^2)(1 - k_1^2 k_2^2) > 0$$

$$\Leftrightarrow \quad 1 + k_1^2 k_2^4 + 2k_1 k_2^2 > k_2^2(1 + k_1^2) + 2k_1 k_2^2$$

$$\Leftrightarrow \quad \left| \frac{1 + k_1 k_2^2}{k_2(1 + k_1)} \right| > 1.$$

which implies that there is no eigenvalue on the imaginary axis. Then, using also (2.34), we can conclude, using a standard deformation argument on k_3, that, when $|k_2| < 1$ and $|k_1 k_2| < 1$, the poles remain stable for every $k_3 > 0$.

Let us now consider the case where $\lambda_1 > \lambda_2$, i.e., $k_2 < -1$ (the case $\lambda_1 = \lambda_2$ is discussed later). In this case, it can be readily verified that

$$\left| \frac{1 + k_1 k_2^2}{k_2(1 + k_1)} \right| < 1.$$

Therefore, from (2.36) and (2.38) with $\sigma = 0$, there is a pair of poles $\pm i\omega$ on the imaginary axis for any positive value of ω such that:

$$\cos(\pm\omega\tau) = \frac{1 + k_1 k_2^2}{k_2(1 + k_1)} \quad \text{and} \quad \omega\sin(\omega\tau) = -\frac{k_3(k_2^2 - 1)}{k_2(1 + k_1)}. \tag{2.40}$$

Let ω_0 be the smallest value of ω such that (2.40) is satisfied. Now, if $i\omega_0$ is a pole on the imaginary axis, the corresponding value of k_3 computed from (2.40) $\omega = \omega_0$ is as follows:

$$k_3^* = \omega_0 \sin(\omega_0\tau)\frac{k_2(1 + k_1)}{1 - k_2^2} > 0.$$

Then, using again (2.34), we can conclude, using a standard deformation argument on k_3, that the poles are stable for any k_3 such that $0 < k_3 < k_3^*$. In order to determine the motion of the pole on the imaginary axis for small variations of k_3 around k_3^*, we consider the root s of the characteristic equation as an explicit function of k_3. Then, by differentiating the characteristic equation (2.33), we have the following expression for the derivative of s with respect to k_3:

$$s' \triangleq \frac{ds}{dk_3} = \frac{k_2 - e^{s\tau}}{e^{s\tau}(1 + \tau(s + k_3)) - k_1 k_2}. \tag{2.41}$$

We now evaluate this expression at $i\omega$:

$$s' = \frac{k_2 - e^{i\omega\tau}}{e^{i\omega\tau}(1 + \tau(i\omega + k_3)) - k_1 k_2}.$$

Using (2.40), after some calculations, we obtain that the real part of s' at $i\omega$ is given by:

$$\Re(s') = \frac{\tau k_3(k_2^2 - 1)}{\left|e^{i\omega\tau}(1 + \tau(i\omega + k_3)) - k_1 k_2\right|^2}.$$

Hence, since $k_2^2 > 1$ and $k_3 > 0$ by assumption, $\Re(s')$ is a positive number. It follows that any pole reaching the imaginary axis from the left when k_3 is increasing will cross the imaginary axis from left to right. This readily implies that, as soon as $k_3 > k_3^*$, there is necessarily at least one pole in the right half plane.

Let us finally consider the case where $\lambda_1 = \lambda_2$ (i.e., $k_2 = -1$). In that case, it follows directly from (2.40) that $\cos(\omega\tau) = -1$ and $\sin(\omega\tau) = 0$ for any pole $i\omega$ on the imaginary axis. Therefore the characteristic equation (2.33) reduces to

$$(k_1 - 1)i\omega = 0$$

Fig. 2.9 Sketch of the root locus for fixed values of k_1 and k_2 and increasing values of k_3 from 0 to $+\infty$

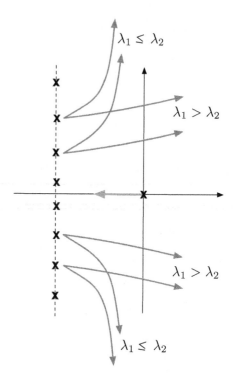

which is impossible if $\omega \neq 0$ because the conditions $k_2 = -1$ and $|k_1 k_2| < 1$ imply that $|k_1| < 1$. Hence there is no imaginary pole when $\lambda_1 = \lambda_2$. This completes the proof of Theorem 2.7.   □

As a matter of illustration, a sketch of the root locus for fixed values of k_1 and k_2 and increasing values of k_3 from 0 to $+\infty$ is given in Fig. 2.9.

In the previous theorem, for the clarity of the proof, we have carried out the analysis in terms of the parameters k_1, k_2 and k_3. However, from a practical viewpoint, it is clearly more relevant and more interesting to express the stability conditions in terms of the control tuning parameters k_P and k_I. Replacing k_1, k_2, and k_3 by their expressions in function of k_P, k_I, λ_1, and λ_2 as given in (2.26), (2.27), the conditions of Theorem 2.7 are translated as follows.

Theorem 2.8. There exist $\delta > 0$ such that the poles of the closed-loop system (2.20),(2.25) are in the half plane $(-\infty, -\delta] \times \mathbb{R}$ if and only if the control tuning parameters k_P, k_I are selected such that:

- when $\lambda_1 < \lambda_2$,

$$k_P > 0 \text{ and } k_I > 0 \quad \text{or} \quad k_P < -\frac{2\lambda_1\lambda_2}{\lambda_2 - \lambda_1} \text{ and } k_I < 0;$$

- when $\lambda_1 = \lambda_2$, $k_P > 0$ and $k_I > 0$;

- when $\lambda_1 > \lambda_2$,

$$0 < k_P < \frac{2\lambda_1\lambda_2}{\lambda_1 - \lambda_2} \text{ and } 0 < k_I < \omega_0 \frac{(2k_P + \lambda_1 - \lambda_2)\lambda_1\lambda_2}{\lambda_1^2 - \lambda_2^2} \cdot \sin(\omega_0\tau)$$

where ω_0 is the smallest positive ω such that

$$\cos(\omega\tau) = \frac{\lambda_2^2(k_P + \lambda_1) + \lambda_1^2(k_P - \lambda_2)}{\lambda_1\lambda_2(\lambda_2 - \lambda_1 - 2k_P)}.$$

$\square$

2.2.4.2 Lyapunov Stability Analysis

Up to now, in this subsection, we have taken the frequency domain viewpoint for the stability analysis of PI control of density-flow systems. We now move to the Lyapunov approach. In the next two theorems, we show how a quadratic Lyapunov function can be used to establish sufficient stability conditions which are similar but slightly more restrictive.

Theorem 2.9. The solution $R_1(t, x)$, $R_2(t, x)$ of the system (2.1), (2.26), (2.27) exponentially converges to zero for the L^2-norm if

$$|k_2| < 1, \quad |k_1k_2| < 1, \quad k_3 > 0. \tag{2.42}$$

Proof. We define the candidate Lyapunov function

$$\mathbf{V}(t) = \int_0^L \left[\frac{p_1}{\lambda_1} R_1^2(t, x) \exp(-\frac{\mu x}{\lambda_1}) + \frac{p_2}{\lambda_2} R_2^2(t, x) \exp(+\frac{\mu x}{\lambda_2}) \right] dx + bX^2(t)$$

with positive constant coefficients p_1, p_2, b, and μ. The time derivative of $\mathbf{V}(t)$ along the system trajectories is

$$\frac{d\mathbf{V}}{dt} = -\mu \int_0^L \left[\frac{p_1}{\lambda_1} R_1^2 \exp(-\frac{\mu x}{\lambda_1}) + \frac{p_2}{\lambda_2} R_2^2 \exp(+\frac{\mu x}{\lambda_2}) \right] dx$$

$$- \left[p_1 R_1^2 \exp(-\frac{\mu x}{\lambda_1}) \right]_0^L + \left[p_2 R_2^2 \exp(+\frac{\mu x}{\lambda_2}) \right]_0^L + 2bX\dot{X},$$

Using the boundary conditions (2.26), (2.27), we have:

$$\frac{d\mathbf{V}}{dt} = -\mu\mathbf{V} + \mathbf{W}_1 + \mathbf{W}_2$$

with

$$\mathbf{W}_1 \triangleq \left[p_2 \frac{\lambda_1^2}{\lambda_2^2} \exp(+\frac{\mu L}{\lambda_2}) - p_1 \exp(-\frac{\mu L}{\lambda_1}) \right] R_1^2(t, L), \tag{2.43}$$

$$\mathbf{W}_2 \triangleq \left[p_1 k_1^2 - p_2 \right] R_2^2(t, 0) + \left[p_1 k_3^2 - 2bk_3 + \mu b \right] X^2 \tag{2.44}$$

$$+ \left[2p_1 k_1 k_3 + 2b(1 - k_1) \right] XR_2(t, 0).$$

We have to prove that there exist positive $\mu > 0$, $p_1 > 0$, $p_2 > 0$, $b > 0$ such that (2.43) and (2.44) are negative definite (ND) quadratic forms in $R_1(t, L)$ and in $R_2(t, 0)$ and X respectively. We set $p_2 = 1$. Then, it follows that there exists $\mu > 0$ such that (2.43) is ND if and only if

$$p_1 > \left(\frac{\lambda_1}{\lambda_2} \right)^2 = k_2^2. \tag{2.45}$$

We now consider the special case where $\mu = 0$. Then the right-hand side of (2.44) is:

$$\left[p_1 k_1^2 - 1 \right] R_2^2(t, 0) + \left[p_1 k_3^2 - 2bk_3 \right] X^2 + \left[2p_1 k_1 k_3 + 2b(1 - k_1) \right] XR_2(t, 0).$$

This is a ND quadratic form if and only if

$$p_1 < \frac{1}{k_1^2} \tag{2.46}$$

and

$$P \triangleq (1 - k_1)^2 b^2 + 2k_3(p_1 k_1 - 1)b + p_1 k_3^2 < 0. \tag{2.47}$$

We consider P as a degree-2 polynomial in b. The discriminant Δ of P is

$$\Delta = 4k_3^2 (p_1 k_1^2 - 1)(p_1 - 1),$$

which, if (2.46) holds, is positive if

$$p_1 < 1. \tag{2.48}$$

In this case, the two roots of P (which have the sign) are positive if

$$p_1 k_1 < 1. \tag{2.49}$$

Since $|k_1 k_2| < 1$ and $|k_2| < 1$, there exists $p_1 > 0$ such that (2.45), (2.48), and (2.49) hold. We choose such a p_1. Then, it follows from our analysis that there exists $b > 0$ such that (2.43), (2.44) are ND quadratic forms when $\mu = 0$.

By continuity with respect to μ, under the same conditions, there exist $\mu > 0$, $p_1 > 0$, $p_2 > 0$, $b > 0$ such that (2.43), (2.44) are ND quadratic forms. $\qquad\square$

Theorem 2.10. The solution $R_1(t, x)$, $R_2(t, x)$ of the system (2.1), (2.26), (2.27) exponentially converges to zero for the L^2-norm if

$$|k_1 k_2| < 1, \quad \text{and } k_3 > 0 \text{ is sufficiently small.}$$

Proof. We introduce the following change of coordinates:

$$S_1 \triangleq R_1 - \frac{k_3}{1 - k_1 k_2} X, \quad S_2 \triangleq R_2 - \frac{k_2 k_3}{1 - k_1 k_2} X.$$

In these coordinates, the system (2.22) is rewritten

$$\partial_t S_1 + \lambda_1 \partial_x S_1 + \frac{k_3}{1 - k_1 k_2} \dot{X} = 0,$$

$$\partial_t S_2 - \lambda_2 \partial_x S_2 + \frac{k_2 k_3}{1 - k_1 k_2} \dot{X} = 0$$

and the boundary conditions (2.26), (2.27) are respectively

$$S_1(t, 0) = k_1 S_2(t, 0), \quad S_2(t, L) = k_2 S_1(t, L).$$

Under these boundary conditions, we have

$$\dot{X} = (1 - k_1) S_2(t, 0) - (1 - k_2) \frac{k_3}{1 - k_1 k_2} X. \tag{2.50}$$

We define the candidate Lyapunov function

$$\mathbf{V}(t) \triangleq \int_0^L \left[\frac{p_1}{\lambda_1} S_1^2(t, x) \exp(-\frac{\mu x}{\lambda_1}) + \frac{p_2}{\lambda_2} S_2^2(t, x) \exp(+\frac{\mu x}{\lambda_2}) \right] dx + b X^2(t).$$

with positive constant coefficients p_1, p_2, b, and μ. The time derivative of $\mathbf{V}(t)$ along the system trajectories is

$$\frac{d\mathbf{V}}{dt} = -\mu \int_0^L \left[\frac{p_1}{\lambda_1} S_1^2 \exp(-\frac{\mu x}{\lambda_1}) + \frac{p_2}{\lambda_2} S_2^2 \exp(+\frac{\mu x}{\lambda_2}) \right] dx$$

$$- \left[p_1 \exp(-\frac{\mu L}{\lambda_1}) - p_2 k_2^2 \exp(+\frac{\mu L}{\lambda_2}) \right] S_1^2(t, L) - [p_2 - p_1 k_1^2] S_2^2(t, 0)$$

$$+ 2bX\dot{X} - \int_0^L \left[\left(\frac{p_1}{\lambda_1} S_1 \exp(-\frac{\mu x}{\lambda_1}) + 2\frac{p_2}{\lambda_2} S_2 k_2 \exp(+\frac{\mu x}{\lambda_2}) \right) \frac{k_3}{1 - k_1 k_2} \dot{X} \right] dx.$$

Since $|k_1 k_2| < 1$, we can select $\mu > 0$ such that

$$\exp(\frac{\mu L}{\lambda_1} + \frac{\mu L}{\lambda_2}) k_1^2 k_2^2 < 1.$$

Then, we can select p_1 and p_2 such that

$$\exp(\frac{\mu L}{\lambda_1} + \frac{\mu L}{\lambda_2}) k_2^2 < \frac{p_1}{p_2} < \frac{1}{k_1^2}$$

which implies that

$$p_1 \exp(-\frac{\mu L}{\lambda_1}) - p_2 k_2^2 \exp(+\frac{\mu L}{\lambda_2}) > 0 \quad \text{and} \quad p_2 - p_1 k_1^2 > 0.$$

Then, using also (2.50), there are $\delta > 0$ and $\kappa > 0$ such that

$$\frac{d\mathbf{V}(t)}{dt} \leqslant -\delta \int_0^L (S_1^2 + S_2^2) dx - \delta (S_1^2(t, L) + S_2^2(t, 0))$$

$$- 2bk_3 \frac{1 - k_2}{1 - k_1 k_2} X^2 + \kappa b |S_2(t, 0)| |X|$$

$$+ \kappa k_3 \left[\int_0^L (S_1^2 + S_2^2) dx \right]^{\frac{1}{2}} (|S_2(t, 0)| + k_3 |X|).$$

We introduce the following notations:

$$A \triangleq \frac{1 - k_2}{1 - k_1 k_2} > 0, \qquad \vartheta \triangleq \left[\int_0^L (S_1^2 + S_2^2) dx \right]^{\frac{1}{2}}.$$

Then

$$\frac{d\mathbf{V}(t)}{dt} \leqslant - \delta S_1^2(t, L)$$

$$- (|X| \; |S_2(t,0)| \; \vartheta) \underbrace{\begin{pmatrix} 2bk_3 A & -bk/2 & -\kappa k_3^2/2 \\ -bk/2 & \delta & -\kappa k_3/2 \\ -\kappa k_3^2/2 & -\kappa k_3/2 & \delta \end{pmatrix}}_{Q} \begin{pmatrix} |X| \\ |S_2(t,0)| \\ \vartheta \end{pmatrix}.$$

We set $b = k_3^2$ and we compute the three principal minors of the matrix Q:

$$2k_3^3 A, \quad (2A\delta) k_3^3 - (\kappa^4/4) k_3^4, \quad (2A\delta^2) k_3^3 - \delta \kappa^2 k_3^4/2 - \left(\frac{A\kappa^2}{2} + \frac{\kappa^3}{4}\right) k_3^5.$$

Clearly, we have that the three minors are strictly positive if $k_3 > 0$ is chosen sufficiently small and therefore that Q is positive definite. We conclude that, along the system trajectories, there exists $\nu > 0$ such that

$$\frac{d\mathbf{V}(t)}{dt} \leqslant -\nu \mathbf{V} \quad \forall t.$$

Consequently the equilibrium is asymptotically stable and the solutions exponentially converge to zero for the L^2-norm. $\qquad\qquad\qquad\qquad\qquad\qquad\qquad$ □

2.3 The Nonuniform Case

In this section, we examine how the previous results can be extended to the nonuniform case where the characteristic velocities λ_i are function of the space coordinate x. More precisely, we now consider the following system:

$$\partial_t R_1 + \lambda_1(x)\partial_x R_1 = 0,$$
$$\lambda_1(x) > 0 > -\lambda_2(x), \quad \forall x \in [0, L], \qquad (2.51)$$
$$\partial_t R_2 - \lambda_2(x)\partial_x R_2 = 0,$$

under boundary conditions in canonical form

$$R_1(t, 0) = k_1 R_2(t, 0), \qquad R_2(t, L) = k_2 R_1(t, L), \qquad (2.52)$$

and an initial condition

$$R_1(0, x) = R_{10}(x) \in L^2([0, L], \mathbb{R}), \qquad R_2(0, x) = R_{20}(x) \in L^2([0, L], \mathbb{R}). \qquad (2.53)$$

The well-posedness of the Cauchy problem (2.51), (2.52), (2.53) in L^2 follows, as a special case, from Theorem A.4 in Appendix A. In order to analyze the exponential stability of the system we introduce the following tentative Lyapunov function:

$$\mathbf{V}(t) = \int_0^L \left(q_1(x)R_1^2(t, x) + q_2(x)R_2^2(t, x)\right)dx$$

where $q_1 \in C^1([0, L]; (0, +\infty))$ and $q_2 \in C^1([0, L]; (0, +\infty))$ have to be determined. The time derivative of $\mathbf{V}$ along the trajectories of (2.51), (2.52), (2.53) is

$$\frac{d\mathbf{V}}{dt} = \int_0^L \left(2q_1 R_1 \partial_t R_1 + 2q_2 R_2 \partial_t R_2\right)dx = -\int_0^L \left(2q_1 R_1 \lambda_1 \partial_x R_1 - 2q_2 R_2 \lambda_2 \partial_x R_2\right)dx$$

$$= \int_0^L \left[((\lambda_1 q_1)_x)R_1^2 - ((\lambda_2 q_2)_x)R_2^2\right]dx - \left[\lambda_1(L)q_1(L) - \lambda_2(L)q_2(L)k_2^2\right]R_1^2(t, L)$$

$$- \left[\lambda_2(0)q_2(0) - \lambda_1(0)q_1(0)k_1^2\right]R_2^2(t, 0).$$

It follows that $\mathbf{V}$ is a Lyapunov function if q_1 and q_2 are such that

$$(\lambda_1 q_1)_x > 0, \quad (\lambda_2 q_2)_x < 0, \quad \forall x \in [0, L],$$

$$\text{and } k_1^2 \le \frac{\lambda_2(0)q_2(0)}{\lambda_1(0)q_1(0)}, \quad k_2^2 \le \frac{\lambda_1(L)q_1(L)}{\lambda_2(L)q_2(L)}.$$

On the basis of our previous results in this chapter, a natural and convenient choice for the functions q_1 and q_2 is:

$$q_1(x) = \frac{p_1}{\lambda_1(x)} \exp\left(-\int_0^x \frac{\mu}{\lambda_1(\sigma)} d\sigma\right),$$

$$q_2(x) = \frac{p_2}{\lambda_2(x)} \exp\left(+\int_0^x \frac{\mu}{\lambda_2(\sigma)} d\sigma\right)$$

(2.54)

where p_1, p_2, μ are positive constants.

With this choice, we have the following stability theorem.

Theorem 2.11. The solution $R_1(t, x), R_2(t, x)$ of the Cauchy problem (2.51), (2.52), (2.53) exponentially converges to zero for the L^2-norm if $|k_1 k_2| < 1$.

Proof. With the functions (2.54), the derivative of the Lyapunov function becomes

$$\frac{d\mathbf{V}}{dt} = -\mu\mathbf{V}$$

$$-\left[p_1 \exp\left(-\int_0^L \frac{\mu}{\lambda_1(\sigma)} d\sigma\right) - p_2 k_2^2 \exp\left(+\int_0^L \frac{\mu}{\lambda_2(\sigma)} d\sigma\right)\right] R_1^2(t, L)$$

$$-\left[p_2 - p_1 k_1^2\right] R_2^2(t, 0).$$

Since $|k_1 k_2| < 1$, we can select μ sufficiently small such that

$$k_1^2 k_2^2 \exp\left(\mu \int_0^L \left[\frac{1}{\lambda_1(\sigma)} + \frac{1}{\lambda_2(\sigma)}\right] d\sigma\right) < 1.$$

Then, we can select p_1 and p_2 such that

$$k_2^2 \exp\left(\mu \int_0^L \left[\frac{1}{\lambda_1(\sigma)} + \frac{1}{\lambda_2(\sigma)}\right] d\sigma\right) < \frac{p_1}{p_2} < \frac{1}{k_1^2}$$

which implies that

$$\left[p_1 \exp\left(-\int_0^L \frac{\mu}{\lambda_1(\sigma)} d\sigma\right) - p_2 k_2^2 \exp\left(+\int_0^L \frac{\mu}{\lambda_2(\sigma)} d\sigma\right)\right] > 0$$

and

$$\left[p_2 - p_1 k_1^2\right] > 0.$$

Hence, we have

$$\frac{d\mathbf{V}}{dt} \leqslant -\mu\mathbf{V}.$$

Therefore $\mathbf{V}$ is a strict Lyapunov function and the solutions of the system (2.51), (2.52), (2.53) exponentially converge to zero for the L^2-norm. $\quad\square$

2.4 Conclusions

In Section 2.1, this chapter was first devoted to a comprehensive treatment of the exponential stability of a system of two linear conservation laws under local linear static boundary conditions. A necessary and sufficient stability condition has been given. This condition ensures that the L^2-norm is an exponentially decaying Lyapunov function and, equivalently, it guarantees that the poles of the system are located in the left-half complex plane and bounded away from the imaginary axis.

In the subsequent chapters, the main concern will be to examine how such static boundary conditions can be generalized for the exponential stability analysis of hyperbolic systems of conservation and balance laws.

Then, Section 2.2 was devoted the case when the stability boundary conditions are obtained by using boundary feedback control with actuators and sensors located at the boundaries. The stabilization of density-flow systems which are open-loop unstable and subject to unknown disturbances was investigated in detail. Using *Proportional-Integral* (PI) control the stabilization of the two conservation laws is achieved under dynamic boundary conditions. A necessary and sufficient condition on the values of the control tuning parameters is given (see also Bastin et al. (2015)). Let us also mention that, in this case, adding a derivative action to the PI controller is known to always produce an unstable closed-loop (see, e.g., Coron and Tamasoiu (2015)).

Chapter 3
Systems of Linear Conservation Laws

$\mathbf{T}$HIS CHAPTER mainly deals with the stability of general systems of linear conservation laws under static linear boundary conditions. Depending on whether the issue is examined in the time or in the frequency domain, different stability criteria emerge and are compared, namely from the viewpoint of robustness against uncertainties in the characteristic velocities. The chapter ends with the study of the stability of linear conservation laws under more general boundary conditions that may be dynamic, nonlinear, or switching.

We consider hyperbolic systems of linear conservation laws in Riemann coordinates

$$\mathbf{R}_t + \Lambda \mathbf{R}_x = \mathbf{0}, \qquad t \in [0, +\infty), \ x \in [0, L], \tag{3.1}$$

where $\mathbf{R} : [0, +\infty) \times [0, L] \to \mathbb{R}^n$. As we have already explained in Chapter 1, the matrix Λ is diagonal and defined as

$$\Lambda \triangleq \begin{pmatrix} \Lambda^+ & 0 \\ 0 & -\Lambda^- \end{pmatrix} \quad \text{with} \quad \begin{cases} \Lambda^+ = \text{diag}\{\lambda_1, \ldots, \lambda_m\}, \\ \Lambda^- = \text{diag}\{\lambda_{m+1}, \ldots, \lambda_n\}, \end{cases} \quad \lambda_i > 0 \ \forall i. \tag{3.2}$$

With the notations

$$\mathbf{R}^+ = \begin{pmatrix} R_1 \\ \vdots \\ R_m \end{pmatrix} \quad \text{and} \quad \mathbf{R}^- = \begin{pmatrix} R_{m+1} \\ \vdots \\ R_n \end{pmatrix} \quad \text{such that} \quad \mathbf{R} = \begin{pmatrix} \mathbf{R}^+ \\ \mathbf{R}^- \end{pmatrix},$$

© Springer International Publishing Switzerland 2016

G. Bastin, J.-M. Coron, *Stability and Boundary Stabilization of 1-D Hyperbolic Systems*, Progress in Nonlinear Differential Equations and Their Applications 88, DOI 10.1007/978-3-319-32062-5_3

the system (3.1) is also written

$$\partial_t \begin{pmatrix} \mathbf{R}^+ \\ \mathbf{R}^- \end{pmatrix} + \begin{pmatrix} \Lambda^+ & \mathbf{0} \\ \mathbf{0} & -\Lambda^- \end{pmatrix} \partial_x \begin{pmatrix} \mathbf{R}^+ \\ \mathbf{R}^- \end{pmatrix} = \mathbf{0}. \tag{3.3}$$

Our concern is to analyze the exponential stability of this system under linear boundary conditions in canonical form

$$\begin{pmatrix} \mathbf{R}^+(t,0) \\ \mathbf{R}^-(t,L) \end{pmatrix} = \mathbf{K} \begin{pmatrix} \mathbf{R}^+(t,L) \\ \mathbf{R}^-(t,0) \end{pmatrix} \quad \text{with } \mathbf{K} \triangleq \begin{pmatrix} K_{00} & K_{01} \\ K_{10} & K_{11} \end{pmatrix}, \quad t \in [0,+\infty), \tag{3.4}$$

and an initial condition

$$\mathbf{R}(0,x) = \mathbf{R}_0(x), \quad x \in (0,L). \tag{3.5}$$

3.1 Exponential Stability for the L^2-Norm

In this section, using a Lyapunov approach, we give an explicit condition on the matrix $\mathbf{K}$ under which the steady state $\mathbf{R}(t,x) \equiv 0$ of the system (3.3), (3.4) is globally exponentially stable for the L^2-norm according to the following definition.

Definition 3.1. The system (3.3), (3.4) is exponentially stable for the L^2-norm if there exist $\nu > 0$ and $C > 0$ such that, for every $\mathbf{R}_0 \in L^2((0,L); \mathbb{R}^n)$, the L^2-solution of the Cauchy problem (3.3), (3.4), (3.5) satisfies

$$\|\mathbf{R}(t,.)\|_{L^2((0,L);\mathbb{R}^n)} \leq Ce^{-\nu t} \|\mathbf{R}_0\|_{L^2((0,L);\mathbb{R}^n)}, \quad \forall t \in [0,+\infty).$$

$\square$

The definition of the L^2-solutions and the well-posedness of the Cauchy problem (3.3), (3.4), (3.5) are given in Appendix A, see Definition A.3 and Theorem A.4.

In order to state the stability condition, we first introduce the functions ρ_p : $\mathcal{M}_{n,n}(\mathbb{R}) \to \mathbb{R}$ defined by

$$\rho_p(M) \triangleq \inf \left\{ \|\Delta M \Delta^{-1}\|_p, \ \Delta \in \mathcal{D}_n^+ \right\}, \quad 1 \leq p \leq \infty, \tag{3.6}$$

where $\mathcal{D}_n^+$ denotes the set of diagonal $n \times n$ real matrices with strictly positive diagonal entries and

for $\boldsymbol{\xi} \triangleq (\xi_1, \dots, \xi_n)^\mathsf{T} \in \mathbb{R}^n$, $\|\boldsymbol{\xi}\|_p \triangleq \left[\sum_{i=1}^n |\xi_i|^p \right]^{\frac{1}{p}}$, $\|\boldsymbol{\xi}\|_\infty \triangleq \max\{|\xi_1|, \dots, |\xi_n|\}$,

for $M \in \mathcal{M}_{n,n}(\mathbb{R})$, $\|M\|_p \triangleq \max_{\|\boldsymbol{\xi}\|_p=1} \|M\boldsymbol{\xi}\|_p$.

We have the following stability theorem.

Theorem 3.2. The system (3.3), (3.4) is exponentially stable for the L^2-norm if $\rho_2(\mathbf{K}) < 1$.

Proof. We introduce the following candidate Lyapunov function, which is a direct extension of the function used in Chapter 2:

$$\mathbf{V} = \int_0^L \left[\sum_{i=1}^m \frac{p_i}{\lambda_i} R_i^2(t,x) \exp(-\frac{\mu x}{\lambda_i}) + \sum_{i=m+1}^n \frac{p_i}{\lambda_i} R_i^2(t,x) \exp(+\frac{\mu x}{\lambda_i}) \right] dx \tag{3.7}$$

$$= \int_0^L \left[(\mathbf{R}^{+\mathsf{T}}(\Lambda^+)^{-1} P^+(\mu x)\mathbf{R}^+) + (\mathbf{R}^{-\mathsf{T}}(\Lambda^-)^{-1} P^-(\mu x)\mathbf{R}^-) \right] dx$$

with

$$P^+(\mu x) \triangleq \mathrm{diag}\left\{ p_1 \exp(-\frac{\mu x}{\lambda_1}), \dots, p_m \exp(-\frac{\mu x}{\lambda_m}) \right\}, \quad p_i > 0, \tag{3.8}$$

$$P^-(\mu x) \triangleq \mathrm{diag}\left\{ p_{m+1} \exp(+\frac{\mu x}{\lambda_{m+1}}), \dots, p_n \exp(+\frac{\mu x}{\lambda_n}) \right\}, \quad p_i > 0. \tag{3.9}$$

The time derivative of $\mathbf{V}$ along the C^1-solutions of (3.3), (3.4) is

$$\frac{d\mathbf{V}}{dt} = -\mu\mathbf{V} + \mathbf{W}$$

with

$$\mathbf{W} \triangleq -\left[\mathbf{R}^{+\mathsf{T}} P^+(\mu x)\mathbf{R}^+\right]_0^L + \left[\mathbf{R}^{-\mathsf{T}} P^-(\mu x)\mathbf{R}^-\right]_0^L.$$

First we will show that the parameters p_i and μ can be selected such that, under the condition $\rho_2(\mathbf{K}) < 1$, $\mathbf{W}$ is a negative definite quadratic form in $\mathbf{R}^-(t,0)$ and $\mathbf{R}^+(t,L)$. For this analysis, we introduce the following notations:

$$\mathbf{R}_0^-(t) \triangleq \mathbf{R}^-(t,0), \quad \mathbf{R}_L^+(t) \triangleq \mathbf{R}^+(t,L).$$

Using the boundary condition (3.4), we have

$$\mathbf{W} = -\left[\mathbf{R}^{+\mathsf{T}} P^+(\mu x)\mathbf{R}^+\right]_0^L + \left[\mathbf{R}^{-\mathsf{T}} P^-(\mu x)\mathbf{R}^-\right]_0^L$$

$$= -\left(\mathbf{R}_L^{+\mathsf{T}} P^+(\mu L)\mathbf{R}_L^+ + \mathbf{R}_0^{-\mathsf{T}} P^-(0)\mathbf{R}_0^- \right)$$

$$+ \left(\mathbf{R}_L^{+\mathsf{T}} K_{00}^\mathsf{T} + \mathbf{R}_0^{-\mathsf{T}} K_{01}^\mathsf{T} \right) P^+(0) \left(K_{00}\mathbf{R}_L^+ + K_{01}\mathbf{R}_0^- \right) \tag{3.10}$$

$$+ \left(\mathbf{R}_L^{+\mathsf{T}} K_{10}^\mathsf{T} + \mathbf{R}_0^{-\mathsf{T}} K_{11}^\mathsf{T} \right) P^-(\mu L) \left(K_{10}\mathbf{R}_L^+ + K_{11}\mathbf{R}_0^- \right).$$

Since $\rho_2(\mathbf{K}) < 1$ by assumption, there exist $D_0 \in \mathcal{D}_m^+$, $D_1 \in \mathcal{D}_{n-m}^+$ and $\Delta \triangleq$ diag$\{D_0, D_1\}$ such that

$$\|\Delta \mathbf{K} \Delta^{-1}\| < 1. \tag{3.11}$$

The parameters p_i are selected such that $P^+(0) = D_0^2$ and $P^-(0) = D_1^2$. With these definitions, regarding $\mathbf{W}$ as a function of μ, we have

$$\mathbf{W}(\mu) = -\left(\mathbf{R}_L^{+\mathsf{T}} D_0 \ \mathbf{R}_0^{-\mathsf{T}} D_1\right) \mathbf{\Omega}(\mu) \begin{pmatrix} D_0 \mathbf{R}_L^+ \\ D_1 \mathbf{R}_0^- \end{pmatrix}$$

with

$$\Omega(\mu) \triangleq \begin{pmatrix} P^+(\mu L)D_0^{-2} & 0 \\ 0 & I \end{pmatrix} -$$

$$\begin{pmatrix} D_0 K_{00} D_0^{-1} & D_0 K_{01} D_1^{-1} \\ D_1 K_{10} D_0^{-1} & D_1 K_{11} D_1^{-1} \end{pmatrix}^{\mathsf{T}} \begin{pmatrix} D_0 K_{00} D_0^{-1} & D_0 K_{01} D_1^{-1} \\ P^-(\mu L)D_1^{-1} K_{10} D_0^{-1} & P^-(\mu L)D_1^{-1} K_{11} D_1^{-1} \end{pmatrix}$$

and, for $\mu = 0$,

$$\mathbf{W}(0) = -\left(\mathbf{R}_L^{+\mathsf{T}} D_0 \ \mathbf{R}_0^{-\mathsf{T}} D_1\right)\left(\mathbf{I} - (\Delta \mathbf{K} \Delta^{-1})^{\mathsf{T}}(\Delta \mathbf{K} \Delta^{-1})\right)\begin{pmatrix} D_0 \mathbf{R}_L^+ \\ D_1 \mathbf{R}_0^- \end{pmatrix}.$$

Since $\|\Delta \mathbf{K} \Delta^{-1}\| < 1$, it follows that $\mathbf{W}(0)$ is a strictly negative definite quadratic form in $\mathbf{R}_L^+$ and $\mathbf{R}_0^-$. Then, by continuity, $\mathbf{W}(\mu)$ remains a strictly negative definite quadratic form for $\mu > 0$ sufficiently small.

Hence, we have

$$\frac{d\mathbf{V}}{dt} = -\mu \mathbf{V} + \mathbf{W} \leqslant -\mu \mathbf{V}.$$

along the system trajectories.

Therefore $\mathbf{V}$ is a strict Lyapunov function and the solutions of the system (3.3), (3.4), (3.5) exponentially converge to zero for the L^2-norm. $\quad\square$

3.1.1 Dissipative Boundary Conditions

It is remarkable that the stability condition $\rho_2(\mathbf{K}) < 1$ depends on the value of $\mathbf{K}$ but not on the values of the characteristic velocities λ_i. In other words, the stability condition is independent of the system dynamics (3.3) and depends only on the

Fig. 3.1 The linear
hyperbolic system (3.3), (3.4)
viewed as a closed loop
interconnection of two causal
input-output systems

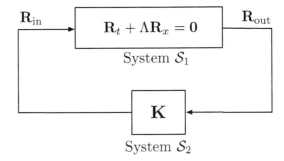

boundary conditions (3.4). When the matrix $\mathbf{K}$ satisfies such a stability condition, the boundary conditions are said to be **dissipative** and the stability is guaranteed whatever the length L and the time required for solutions to cross the system. Intuitively, this is understood as follows: the solutions, which are moving back and forth between the two boundaries, remain constant along the characteristic lines and are exponentially damped at the boundaries only. This can be also understood using a *small gain principle*. Indeed we have observed in Chapter 1, Section 1.1, that the hyperbolic systems (1.3) under the boundary conditions (1.11) can be regarded as a closed loop interconnection of two *causal* input-output systems as represented in Fig. 3.1. It is therefore natural that the stability requires a 'small gain' of the feedback loop. It is not surprising that the condition relies only on the gain $\rho_2(\mathbf{K})$ of the system $\mathcal{S}_2$ since the system $\mathcal{S}_1$ has a unit gain by definition.

3.2 Exponential Stability for the C^0-Norm: Analysis in the Frequency Domain

In this section, we now take the frequency domain viewpoint to analyze the exponential stability of the system (3.3), (3.4) for the C^0-norm according to the following definition.

Definition 3.3. The system (3.3), (3.4) is exponentially stable for the C^0-norm if there exist $v > 0$ and $C > 0$ such that, for every $\mathbf{R}_o \in C^0([0, L], \mathbb{R}^n)$ satisfying the compatibility condition

$$\begin{pmatrix} \mathbf{R}_o^+(0) \\ \mathbf{R}_o^-(L) \end{pmatrix} = \mathbf{K} \begin{pmatrix} \mathbf{R}_o^+(L) \\ \mathbf{R}_o^-(0) \end{pmatrix}, \tag{3.12}$$

the solution of the Cauchy problem (3.3), (3.4), (3.5) satisfies

$$\|\mathbf{R}(t, .)\|_{C^0((0,L);\mathbb{R}^n)} \le Ce^{-vt}\|\mathbf{R}_o\|_{C^0((0,L);\mathbb{R})}, \quad \forall t \in [0, +\infty). \qquad \square$$

As we have already emphasized in Chapter 2, the system (3.3) can be regarded as a set of scalar delay systems

$$R_i(t, L) = R_i(t - \tau_i, 0) \quad i = 1, \ldots, m,$$
$$\tau_k \triangleq \frac{L}{\lambda_k}, \quad k = 1, \ldots, n,$$
$$R_j(t, 0) = R_j(t - \tau_j, L) \quad j = m + 1, \ldots, n,$$

which are interconnected by the boundary conditions (3.4). Taking the Laplace transform, it follows that the characteristic function of the system (3.3), (3.4) is:

$$\det \left[\mathbf{I}_n - \mathrm{diag}\{e^{-s\tau_1}, \ldots, e^{-s\tau_n}\} \, \mathbf{K} \right], \tag{3.13}$$

where $\mathbf{I}_n$ is the identity matrix of $\mathcal{M}_{n,n}(\mathbb{R})$. The roots of this function are called the *poles* of the system.

Definition 3.4. The poles of the system (3.3), (3.4) are stable if there exists $\delta > 0$ such that the poles are located in the half plane $(-\infty, -\delta] \times \mathbb{R}$.

A fundamental property is given in the following theorem.

Theorem 3.5. The system (3.3), (3.4) is exponentially stable for the C^0-norm if and only if the poles of the system are stable.

Proof. See (Hale and Verduyn-Lunel 1993, Chapter 9, Theorem 3.5). $\qquad\square$

Remark 3.6. Theorem 3.5 deals with the C^0-norm. However, it must be pointed out that the proof, as it is given by Hale and Verduyn-Lunel, also works for the L^p-norm for every $p \in [1, +\infty]$.

Hence the stability analysis does not require to know the actual location of the poles. It is sufficient to know that they have a negative real part which is bounded away from zero. From the viewpoint of boundary control design, it is obviously of major interest to predict the stability, and therefore the sign of the real parts of the poles, directly from the coefficients of the matrix $\mathbf{K}$. Two stability conditions are presented below. The first one is the same as in the previous section.

Theorem 3.7. The poles of the system (3.3), (3.4) are stable if $\rho_2(\mathbf{K}) < 1$.

Proof. If $\rho_2(\mathbf{K}) < 1$ there exists $\eta \in (0, 1)$ and $\Delta \in \mathcal{D}_n$ such that

$$\|\Delta \mathbf{K} \Delta^{-1}\| \leq \eta. \tag{3.14}$$

Let us assume that s is a pole of the system. Then

$$\det \left[\mathbf{I}_n - \mathrm{diag}\{e^{-s\tau_1}, \ldots, e^{-s\tau_n}\} \, \Delta \mathbf{K} \Delta^{-1} \right]$$

$$= \det \left[\Delta (\mathbf{I}_n - \mathrm{diag}\{e^{-s\tau_1}, \ldots, e^{-s\tau_n}) \mathbf{K} \} \, \Delta^{-1} \right]$$

$$= \det \left[\mathbf{I}_n - \mathrm{diag}\{e^{-s\tau_1}, \ldots, e^{-s\tau_n}\} \, \mathbf{K}) \right]$$

$$= 0,$$

which implies that

$$\|\text{diag}\{e^{-s\tau_1},\ldots,e^{-s\tau_n}\}\Delta \mathbf{K}\Delta^{-1}\| \geq 1. \tag{3.15}$$

Since

$$\|\text{diag}\{e^{-s\tau_1},\ldots,e^{-s\tau_n}\}\Delta \mathbf{K}\Delta^{-1}\| \leq \|\text{diag}\{e^{-s\tau_1},\ldots,e^{-s\tau_n}\}\|\|\Delta \mathbf{K}\Delta^{-1}\|$$

$$\leq \exp(-\min\{\tau_1\Re(s),\ldots,\tau_n\Re(s)\})\|\Delta \mathbf{K}\Delta^{-1}\|$$

where $\Re(s)$ denotes the real part of the pole s, we have, using also (3.14) and (3.15),

$$\exp(-\min\{\tau_1\Re(s),\ldots,\tau_n\Re(s)\})\eta \geq 1. \tag{3.16}$$

Inequality (3.16) implies that

$$\Re(s) \leq \delta \triangleq \frac{\ln(\eta)}{\max\{\tau_1,\ldots,\tau_n\}} < 0.$$

$\square$

Another stability condition is stated in the following theorem by Silkowski (1976) which relies on the *Kronecker density theorem* (e.g., Bridges and Schuster (2006)).

Theorem 3.8. Let

$$\bar{\rho}(\mathbf{K}) \triangleq \max\{\rho(\text{diag}\{e^{-i\theta_1},\ldots,e^{-i\theta_n}\}\mathbf{K}); (\theta_1,\ldots,\theta_n)^{\mathsf{T}} \in \mathbb{R}^n\} \tag{3.17}$$

where $\rho(M)$ denotes the spectral radius of the matrix M. If the time delays $(\tau_1,\ldots,\tau_n)$ are rationally independent, the poles of the system (3.3), (3.4) are stable if and only if $\bar{\rho}(\mathbf{K}) < 1$.

Proof. See (Hale and Verduyn-Lunel 1993, Chapter 9, Theorem 6.1). $\square$

The statement of this theorem includes the rather unexpected feature that the time delays have to be 'rationally independent' which is a generic property. In fact, when the τ_i's are rationally dependent the condition $\bar{\rho}(\mathbf{K}) < 1$ is no longer necessary and can be violated while keeping the exponential stability as we shall illustrate with a simple example below. In Michiels et al. (2001), it is explained how "when approaching rational dependence of the delays, the supremum of the real parts of the poles can have a discontinuity (…) compatible with the continuous movement of individual roots" in the complex plane.

3.2.1 A Simple Illustrative Example

Let us now present an example that illustrates the conditions of Theorems 3.7 and 3.8. We consider the most simple case of a system of two linear conservation laws with a full matrix $\mathbf{K}$. More precisely, we have the system

$$\partial_t \begin{pmatrix} R_1 \\ R_2 \end{pmatrix} + \begin{pmatrix} \lambda_1 & 0 \\ 0 & -\lambda_2 \end{pmatrix} \partial_x \begin{pmatrix} R_1 \\ R_2 \end{pmatrix} = \mathbf{0}, \quad -\lambda_2 < 0 < \lambda_1, \tag{3.18}$$

with the boundary condition

$$\begin{pmatrix} R_1(t,0) \\ R_2(t,L) \end{pmatrix} = \overbrace{\begin{pmatrix} k_0 & k_1 \\ k_2 & k_3 \end{pmatrix}}^{\mathbf{K}} \begin{pmatrix} R_1(t,L) \\ R_2(t,0) \end{pmatrix}. \tag{3.19}$$

Taking the Laplace transform of system (3.18), (3.19), the characteristic equation is

$$(e^{s\tau_1} - k_0)(e^{s\tau_2} - k_3) - k_1 k_2 = 0. \tag{3.20}$$

Let us consider the very special case $\tau_1 = 1$, $\tau_2 = 2$ which allows a simple and explicit computation of the poles. In this case, the characteristic equation is

$$e^{3s} - k_0 e^{2s} - k_3 e^s + k_0 k_3 - k_1 k_2 = 0. \tag{3.21}$$

Defining $z \triangleq e^s$, we get the third-order polynomial equation

$$z^3 - k_0 z^2 - k_3 z + k_0 k_3 - k_1 k_2 = 0. \tag{3.22}$$

Let z_ℓ ($\ell = 1, 2, 3$) denote the three roots of this polynomial. Then, for each $z_\ell \neq 0$, there is an infinity of system poles $s_n = \sigma_n + j\omega_n$ lying on a vertical line in the complex plane:

$$\sigma_n = \ln|z_i|, \quad \omega_n = 2\pi n + \arg(z_i), \quad n = 0, \pm 1, \pm 2, \ldots. \tag{3.23}$$

The poles are stable if and only if $|z_\ell| < 1$, $\ell = 1, 2, 3$.

For simplicity, let us now address the special case where $k_0 k_3 = k_1 k_2$. In that case, it can be shown after a few calculations that the stability condition of Theorems 3.7 and 3.8 is

$$\bar{\rho}(\mathbf{K}) = \rho_2(\mathbf{K}) = |k_0| + |k_3| < 1. \tag{3.24}$$

The region of stability corresponding to this condition is thus the square represented in Fig. 3.2. From Theorem 3.8 we know that the condition is necessary and sufficient when τ_1/τ_2 is an irrational number. But, when τ_1/τ_2 is rational, the stability region may be larger as we shall now illustrate by computing the poles of the system.

Using the condition $k_0 k_3 = k_1 k_2$ the polynomial equation (3.22) becomes

$$z(z^2 - k_0 z - k_3) = 0 \qquad (3.25)$$

and we can compute the roots explicitly

$$z_1 = 0, \qquad z_{2,3} = \frac{k_0 \pm \sqrt{k_0^2 + 4k_3}}{2}. \qquad (3.26)$$

Remark that $e^s = z_1 = 0$ has no solution. The system poles corresponding to z_2 and z_3 are stable if and only if the parameters k_0 and k_3 satisfy one of the following two conditions:

$$k_0^2 + 4k_3 \geqslant 0 \quad \text{and} \quad |k_0 \pm \sqrt{k_0^2 + 4k_3}| < 2 \qquad (3.27)$$

or

$$k_0^2 + 4k_3 < 0 \quad \text{and} \quad \sqrt{k_0^2 + |k_0^2 + 4k_3|} < 2. \qquad (3.28)$$

The region of stability in the (k_0, k_3) plane is the triangular region shown in Fig. 3.2. It can be easily checked that the stability conditions (3.27), (3.28) can be in fact formulated in the following simpler way:

$$k_3 > -1, \quad k_0 + k_3 < 1, \quad k_0 - k_3 > -1. \qquad (3.29)$$

A numerical illustration is given in Fig. 3.3(a) for parameter values $k_0 = k_1 = k_2 = k_3 = -0.6$ which satisfy inequalities (3.29) but not inequality (3.24). As expected from the above analysis, the poles are located on a vertical line in the left half complex plane.

Fig. 3.2 Regions of stability

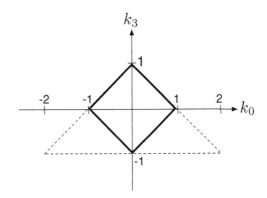

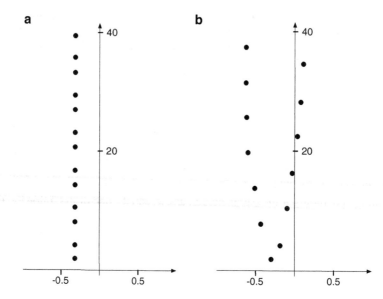

Fig. 3.3 Pole configuration for the characteristic equation (3.20) with $k_0 = k_1 = k_2 = k_3 = -0.6$: (a) $\tau_1 = 1, \tau_2 = 2$, (b) $\tau_1 = 1, \tau_2 = 2.1$

3.2.2   Robust Stability

From this simple example, it appears that the strict negativity of the pole real parts is not a robust stability condition as far as robustness with respect to small variations of the characteristic velocities is concerned. A numerical illustration of this fact is given in Fig. 3.3(b) where the τ_2 value of the previous example is slightly perturbed: τ_2 is set to 2.1 instead of 2. The consequence of this small variation of τ_2 is that half of the poles are progressively shifted to the right up to instability.

In contrast, we may observe that the stability condition $\bar{\rho}(\mathbf{K}) < 1$ is robust with respect to small changes on $\mathbf{K}$ since $\bar{\rho}(\mathbf{K})$ depends continuously on $\mathbf{K}$. We then introduce the following definition for the robustness with respect to the characteristic velocities.

Definition 3.9. The system (3.3), (3.4) is robustly exponentially stable with respect to the characteristic velocities if there exists $\varepsilon > 0$ such that the perturbed system

$$
\partial_t \begin{pmatrix} \mathbf{R}^+ \\ \mathbf{R}^- \end{pmatrix} + \begin{pmatrix} \tilde{\Lambda}^+ & \mathbf{0} \\ \mathbf{0} & -\tilde{\Lambda}^- \end{pmatrix} \partial_x \begin{pmatrix} \mathbf{R}^+ \\ \mathbf{R}^- \end{pmatrix} = \mathbf{0}.
$$

is exponentially stable for every $\tilde{\Lambda}$ such that $|\tilde{\lambda}_i - \lambda_i| \leq \varepsilon \;\; \forall i \in 1, \ldots, n$.

It is then evident that the following robust stability condition follows as a simple corollary of Theorem 3.8.

Corollary 3.10. The system (3.3), (3.4) is robustly exponentially stable with respect to the characteristic velocities if and only if $\bar{\rho}(\mathbf{K}) < 1$. □

3.2.3 Comparison of the Two Stability Conditions

Another interesting observation is that we have $\bar{\rho}(\mathbf{K}) = \rho_2(\mathbf{K})$ in the above simple example. This observation suggests to further investigate the comparison between $\bar{\rho}$ and ρ_2 and to determine to which extent they can be equal. This is done in the following theorems.

Theorem 3.11. For every integer n and for every real $n \times n$ matrix $\mathbf{K}$, $\bar{\rho}(\mathbf{K}) \leqslant \rho_2(\mathbf{K})$.

Proof. For every $(\theta_1, \ldots, \theta_n)^\top \in \mathbb{R}^n$ and for every $D \in \mathcal{D}_n$

$$\rho(\mathrm{diag}\{e^{\iota\theta_1}, \ldots, e^{\iota\theta_n}\} K) = \rho(D \, \mathrm{diag}\{e^{\iota\theta_1}, \ldots, e^{\iota\theta_n}\} KD^{-1})$$

$$= \rho(\mathrm{diag}\{e^{\iota\theta_1}, \ldots, e^{\iota\theta_n}\} DKD^{-1})$$

$$\leqslant \|\mathrm{diag}\{e^{\iota\theta_1}, \ldots, e^{\iota\theta_n}\} DKD^{-1}\|$$

$$\leqslant \|\mathrm{diag}\{e^{\iota\theta_1}, \ldots, e^{\iota\theta_n}\}\| \|DKD^{-1}\| = \|DKD^{-1}\|.$$

□

Theorem 3.12.

(a) For every $n \in \{1, 2, 3, 4, 5\}$ and for every real $n \times n$ matrix $\mathbf{K}$, $\bar{\rho}(\mathbf{K}) = \rho_2(\mathbf{K})$.
(b) For every integer $n > 5$, there exist a real $n \times n$ matrix $\mathbf{K}$ such that $\bar{\rho}(\mathbf{K}) < \rho_2(\mathbf{K})$.

The proof of this theorem can be found in Appendix C. The following corollary follows trivially.

Corollary 3.13. If there exist a permutation matrix P such that the matrix $\widetilde{\mathbf{K}} = PKP^{-1}$ is a block diagonal matrix

$$\widetilde{\mathbf{K}} = \mathrm{diag}\{\widetilde{\mathbf{K}}_1, \widetilde{\mathbf{K}}_2, \ldots, \widetilde{\mathbf{K}}_p\}$$

where each block $\widetilde{\mathbf{K}}_i$ is a real $n_i \times n_i$ matrix with $n_i \in \{1, 2, 3, 4, 5\}$, then $\bar{\rho}(\mathbf{K}) = \rho_2(\mathbf{K})$. □

3.3   The Rate of Convergence

In the previous two sections, we have given explicit conditions on the matrix **K** that guarantee the exponential convergence to zero of the solutions of the system (3.3), (3.4). From a control design viewpoint it is also of major interest to be able to quantify the rate of convergence. For that purpose, we define the following change of state variables:

$$\mathbf{S}(t,x) \triangleq e^{\mu t}P(\mu x)\mathbf{R}(t,x), \;\; 0 < \mu \in \mathbb{R}, \;\; P(\mu x) \triangleq \begin{pmatrix} P^+(\mu x) & 0 \\ 0 & P^-(\mu x) \end{pmatrix},$$
(3.30)

where P^+ and P^- are defined in (3.8) and (3.9) respectively. The dynamics of these new coordinates can be shown to be governed by the same hyperbolic system as is **R**:

$$\mathbf{S}_t + \Lambda\mathbf{S}_x = 0,$$

with adequately adapted boundary conditions:

$$\begin{pmatrix} \mathbf{S}^+(t,0) \\ \mathbf{S}^-(t,L) \end{pmatrix} = \widetilde{\mathbf{K}}(\mu) \begin{pmatrix} \mathbf{S}^+(t,L) \\ \mathbf{S}^-(t,0) \end{pmatrix},$$

$$\widetilde{\mathbf{K}}(\mu) \triangleq \begin{pmatrix} D_0^2 K_{00}P^+(-\mu L)D_0^{-4} & D_0^2 K_{01}D_1^{-2} \\ P^-(\mu L)K_{10}P^+(-\mu L)D_0^{-4} & P^-(\mu L)K_{11}D_1^{-2} \end{pmatrix}.$$

It follows that, for a given **K** such that $\bar{\rho}(\mathbf{K}) < 1$, the robust convergence is guaranteed with any μ such that

$$\bar{\rho}(\widetilde{\mathbf{K}}(\mu)) < 1.$$

Let us define

$$\mu_c \triangleq \sup\{\mu : \bar{\rho}(\widetilde{\mathbf{K}}(\mu)) < 1\}.$$

It follows that, for any $\nu \in (0, \mu_c)$, there exists $C > 0$ such that, for every solution of (3.3), (3.4),

$$\|\mathbf{R}(t,.)\|_{L^2((0,L);\mathbb{R}^n)} \leq Ce^{-\nu t}\|\mathbf{R}_o\|_{L^2((0,L);\mathbb{R}^n)}, \;\; \forall t \in [0, +\infty).$$

3.3.1 Application to a System of Two Conservation Laws

Let us consider again the system of two conservation laws

$$\partial_t \begin{pmatrix} R_1 \\ R_2 \end{pmatrix} + \begin{pmatrix} \lambda_1 & 0 \\ 0 & -\lambda_2 \end{pmatrix} \partial_x \begin{pmatrix} R_1 \\ R_2 \end{pmatrix} = \mathbf{0}, \qquad \lambda_1 > 0, \ \lambda_2 > 0, \tag{3.31}$$

under the boundary condition

$$\begin{pmatrix} R_1(t,0) \\ R_2(t,L) \end{pmatrix} = \begin{pmatrix} k_0 & k_1 \\ k_2 & k_3 \end{pmatrix} \begin{pmatrix} R_1(t,L) \\ R_2(t,0) \end{pmatrix}. \tag{3.32}$$

In this case, the change of coordinates (3.30) is

$$S_1 = p_1 e^{\mu t} e^{-\mu x/\lambda_1} R_1, \qquad S_2 = p_2 e^{\mu t} e^{\mu x/\lambda_2} R_2,$$

and the matrix $\widetilde{\mathbf{K}}(\mu)$ is

$$\widetilde{\mathbf{K}}(\mu) \triangleq \begin{pmatrix} k_0 e^{\mu \tau_1} & k_1 \dfrac{D_0^2}{D_1^2} \\ k_2 e^{\mu(\tau_1 + \tau_2)} \dfrac{D_1^2}{D_0^2} & k_3 e^{\mu \tau_2} \end{pmatrix}, \qquad \text{with } \tau_i \triangleq \dfrac{L}{\lambda_i}, \ i = 1, 2.$$

In the special case of local boundary conditions where $k_0 = k_3 = 0$ and $|k_1 k_2| < 1$, the value of μ_c is explicitly given by

$$\mu_c = \sup\{\mu : \bar{\rho}(\widetilde{\mathbf{K}}(\mu)) = 1\} = \frac{1}{\tau} \ln \left(\frac{1}{|k_1 k_2|} \right)$$

which is, as expected, identical to the convergence rate that we have found in Chapter 2.

3.4 Differential Linear Boundary Conditions

Up to now, in this chapter, we have discussed the stability of linear hyperbolic systems under static linear boundary conditions. In this section, we examine how the previous results can be generalized to the case of boundary conditions that are dynamic and represented by linear differential equations. More precisely, we consider the linear hyperbolic system of conservation laws in Riemann coordinates (3.3) under linear differential boundary conditions of the following form:

$$\dot{\mathbf{X}} = A\mathbf{X} + B\mathbf{R}_{\text{out}}(t),$$

$$\mathbf{R}_{\text{in}}(t) = C\mathbf{X} + K\mathbf{R}_{\text{out}}(t), \tag{3.33}$$

where $A \in \mathcal{M}_{\ell,\ell}(\mathbb{R})$, $B \in \mathcal{M}_{\ell,n}(\mathbb{R})$, $C \in \mathcal{M}_{n,\ell}(\mathbb{R})$, $\mathbf{K} \in \mathcal{M}_{n,n}(\mathbb{R})$, $\mathbf{X} \in \mathbb{R}^{\ell}$, $\ell \leqslant n$. The notations $\mathbf{R}_{\mathrm{in}}$ and $\mathbf{R}_{\mathrm{out}}$ were introduced in Section 1.1 and stand for

$$\mathbf{R}_{\mathrm{in}}(t) \triangleq \begin{pmatrix} \mathbf{R}^+(t,0) \\ \mathbf{R}^-(t,L) \end{pmatrix}, \qquad \mathbf{R}_{\mathrm{out}}(t) \triangleq \begin{pmatrix} \mathbf{R}^+(t,L) \\ \mathbf{R}^-(t,0) \end{pmatrix}.$$

The well-posedness of the Cauchy problem associated with this system is addressed in Appendix A, see Theorem A.6.

3.4.1   Frequency Domain

Using the Laplace transform, the system (3.3), (3.33) is written in the frequency domain as

$$\mathbf{R}_{\mathrm{out}}(s) = D(s)\mathbf{R}_{\mathrm{in}}(s) \quad \text{with} \quad D(s) \triangleq \mathrm{diag}\{e^{-s\tau_1}, \ldots, e^{-s\tau_n}\}, \quad \tau_i = L/\lambda_i,$$

$$(sI - A)\mathbf{X}(s) = B\mathbf{R}_{\mathrm{out}}(s), \qquad \mathbf{R}_{\mathrm{in}}(s) = C\mathbf{X}(s) + \mathbf{K}\mathbf{R}_{\mathrm{out}}(s).$$

Hence the poles of the system are the roots of the characteristic equation

$$\det\left[I - D(s)\big(C(sI - A)^{-1}B + \mathbf{K}\big)\right] = 0.$$

Theorem 3.14. The steady state $\mathbf{R}(t,x) \equiv 0$ of the system (3.3), (3.4) is exponentially stable for the L^{∞}-norm if and only if the poles of the system are stable (i.e., have strictly negative real parts and are bounded away from zero).

Proof. See (Hale and Verduyn-Lunel 2002, Section 3) and (Michiels and Niculescu 2007, Section 1.2).   □

3.4.2   Lyapunov Approach

In the line of the previous results of this chapter, we may also introduce the following Lyapunov function candidate:

$$\mathbf{V} = \int_0^L \left[\sum_{i=1}^m \frac{p_i}{\lambda_i} R_i^2(t,x) \exp(-\frac{\mu x}{\lambda_i}) + \sum_{i=m+1}^n \frac{p_i}{\lambda_i} R_i^2(t,x) \exp(+\frac{\mu x}{\lambda_i}) \right] dx + \sum_{j=1}^{\ell} q_j X_j^2$$

$$\tag{3.34}$$

with $\mathbf{X} \triangleq (X_1, \ldots, X_{\ell})^{\mathsf{T}}$, $p_i > 0$ $(i = 1, \ldots, n)$, $q_j > 0$ $(j = 1, \ldots, \ell)$.

The time derivative of this function along the C^1-solutions of (3.3), (3.33) is

$$\dot{\mathbf{V}} = -\mu\mathbf{V} + (\mathbf{R}_{\mathrm{out}}^{\mathsf{T}}, \mathbf{X}^{\mathsf{T}})\mathcal{M}(\mu)\begin{pmatrix} \mathbf{R}_{\mathrm{out}} \\ \mathbf{X} \end{pmatrix},$$

with the matrix

$$\mathcal{M}(\mu) \triangleq \begin{pmatrix} \mathbf{K}^{\mathsf{T}} P_1(\mu)\mathbf{K} - P_2(\mu) & \mathbf{K}^{\mathsf{T}} P_1(\mu)C + B^{\mathsf{T}}Q \\ C^{\mathsf{T}} P_1(\mu)\mathbf{K} + QB & \mu Q + C^{\mathsf{T}} P_1(\mu)C + (A^{\mathsf{T}}Q + QA) \end{pmatrix},$$

and

$$P_1(\mu) \triangleq \mathrm{diag}\left\{ p_1, \ldots, p_m, p_{m+1}\exp(\frac{\mu L}{\lambda_{m+1}}), \ldots, p_n\exp(\frac{\mu L}{\lambda_n}) \right\}$$

$$P_2(\mu) \triangleq \mathrm{diag}\left\{ p_1\exp(-\frac{\mu L}{\lambda_1}), \ldots, p_m\exp(-\frac{\mu L}{\lambda_m}), p_{m+1}, \ldots, p_n \right\}$$

$$Q \triangleq \mathrm{diag}\{q_1, \ldots, q_\ell\}.$$

Exponential stability holds if there exist $p_i > 0$ and $q_j > 0$ such that the matrix $\mathcal{M}(0)$ is negative definite (see Castillo et al. (2012) for a related reference). A simple example of this Lyapunov approach can be found in Theorem 2.9 for the stability analysis of a density-flow system under Proportional-Integral control.

3.4.3 Example: A Lossless Electrical Line Connecting an Inductive Power Supply to a Capacitive Load

Let us come back to the example of a lossless electrical line that we have presented in Section 2.1. We now consider the case where the line connects an inductive power supply to a capacitive load as shown in Fig. 3.4.

The dynamics of the line are described by the following system of two conservation laws:

$$\partial_t I + \frac{1}{L_\ell}\partial_x V = 0, \qquad \partial_t V + \frac{1}{C_\ell}\partial_x I = 0, \tag{3.35}$$

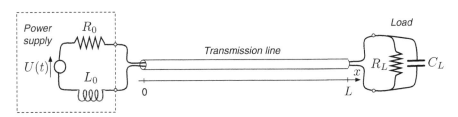

Fig. 3.4 Transmission line connecting an inductive power supply to a capacitive load

with the differential boundary conditions:

$$L_0 \frac{dI(t,0)}{dt} + R_0 I(t,0) + V(t,0) = U(t), \qquad C_L \frac{dV(t,L)}{dt} + \frac{V(t,L)}{R_L} = I(t,L).$$

$$(3.36)$$

For a given constant input voltage $U(t) = U^*$, the system has a unique constant steady state

$$I^* = \frac{U^*}{R_0 + R_\ell}, \qquad V^* = \frac{R_\ell U^*}{R_0 + R_\ell}.$$

The Riemann coordinates are defined as

$$R_1 \triangleq (V - V^*) + (I - I^*) \sqrt{L_\ell/C_\ell},$$

$$R_2 \triangleq (V - V^*) - (I - I^*) \sqrt{L_\ell/C_\ell},$$

with the inverse coordinates

$$I = I^* + \frac{R_1 - R_2}{2} \sqrt{L_\ell/C_\ell},$$

$$V = V^* + \frac{R_1 + R_2}{2}.$$

Then, expressing the dynamics (3.35) and the boundary conditions (3.36) in Riemann coordinates, we have:

$$\partial_t R_1 + \lambda_1 \partial_x R_1 = 0, \qquad \partial_t R_2 - \lambda_2 \partial_x R_2 = 0, \qquad \lambda_1 = \lambda_2 \triangleq \frac{1}{\sqrt{L_\ell C_\ell}},$$

$$\begin{pmatrix} \dot{X}_1 \\ \dot{X}_2 \end{pmatrix} = \underbrace{\begin{pmatrix} -\alpha_1 & 0 \\ 0 & -\alpha_2 \end{pmatrix}}_{A} \begin{pmatrix} X_1 \\ X_2 \end{pmatrix} + \underbrace{\begin{pmatrix} 0 & -\beta_1 \\ \beta_2 & 0 \end{pmatrix}}_{B} \begin{pmatrix} R_1(t,L) \\ R_2(t,0) \end{pmatrix},$$

$$\begin{pmatrix} R_1(t,0) \\ R_2(t,L) \end{pmatrix} = \underbrace{\begin{pmatrix} 1 & 0 \\ 0 & 1 \end{pmatrix}}_{C} \begin{pmatrix} X_1 \\ X_2 \end{pmatrix} + \underbrace{\begin{pmatrix} 0 & 1 \\ -1 & 0 \end{pmatrix}}_{K} \begin{pmatrix} R_1(t,L) \\ R_2(t,0) \end{pmatrix},$$

with

$$\alpha_1 = \frac{1}{L_0} \sqrt{\frac{C_\ell}{L_\ell}} + \frac{R_0}{L_0}, \qquad \alpha_2 = \frac{1}{C_L} \sqrt{\frac{L_\ell}{C_\ell}} + \frac{1}{R_L C_L},$$

$$\beta_1 = \frac{2}{L_0} \sqrt{\frac{C_\ell}{L_\ell}}, \qquad \beta_2 = \frac{2}{C_L} \sqrt{\frac{L_\ell}{C_\ell}}.$$

The characteristic equation is

$$(s + \alpha_1)(s + \alpha_2) + \underbrace{(s + \alpha_1 - \beta_1)(s + \alpha_2 - \beta_2)\, e^{-s\tau}}_{n(s)} = 0, \qquad \tau \overset{\Delta}{=} 2L\sqrt{L_\ell C_\ell}.$$

$$\underbrace{}_{d(s)}$$

In order to analyze the stability of the poles in function of the length L of the line, we follow the Walton and Marshall procedure as it is described in (Silva et al. 2005, Section 5.6).

The first step is to examine the stability when $L = 0$ (i.e., $\tau = 0$) where the characteristic equation reduces to the following second order polynomial with positive coefficients:

$$\tau = 0 \implies s^2 + \left(\frac{R_0}{L_0} + \frac{1}{R_L C_L}\right) s + \frac{1}{L_0 C_L}\left(1 + \frac{R_0}{R_L}\right) = 0.$$

Obviously, in that case the two poles are stable.

In the second step, we compute the following polynomial in ω^2:

$$W(\omega^2) \overset{\Delta}{=} d(j\omega)d(-j\omega) - n(j\omega)n(-j\omega)$$

$$= (-\omega^2 + (\alpha_1 + \alpha_2)j\omega + \alpha_1\alpha_2)(-\omega^2 - (\alpha_1 + \alpha_2)j\omega + \alpha_1\alpha_2)$$

$$- (-\omega^2 + (\alpha_1 + \alpha_2 - \beta_1 - \beta_2)j\omega + (\alpha_1 - \beta_1)(\alpha_2 - \beta_2))$$

$$(-\omega^2 - (\alpha_1 + \alpha_2 - \beta_1 - \beta_2)j\omega + (\alpha_1 - \beta_1)(\alpha_2 - \beta_2))$$

$$= (\alpha_1\alpha_2 - \omega^2)^2 + (\alpha_1 + \alpha_2)^2\omega^2 - \left((\alpha_1 - \beta_1)(\alpha_2 - \beta_2) - \omega^2\right)^2$$

$$- (\alpha_1 + \alpha_2 - \beta_1 - \beta_2)^2\omega^2.$$

After a few computations, we get

$$W(\omega^2) = (\gamma_1 + \gamma_2)\omega^2 - \gamma_1\gamma_2 + \gamma_1\alpha_2^2 + \gamma_2\alpha_1^2, \tag{3.37}$$

with

$$\gamma_1 \overset{\Delta}{=} \alpha_1^2 - (\alpha_1 - \beta_1)^2 = \frac{4R_0}{L_0^2}\sqrt{\frac{C_\ell}{L_\ell}}, \qquad \gamma_2 \overset{\Delta}{=} \alpha_2^2 - (\alpha_2 - \beta_2)^2 = \frac{4}{R_L C_L^2}\sqrt{\frac{L_\ell}{C_\ell}}.$$

It follows that the sign of $W(\omega^2)$ for large ω is positive. This means that all the system poles have strictly negative real parts for sufficiently small nonzero values of L.

In the third step, we observe that the polynomial (3.37) has a single root:

$$\omega^2 = \frac{\gamma_1\gamma_2 - \gamma_1\alpha_2^2 - \gamma_2\alpha_1^2}{(\gamma_1 + \gamma_2)}$$

which is negative for all positive values of the physical parameters R_0, R_L, L_0, L_ℓ, C_L, C_ℓ. In accordance with the physical intuition, we conclude that, whatever the length of the line, the poles of the system are stable for any line length L.

3.4.4 Example: A Network of Density-Flow Systems Under PI Control

In Section 1.15, we have emphasized that many physical networks of interest are represented by hyperbolic systems of conservation or balance laws. In this section, we consider the special case of acyclic networks of density-flow conservation laws under PI control which is a typical example of a hyperbolic system of conservation laws with differential boundary conditions. We examine how the stability conditions of Section 2.2 can be extended.

Depending on the concerned application, there are different ways of designing such networks. Here, as a matter of example, we consider a specific structure which leads to a natural generalization of Theorem 2.8. But other structures could be dealt with in a similar way, see, e.g., Marigo (2007) or Engel et al. (2008) for relevant related references.

The network has a compartmental structure illustrated in Fig. 3.5. The nodes of the network are n storage compartments having the dynamics of density-flow systems (e.g., the pipes of an hydraulic network):

$$\begin{cases} \partial_t H_j + \partial_x Q_j = 0, \\ \partial_t Q_j + \lambda_j \lambda_{n+j} \partial_x H_j + (\lambda_j - \lambda_{n+j}) \partial_x Q_j = 0, \end{cases} \quad j = 1, \ldots, n. \quad (3.38)$$

Without loss of generality and for simplicity, it can always be assumed that, by an appropriate scaling, all the systems have exactly the same length L.

The directed arcs $i \to j$ of the network represent instantaneous transfer flows between the compartments. Additional input and output arcs represent interactions with the surroundings: either inflows injected from the outside into some

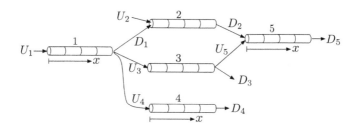

Fig. 3.5 Physical network of density-flow systems

compartments or outflows from some compartments to the outside. We assume that there is exactly one and only one control flow, denoted U_i, at the input of each compartment. All the other flows are assumed to be disturbances and denoted D_k ($k = 1, \ldots, m$). The set of $2n$ PDEs (3.38) is therefore subject to $2n$ boundary flow balance conditions of the form:

$$Q_i(t, 0) = U_i(t) + \sum_{k=1}^{m} \beta_{ik} D_k(t), \quad i = 1, \ldots, n,$$

$$Q_i(t, L) = \sum_{j=1}^{n} \alpha_{ij} U_j(t) + \sum_{k=1}^{m} \gamma_{ik} D_k(t), \quad i = 1, \ldots, n.$$

(3.39)

In the summations, only the terms corresponding to actual links between adjacent compartments of the network are taken into account, i.e., the coefficients β_{ik} and γ_{ik} are equal to 1 for the existing links and 0 for the others (see Fig. 3.5 for illustration).

With the matrix notations

$$\mathbf{H} \triangleq \begin{pmatrix} H_1 \\ \vdots \\ H_n \end{pmatrix}, \quad \mathbf{Q} \triangleq \begin{pmatrix} Q_1 \\ \vdots \\ Q_n \end{pmatrix}, \quad \mathbf{U} \triangleq \begin{pmatrix} U_1 \\ \vdots \\ U_n \end{pmatrix}, \quad \mathbf{D} \triangleq \begin{pmatrix} D_1 \\ \vdots \\ D_m \end{pmatrix},$$

$$\Lambda^+ = \mathrm{diag}\{\lambda_1, \ldots, \lambda_n\}, \quad \Lambda^- = \mathrm{diag}\{\lambda_{n+1}, \ldots, \lambda_{2n}\},$$

the system (3.38) is written

$$\partial_t \mathbf{H} + \partial_x \mathbf{Q} = 0,$$

$$\partial_t \mathbf{Q} + \Lambda^+ \Lambda^- \partial_x \mathbf{H} + (\Lambda^+ - \Lambda^-) \partial_x \mathbf{Q} = 0.$$

(3.40)

The boundary conditions (3.39) are written

$$\mathbf{Q}(t, 0) = \mathbf{U}(t) + B_0 \mathbf{D}(t),$$

$$\mathbf{Q}(t, L) = A_L \mathbf{U}(t) + B_L \mathbf{D}(t),$$

(3.41)

where A_L, B_0, and B_L are the matrices with entries α_{ij}, β_{jk}, and γ_{ik} respectively. For example, in the network of Fig. 3.5, we have

$$A_L \triangleq \begin{bmatrix} 0 & 0 & 1 & 1 & 0 \\ 0 & 0 & 0 & 0 & 0 \\ 0 & 0 & 0 & 0 & 1 \\ 0 & 0 & 0 & 0 & 0 \\ 0 & 0 & 0 & 0 & 0 \end{bmatrix}, \quad B_0 \triangleq \begin{bmatrix} 0 & 0 & 0 & 0 & 0 \\ 1 & 0 & 0 & 0 & 0 \\ 0 & 0 & 0 & 0 & 0 \\ 0 & 0 & 0 & 0 & 0 \\ 0 & 1 & 0 & 0 & 0 \end{bmatrix}, \quad B_L \triangleq \begin{bmatrix} 1 & 0 & 0 & 0 & 0 \\ 0 & 1 & 0 & 0 & 0 \\ 0 & 0 & 1 & 0 & 0 \\ 0 & 0 & 0 & 1 & 0 \\ 0 & 0 & 0 & 0 & 1 \end{bmatrix}.$$

Since the network is acyclic, the nodes of the network can be numbered such that the square matrix A_L is strictly upper triangular. Therefore the matrix A_L has the property that

$$A_L^p = 0, \qquad (3.42)$$

where p is the length of the longest path in the network.

A steady state for the system (3.40), (3.41) is a quadruple

$$\{\mathbf{H}^*, \mathbf{Q}^*, \mathbf{U}^*, \mathbf{D}^*\}$$

which satisfies the boundary conditions:

$$\mathbf{Q}^* = \mathbf{U}^* + B_0 \mathbf{D}^*,$$

$$\mathbf{Q}^* = A_L \mathbf{U}^* + B_L \mathbf{D}^*.$$

The network has an infinity of positive steady states which are not asymptotically stable. In order to stabilize the network, each control input is endowed with a PI control law of the form:

$$U_i(t) \triangleq U_R + k_{Pi}(H_i^* - H_i(t,0)) + k_{Ii} \int_0^t (H_i^* - H_i(\tau,0))d\tau, \qquad (3.43)$$

where U_R is an arbitrary scaling constant, H_i^* is the set point for the ith compartment, k_{Pi} and k_{Ii} are the control tuning parameters. In matrix form, the set of control laws (3.43) is written

$$\mathbf{U} = \mathbf{U}_R + K_P\left(\mathbf{H}^* - \mathbf{H}(t,0)\right) + K_I \int_0^t \left(\mathbf{H}^* - \mathbf{H}(\tau,0)\right)d\tau, \qquad (3.44)$$

with $K_P \triangleq \mathrm{diag}\{k_{P1}, \ldots k_{Pn}\}$ and $K_I \triangleq \mathrm{diag}\{k_{I1}, \ldots k_{In}\}$.

We shall now examine how the stability analysis in the frequency domain can be generalized to the closed-loop system (3.40), (3.41), (3.44) for constant unknown disturbances $\mathbf{D}^*$. The Riemann coordinates are defined as follows:

$$\begin{cases} R_i \triangleq Q_i - Q_i^* + \lambda_{n+i}(H_i - H_i^*), \\ R_{n+i} \triangleq Q_i - Q_i^* - \lambda_i(H_i - H_i^*), \end{cases} \quad i = 1, \ldots, n.$$

Using this definition, we have the following system dynamics

$$\begin{cases} \partial_t R_i + \lambda_i \partial_x R_i = 0, \\ \partial_t R_{n+i} - \lambda_{n+i} \partial_x R_{n+i} = 0, \end{cases} \quad i = 1, \ldots, n,$$

and the following equalities hold at the boundaries:

$$(\lambda_i + \lambda_{n+i})(Q_i(t,0) - Q_i^*) = (\lambda_i + \lambda_{n+i})\left[k_{P_i}(H_i^* - H_i(t,0)) + k_{I_i}Z_i(t)\right]$$

$$= \lambda_i R_i(t,0) + \lambda_{n+i}R_{n+i}(t,0)$$

$$= k_{P_i}(R_{n+i}(t,0) - R_i(t,0)) + (\lambda_i + \lambda_{n+i})k_{I_i}Z_i(t),$$

$$(\lambda_i + \lambda_{n+i})(Q_i(t,L) - Q_i^*) = \lambda_i R_i(t,L) + \lambda_{n+i}R_{n+i}(t,L),$$

with $Z_i(t)$ such that

$$\frac{dZ_i}{dt} = H_i^* - H_i(t,0) = \frac{R_{n+i}(t,0) - R_i(t,0)}{\lambda_i + \lambda_{n+i}}.$$

Since $R_i(t,x)$ and $R_{n+i}(t,x)$ are constant along their respective characteristic lines, we have that

$$R_i(t + \frac{L}{\lambda_i}, L) = R_i(t,0) \quad \text{and} \quad R_{n+i}(t + \frac{L}{\lambda_{n+i}}, 0) = R_{n+i}(t,L).$$

Then, combining appropriately these equalities, it can be shown after some computations that, in the frequency domain, the transfer function between $(Q_i(t,L) - Q_i^*)$ and $(Q_i(t,0) - Q_i^*)$ is given by:

$$G_i(s) \triangleq \frac{Q_i(s,0) - Q_i^*}{Q_i(s,L) - Q_i^*} = \frac{1}{\lambda_{n+i}} \frac{s(\lambda_i k_i - \lambda_{n+i}) + c_i(\lambda_i - \lambda_{n+i})}{(e^{s\tau_i} - k_i k_{n+i})s + c_i(e^{s\tau_i} - k_{n+i})} e^{\frac{sL}{\lambda_i}},$$

with the following notations:

$$k_i \triangleq \frac{k_{P_i} - \lambda_{n+i}}{k_{P_i} + \lambda_i}, \qquad k_{n+i} \triangleq -\frac{\lambda_i}{\lambda_{n+i}},$$

$$c_i \triangleq \frac{k_{I_i}}{k_{P_i} + \lambda_i}, \qquad \tau_i \triangleq \frac{L}{\lambda_i} + \frac{L}{\lambda_{n+i}}.$$

It follows that the poles of each transfer function $G_i(s)$ are the roots of the characteristic equation

$$(e^{s\tau_i} - k_i k_{n+i})s + c_i(e^{s\tau_i} - k_{n+i}) = 0,$$

which is, as expected, identical to the characteristic equation of the simple case of Section 2.2.

Let us now consider the closed-loop system (3.40), (3.41), (3.44) as an input-output dynamical system with input **D** and output **U**. Then, by iterating equations (3.41) p times and using property (3.42), it can be shown that the transfer matrix of the system is as follows:

$$H(s) \triangleq \sum_{i=0}^{p-1}(G(s)A_L)^i(G(s)B_L - B_0),$$

with $G(s) \triangleq \mathrm{diag}\{G_1(s), \ldots, G_n(s)\}$. For example, in the network of Fig. 3.5, since $p = 3$, we have

$$H(s) = (I + G(s)A_L + G(s)A_LG(s)A_L)(G(s)B_L - B_0)$$

$$= \begin{pmatrix} G_1 & -G_1G_3 & G_1G_3 & G_1G_4 & G_1G_3G_5 \\ -1 & G_2 & 0 & 0 & 0 \\ 0 & -G_3 & G_3 & 0 & G_3G_5 \\ 0 & 0 & 0 & G_4 & 0 \\ 0 & -1 & 0 & 0 & G_5 \end{pmatrix}.$$

As illustrated with this example, the poles of $H(s)$ are given by the collection of the poles of the individual scalar transfer function $G_i(s)$. Consequently, the system is stable if and only if the conditions of Theorem 2.8 hold for each PI controller of the network, i.e., if and only if for $i = 1, \ldots, n$,

- when $\lambda_i < \lambda_{n+i}$,

$$k_{Pi} > 0 \text{ and } k_{Ii} > 0 \quad \text{or} \quad k_{Pi} < -\frac{2\lambda_i\lambda_{n+i}}{\lambda_{n+i} - \lambda_i} \text{ and } k_{Ii} < 0;$$

- when $\lambda_i = \lambda_{n+i}$, $k_{Pi} > 0 \text{ and } k_{Ii} > 0;$
- when $\lambda_i > \lambda_{n+i}$,

$$0 < k_{Pi} < \frac{2\lambda_i\lambda_{n+i}}{\lambda_i - \lambda_{n+i}} \text{ and } 0 < k_{Ii} < \omega_i \frac{(2k_P + \lambda_i - \lambda_{n+i})\lambda_i\lambda_{n+i}}{\lambda_i^2 - \lambda_{n+i}^2} \sin(\omega_i\tau_i)$$

where ω_i is the smallest positive ω such that

$$\cos(\omega\tau_i) = \frac{\lambda_{n+i}^2(k_{Pi} + \lambda_i) + \lambda_i^2(k_{Pi} - \lambda_{n+i})}{\lambda_i\lambda_{n+i}(\lambda_{n+i} - \lambda_i - 2k_{Pi})}.$$

3.4.5 Example: Stability of Genetic Regulatory Networks

In Section 1.15.3 (p. 50), we have shown how the dynamics of genetic regulatory networks are represented by *linear* hyperbolic systems with *nonlinear differential boundary conditions* of the form

$$\mathbf{R}_t + \Lambda_{\mathbf{R}}\mathbf{R}_x = 0, \qquad \mathbf{S}_t + \Lambda_{\mathbf{S}}\mathbf{S}_x = 0, \qquad t \in [0, +\infty), \qquad x \in [0, 1],$$

$$\frac{d\mathbf{R}(t,0)}{dt} = AS(t,1) - B\mathbf{R}(t,0), \qquad \frac{d\mathbf{S}(t,0)}{dt} = G(\mathbf{R}(t,1))\mathbf{R}(t,1) - DS(t,0),$$

$$\text{(3.45)}$$

with $\Lambda_{\mathbf{R}} \in \mathcal{D}_n^+$, $\Lambda_{\mathbf{S}} \in \mathcal{D}_n^+$, $A \in \mathcal{D}_n^+$, $B \in \mathcal{D}_n^+$, $D \in \mathcal{D}_n^+$, and $G : \mathcal{R}_+^n \to \mathcal{M}_{n,n}(\mathbb{R})$ is a smooth bounded function of $\mathbf{R}$ representing genetic activations or repressions with $\mathcal{R}_+^n$ being the orthant $\{\mathbf{R} \in \mathbb{R}^n; R_i > -P_i^*, i = 1, n\}$. For this system, we see that the solutions are confined in the orthant $\mathcal{R}_+^n \times \mathcal{S}_+^n$ with $\mathcal{S}_+^n \triangleq \{\mathbf{S} \in \mathbb{R}^n; S_i > -M_i^*, i = 1, n\}$. This means that, in accordance with the physical reality, if the initial conditions $\mathbf{R}(0, x)$ and $\mathbf{S}(0, x)$ are in the orthant $\mathcal{R}_+^n \times \mathcal{S}_+^n$, the solutions of the system are guaranteed to stay in the same orthant for all time.

The stability of the system (3.45) is analyzed with the following Lyapunov function candidate:

$$\mathbf{V} = \int_0^1 \left[\mathbf{R}^\top(t,x) Q_{\mathbf{R}} T_{\mathbf{R}} E_{\mathbf{R}}(\mu, x)\mathbf{R}(t,x) + \mathbf{S}^\top(t,x) Q_{\mathbf{S}} T_{\mathbf{S}} E_{\mathbf{S}}(\mu, x)\mathbf{S}(t,x) \right] dx$$

$$+ \tfrac{1}{2}\mathbf{R}^\top(t,1) W_{\mathbf{R}} \mathbf{R}(t,1) + \tfrac{1}{2}\mathbf{S}^\top(t,1) W_{\mathbf{S}} \mathbf{S}(t,1),$$

$$Q_{\mathbf{R}} = \text{diag}\{q_1, \ldots, q_n\},$$

$$Q_{\mathbf{S}} = \text{diag}\{q_{n+1}, \ldots, q_{2n}\},$$

$$T_{\mathbf{R}} = \text{diag}\{\tau_1, \ldots, \tau_n\},$$

$$T_{\mathbf{S}} = \text{diag}\{\tau_{n+1}, \ldots, \tau_{2n}\},$$

$$E_{\mathbf{R}}(\mu, x) = \text{diag}\{e^{-\mu\tau_1 x}, \ldots, e^{-\mu\tau_n x}\},$$

$$E_{\mathbf{S}}(\mu, x) = \text{diag}\{e^{-\mu\tau_{n+1} x}, \ldots, e^{-\mu\tau_{2n} x}\},$$

$$W_{\mathbf{R}} = \text{diag}\{w_1, \ldots, w_n\},$$

$$W_{\mathbf{S}} = \text{diag}\{w_{n+1}, \ldots, w_{2n}\}.$$

Using integration by parts, it can be shown that the time derivative of $\mathbf{V}$, along the system solutions, is

$$\frac{d\mathbf{V}}{dt} = -\mu\mathbf{V} - \left(\mathbf{R}^\top(t,0) \ \mathbf{S}^\top(t,0) \ \mathbf{R}^\top(t,1) \ \mathbf{S}^\top(t,1)\right) \mathcal{M}(\mu, \mathbf{R}) \begin{pmatrix} \mathbf{R}(t,0) \\ \mathbf{S}(t,0) \\ \mathbf{R}(t,1) \\ \mathbf{S}(t,1) \end{pmatrix},$$

with the matrix $\mathcal{M}(\mu, \mathbf{R})$ defined as

$$\mathcal{M}(\mu, \mathbf{R}) = \begin{pmatrix} W_{\mathbf{R}} B - Q_{\mathbf{R}} & 0 & 0 & -W_{\mathbf{R}} A \\ 0 & W_{\mathbf{S}} D - Q_{\mathbf{S}} & -W_{\mathbf{S}} G(\mathbf{R}) & 0 \\ 0 & 0 & Q_{\mathbf{R}} E_{\mathbf{R}}(\mu, 1) & 0 \\ 0 & 0 & 0 & Q_{\mathbf{S}} E_{\mathbf{S}}(\mu, 1) \end{pmatrix}.$$

We have the following proposition.

Proposition 3.15. There exists $\mu > 0$ sufficiently small such that **V** is a strict exponentially decreasing Lyapunov function along the solutions of the system (3.45) if there exist $q_i > 0$, $q_{n+i} > 0$, $w_i > 0$, $w_{n+i} > 0$, and $\delta > 0$ such that $\mathcal{M}(0, \mathbf{R}) + \mathcal{M}^T(0, \mathbf{R}) > \delta I_n$ for all $\mathbf{R} \in \mathcal{R}_+^n$.

Let us now use the 'toggle switch' as an example of how this proposition can be used. A toggle switch is a system of two genes that repress each other (see, e.g., Smits et al. (2008) and the references therein). In the case of the toggle switch, the general model (3.45) is specialized as follows:

$$\partial_t R_1(t, x) + \frac{1}{\tau_1} \partial_x R_1(t, x) = 0,$$

$$\partial_t R_2(t, x) + \frac{1}{\tau_2} \partial_x R_2(t, x) = 0,$$

$$\partial_t S_1(t, x) + \frac{1}{\tau_3} \partial_x S_1(t, x) = 0,$$

$$\partial_t S_2(t, x) + \frac{1}{\tau_4} \partial_x S_2(t, x) = 0,$$

$$\frac{dR_1(t, 0)}{dt} = \alpha_1 S_1(t, 1) - \beta_1 R_1(t, 0),$$

$$\frac{dR_2(t, 0)}{dt} = \alpha_2 S_2(t, 1) - \beta_2 R_2(t, 0),$$

$$\frac{dS_1(t, 0)}{dt} = g_1(R_2(t, 1))R_2(t, 1) - \delta_1 S_1(t, 0),$$

$$\frac{dS_2(t, 0)}{dt} = g_2(R_1(t, 1))R_1(t, 1) - \delta_2 S_2(t, 0).$$

For this system, the matrix $\mathcal{M}(0, \mathbf{R})$ is

$$\mathcal{M}(0, \mathbf{R}) = \begin{pmatrix} \mathcal{M}_{11} & \mathcal{M}_{12} \\ 0 & \mathcal{M}_{22} \end{pmatrix},$$

with

$$\mathcal{M}_{11} = \begin{pmatrix} w_1\beta_1 - q_1 & 0 & 0 & 0 \\ 0 & w_2\beta_2 - q_2 & 0 & 0 \\ 0 & 0 & w_3\delta_1 - q_3 & 0 \\ 0 & 0 & 0 & w_4\delta_2 - q_4 \end{pmatrix},$$

$$\mathcal{M}_{12} = \begin{pmatrix} 0 & 0 & -w_1\alpha_1 & 0 \\ 0 & 0 & 0 & -w_2\alpha_2 \\ 0 & -w_3g_1(R_2) & 0 & 0 \\ -w_4g_2(R_1) & 0 & 0 & 0 \end{pmatrix},$$

$$\mathcal{M}_{22} = \begin{pmatrix} q_1 & 0 & 0 & 0 \\ 0 & q_2 & 0 & 0 \\ 0 & 0 & q_3 & 0 \\ 0 & 0 & 0 & q_4 \end{pmatrix}.$$

This matrix is positive definite if and only if the leading principal minors of the symmetric matrix $\mathcal{M}(0, \mathbf{R}) + \mathcal{M}^T(0, \mathbf{R})$ are all positive. This leads to the following inequalities:

$$0 < w_1\beta_1 - q_1$$
$$0 < w_2\beta_2 - q_2$$
$$0 < w_3\delta_1 - q_3$$
$$0 < w_4\delta_2 - q_4$$
$$0 < (w_4\delta_2 - q_4)q_1 - (1/4)w_4^2g_2^2(R_1)$$
$$0 < (w_3\delta_1 - q_3)q_2 - (1/4)w_3^2g_1^2(R_2)$$
$$0 < (w_1\beta_1 - q_1)q_3 - (1/4)w_1^2\alpha_1^2$$
$$0 < (w_2\beta_2 - q_2)q_4 - (1/4)w_2^2\alpha_2^2.$$

Hence, the system is stable for any τ_i if there exist positive values of q_i and w_i such that these inequalities are satisfied for all (R_1, R_2) in $\mathcal{R}_+^2$.

3.5 The Nonuniform Case

In this section, we explain how the previous results of this chapter are trivially extended to the nonuniform case where the characteristic velocities $\lambda_i(x)$ depend of the spatial coordinate. We consider the linear hyperbolic system:

$$\mathbf{R}_t + \Lambda(x)\mathbf{R}_x = \mathbf{0} \tag{3.46}$$

with the diagonal matrix $\Lambda(x) \triangleq \text{diag}\{\Lambda^+(x), -\Lambda^-(x)\}$ such that

$$\Lambda^+(x) = \text{diag}\{\lambda_1(x), \dots, \lambda_m(x)\},$$
$$\Lambda^-(x) = \text{diag}\{\lambda_{m+1}(x), \dots, \lambda_n(x)\}, \qquad \lambda_i(x) > 0 \ \forall i, \quad \forall x \in [0, L].$$

Our concern is to analyze the exponential stability of this system under linear boundary conditions in canonical form

$$\begin{pmatrix} \mathbf{R}^+(t,0) \\ \mathbf{R}^-(t,L) \end{pmatrix} = \mathbf{K} \begin{pmatrix} \mathbf{R}^+(t,L) \\ \mathbf{R}^-(t,0) \end{pmatrix}, \quad t \in [0,+\infty), \tag{3.47}$$

and an initial condition

$$\mathbf{R}(0,x) = \mathbf{R}_o(x), \quad x \in (0,L). \tag{3.48}$$

The well-posedness of the Cauchy problem (3.46), (3.47), (3.48) in L^2 results, as a special case, from Theorem A.4 in Appendix A.

We have the following stability theorem.

Theorem 3.16. The system (3.46), (3.47) is exponentially stable for the L^2-norm (in the sense of Definition 3.1) if $\rho_2(\mathbf{K}) < 1$.

Proof. We use the following candidate Lyapunov function which is a direct extension of the function used in Section 2.3:

$$V = \int_0^L \left[(\mathbf{R}^{+\mathsf{T}}(\Lambda^+)^{-1}P^+(\mu x)\mathbf{R}^+) + (\mathbf{R}^{-\mathsf{T}}(\Lambda^-)^{-1}P^-(\mu x)\mathbf{R}^-) \right] dx$$

where

$$P^+(\mu x) \triangleq \mathrm{diag}\left\{ \frac{p_1}{\lambda_1(x)} \exp\left(-\int_0^x \frac{\mu}{\lambda_1(\sigma)} d\sigma \right), \ldots, \frac{p_m}{\lambda_m(x)} \exp\left(-\int_0^x \frac{\mu}{\lambda_m(\sigma)} d\sigma \right) \right\},$$

$$P^-(\mu x) \triangleq \mathrm{diag}\left\{ \frac{p_{m+1}}{\lambda_{m+1}(x)} \exp\left(\int_0^x \frac{\mu}{\lambda_1(\sigma)} d\sigma \right), \ldots, \frac{p_n}{\lambda_n(x)} \exp\left(\int_0^x \frac{\mu}{\lambda_n(\sigma)} d\sigma \right) \right\},$$

with positive coefficients μ and p_i, $i = 1, \ldots, n$.

With this definition of the Lyapunov function, the proof of the theorem is a direct extension of the proof of Theorem 2.11 which can be written as a replicate of the proof of Theorem 3.2.   □

3.6   Switching Linear Conservation Laws

In certain practical applications, it is of interest to address situations where the system exhibits periodic time switching between various sets of boundary conditions. From the viewpoint of exponential stability analysis, a system of conservation laws with switching boundary conditions can be viewed as a hybrid system on an infinite dimensional state space. While hybrid systems based on ordinary differential equations are extensively considered in the literature (e.g., Liberzon (2003), De Schutter and Heemels (2011), Shorten et al. (2007)), hybrid systems based on partial differential equations are relatively unexplored.

In this section, through the specific example of SMB chromatography, our purpose is to illustrate how exponential stability (in L_2-norm) can be established by switching between Lyapunov functions. The obtained stability conditions are direct generalizations of the corresponding results for the unswitched case. Other interesting references on the stability analysis for linear hyperbolic switching systems can be found in Sections 3.7 and 5.7.

3.6.1 The Example of SMB Chromatography

As described in Section 1.13, SMB chromatography is a technology where interconnected chromatographic columns are switched periodically. The SMB chromatography model is given by equations (1.61). It exhibits a periodic steady state denoted $\mathbf{C}^*(t, x)$. Here we consider the linear case which is the special case where $b_A = 0$ and $b_B = 0$. The linear system is therefore written:

$$mT \leqslant t < (m+1)T, \quad m = 0, 1, 2, 3, 4, 5, \ldots, \infty, \quad \ell \in \{A, B\},$$

$$(1 + h_\ell)\partial_t \mathbf{C}^\ell + (P^m)\Upsilon(P^m)^\mathsf{T}\partial_x \mathbf{C}^\ell = 0, \tag{3.49}$$

$$\mathbf{C}^\ell(t, 0) = (P^m)\mathbf{K}(P^m)^\mathsf{T}\mathbf{C}^\ell(t, L) + (P^m)U^\ell.$$

In order to put the system in characteristic form, we define the Riemann coordinates:

$$\begin{aligned} R_i^A &= (1 + h_A)(C_i^A - C_i^{*A}), \\ R_i^B &= (1 + h_B)(C_i^B - C_i^{*B}), \end{aligned} \quad i = 1, 2, 3, 4.$$

In these Riemann coordinates, the periodic linear system is written (see Section 1.13)

$$mT \leqslant t < (m+1)T, \quad m = 0, 1, 2, 3, 4, 5, \ldots, \infty, \quad \ell \in \{A, B\},$$

$$\partial_t \mathbf{R}^\ell + \Lambda_m^\ell \partial_x \mathbf{R}^\ell = 0, \tag{3.50}$$

$$\mathbf{R}^\ell(t, 0) = \mathbf{K}_m \mathbf{R}^\ell(t, L),$$

with the following notations:

$$\Lambda^\ell \triangleq \mathrm{diag}\{\lambda_1^\ell, \lambda_2^\ell, \lambda_1^\ell, \lambda_2^\ell\}$$

$$\text{with} \quad \lambda_1^\ell \triangleq \frac{V_\mathrm{I}}{1 + h_\ell}, \quad \lambda_2^\ell \triangleq \frac{V_\mathrm{II}}{1 + h_\ell},$$

$$\Lambda_m^\ell \triangleq (P^m)\Lambda^\ell(P^m)^\mathsf{T}, \quad \mathbf{K}_m = (P^m)\mathbf{K}(P^m)^\mathsf{T}.$$

We have the following stability property.

Theorem 3.17. The periodic solution $\mathbf{C}^*(t, x)$ of the system (3.50) is exponentially stable if

$$T > \frac{L}{\lambda_2^\ell} - \frac{L}{\lambda_1^\ell}, \quad \ell \in \{A, B\}.$$

Proof. As advocated in Branicky (1998) for the analysis of hybrid systems, we follow a so-called "multiple Lyapunov function" approach with the two following candidate quadratic Lyapunov functions:

$$\mathbf{V}_1 \triangleq \sum_{\ell \in \{A,B\}} \int_0^L \left\{ \frac{p_1}{\lambda_1^\ell} \left([R_1^\ell(t, x)]^2 + [R_3^\ell(t, x)]^2 \right) \exp\left(-\frac{\mu x}{\lambda_1^\ell} \right) \right.$$

$$\left. + \frac{p_2}{\lambda_2^\ell} \left([R_2^\ell(t, x)]^2 + [R_4^\ell(t, x)]^2 \right) \exp\left(-\frac{\mu x}{\lambda_2^\ell} \right) \right\} dx,$$

$$\mathbf{V}_2 \triangleq \sum_{\ell \in \{A,B\}} \int_0^L \left\{ \frac{p_2}{\lambda_2^\ell} \left([R_1^\ell(t, x)]^2 + [R_3^\ell(t, x)]^2 \right) \exp\left(-\frac{\mu x}{\lambda_2^\ell} \right) \right.$$

$$\left. + \frac{p_1}{\lambda_1^\ell} \left([R_2^\ell(t, x)]^2 + [R_4^\ell(t, x)]^2 \right) \exp\left(-\frac{\mu x}{\lambda_1^\ell} \right) \right\} dx,$$

with positive constant coefficients p_1, p_2 and μ.

Now, until the end of the proof, we consider only even values of m: $m \in \{0, 2, 4, \dots\}$. The time derivatives of $\mathbf{V}_1$ and $\mathbf{V}_2$ along the trajectories of the system (3.50) are

For $mT \leqslant t < (m + 1)T$,

$$\frac{d\mathbf{V}_1}{dt} = -\mu \mathbf{V}_1 - \sum_{\ell \in \{A,B\}} \left\{ \left(p_1 \exp\left(-\frac{\mu L}{\lambda_1^\ell} \right) - p_2 \right) \left([R_1^\ell(t, L)]^2 + [R_3^\ell(t, L)]^2 \right) \right\}$$

$$- \sum_{\ell \in \{A,B\}} \left\{ \left(p_2 \exp\left(-\frac{\mu L}{\lambda_2^\ell} \right) - \left(\frac{\lambda_2^\ell}{\lambda_1^\ell} \right)^2 p_1 \right) \left([R_2^\ell(t, L)]^2 + [R_4^\ell(t, L)]^2 \right) \right\}$$

For $(m + 1)T \leqslant t < (m + 2)T$,

$$\frac{d\mathbf{V}_2}{dt} = -\mu \mathbf{V}_2 - \sum_{\ell \in \{A,B\}} \left\{ \left(p_1 \exp\left(-\frac{\mu L}{\lambda_1^\ell} \right) - p_2 \right) \left([R_2^\ell(t, L)]^2 + [R_4^\ell(t, L)]^2 \right) \right\}$$

$$- \sum_{\ell \in \{A,B\}} \left\{ \left(p_2 \exp\left(-\frac{\mu L}{\lambda_2^\ell} \right) - \left(\frac{\lambda_2^\ell}{\lambda_1^\ell} \right)^2 p_1 \right) \left([R_1^\ell(t, L)]^2 + [R_3^\ell(t, L)]^2 \right) \right\}.$$

Since $V_I > V_{II}$ (see Section 1.13), the parameters p_1 and p_2 can be selected such that

$$1 < \frac{p_1}{p_2} < \left(\frac{\lambda_1^\ell}{\lambda_2^\ell}\right)^2 = \left(\frac{V_I}{V_{II}}\right)^2,$$

then $\mu > 0$ can be selected such that

$$\frac{p_1}{p_2}\exp\left(-\frac{\mu L}{\lambda_1^\ell}\right) > 1, \quad \frac{p_2}{p_1}\left(\frac{\lambda_1^\ell}{\lambda_2^\ell}\right)^2\exp\left(-\frac{\mu L}{\lambda_2^\ell}\right) > 1, \tag{3.51}$$

which imply

$$mT \leqslant t < (m+1)T,$$

$$\frac{d\mathbf{V}_1}{dt} \leqslant -\mu\mathbf{V}_1 \text{ and therefore } \mathbf{V}_1((m+1)T) \leqslant \mathbf{V}_1(mT)e^{-\mu T}, \tag{3.52}$$

$$(m+1)T \leqslant t < (m+2)T,$$

$$\frac{d\mathbf{V}_2}{dt} \leqslant -\mu\mathbf{V}_2 \text{ and therefore } \mathbf{V}_2((m+2)T) \leqslant \mathbf{V}_2((m+1)T)e^{-\mu T}. \tag{3.53}$$

Let us select p_1/p_2 as follows:

$$\frac{p_1}{p_2} = \frac{V_I}{V_{II}} > 1. \tag{3.54}$$

Let us define a parameter $\alpha > 1$ selected such that

$$\frac{1}{T}\left(\frac{L}{\lambda_2^\ell} - \frac{L}{\lambda_1^\ell}\right) < \frac{\ln\alpha}{\mu T} < 1. \tag{3.55}$$

Using inequalities (3.54) and (3.55), we then have, for every $x \in [0, L]$,

$$\alpha \geqslant \frac{\exp\left(-\frac{\mu L}{\lambda_1^\ell}\right)}{\exp\left(-\frac{\mu L}{\lambda_2^\ell}\right)} \geqslant \frac{\frac{p_1}{\lambda_1^\ell}\exp\left(-\frac{\mu x}{\lambda_1^\ell}\right)}{\frac{p_2}{\lambda_2^\ell}\exp\left(-\frac{\mu x}{\lambda_2^\ell}\right)} \geqslant 1.$$

By combining this inequality with the definitions of $\mathbf{V}_1$ and $\mathbf{V}_2$, it can be checked that

$$\frac{1}{\alpha}\mathbf{V}_2 \leqslant \mathbf{V}_1 \leqslant \alpha\mathbf{V}_2, \quad \forall t, \forall x. \tag{3.56}$$

From (3.52), (3.53), and (3.56), we then have

$$\mathbf{V}_1((m+2)T) \leqslant \alpha\mathbf{V}_2((m+2)T) \leqslant \alpha e^{-\mu T}\mathbf{V}_2((m+1)T)$$

$$\leqslant \alpha^2 e^{-\mu T}\mathbf{V}_1((m+1)T) \leqslant \left(\alpha e^{-\mu T}\right)^2 \mathbf{V}_1(mT).$$

Mutatis mutandis, obviously we also have

$$\mathbf{V}_2((m+3)T) \leqslant \left(\alpha e^{-\mu T}\right)^2 \mathbf{V}_2((m+1)T).$$

Now, from (3.55) we have:

$$\alpha e^{-\mu T} < 1.$$

Therefore, $\mathbf{V}_1(t)$ and $\mathbf{V}_2(t)$ exponentially converge to zero and the periodic time solution $\mathbf{C}^*$ is exponentially stable. □

3.6.1.1 A Simulation Experiment

As a matter of illustration, we present a simulation experiment of an SMB process implemented under the operating conditions reported in Nobre et al. (2013) for the separation of fructo-oligosaccharides.
The parameter values are

$$L = 0.248 \text{ m}, \quad T = 114.95 \text{ s}, \quad V_{\mathrm{I}} = 0.036 \text{ m/s}, \quad V_{\mathrm{II}} = 0.022 \text{ m/s},$$

$$h_A = 0.3954, \quad h_B = 0.0251, \quad C_F^A = 64 \text{ mg/ml}, \quad C_F^B = 85 \text{ mg/ml},$$

$$V_F = V_{\mathrm{I}} - V_{\mathrm{II}} = 0.014 \text{ m/s}.$$

From these values, we verify that the stability condition of Theorem 3.17 is satisfied since

$$T > \frac{L}{\lambda_2^A} - \frac{L}{\lambda_1^B} \approx 56 \text{ s} > \frac{L}{\lambda_2^B} - \frac{L}{\lambda_1^B} \approx 44 \text{ s}.$$

We simulate the start-up of the process from zero initial conditions (i.e., zero initial concentrations in the columns). The simulation results are shown in Figs. 3.6 and 3.7. We see in Fig. 3.6 that the steady state periodic regime is reached within about 10 column shifts (i.e., 2.5 rounds).

It may also be observed in Fig. 3.7 that the separation between species A and B is effective but not perfect. This is an inherent limitation of SMB processes implemented with four columns as considered here for simplicity. In order to reach a total purity of the separation, industrial SMB processes are generally implemented with eight (e.g., Suvarov et al. (2012)) or even twelve columns (e.g., Lorenz et al. (2001)).

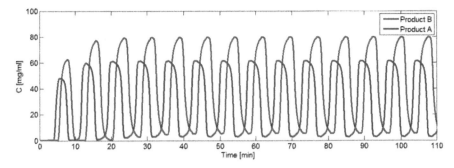

Fig. 3.6 Time evolution of the concentrations inside column 1: exponential convergence towards the periodic regime

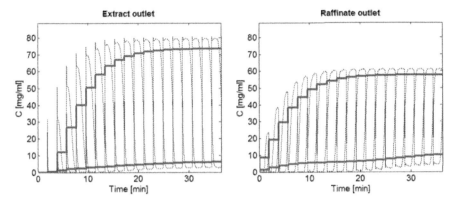

Fig. 3.7 Time evolution of the outlet concentrations (dotted line: actual concentration, solid line: average concentration)

3.7 References and Further Reading

Tang et al. (2015b) address the issue of singular perturbations in linear systems of conservation laws. For systems having a slow-fast behavior, they show that the stability in L^2-norm of the full system implies necessarily the stability for the L^2-norm of both the slow reduced system and the fast boundary-layer system, while the converse is not true. Furthermore, they establish a generalization of Tikhonov's theorem for this class of infinite-dimensional systems. An application to the control of gas transport systems described by Euler equations where the small parameter is the ratio between gas and sound velocities illustrates the theory.

In Section 3.4, we have illustrated the stability of linear hyperbolic systems under differential boundary condition with the example of a power source connected to a load through a lossless electrical line. In Daafouz et al. (2014), the authors address the nonlinear control of this system in the case where the source is a switched power converter.

In Section 3.4, we have given the necessary and sufficient stability conditions for density-flow systems under PI control. Related references dealing with PI control of hyperbolic systems are Xu and Sallet (1999), Dos Santos et al. (2008), Dos Santos Martins and Rodrigues (2011) where sufficient stability conditions are given using respectively spectral, Lyapunov, and LMI approaches. Experimental validations on a mini-channel set-up are also reported in Dos Santos et al. (2008).

In Section 3.6, we have presented a simple example of the stability analysis under switching boundary conditions. The exponential stability in L^2-norm for a class of switched linear systems of conservation laws is further investigated by Lamare et al. (2013) and Lamare et al. (2015b) in the case where the state equations and the boundary conditions are both subject to switching. The authors consider the problem of synthesizing stabilizing switching controllers. By means of Lyapunov techniques, three control strategies are developed based on steepest descent selection, possibly combined with a hysteresis and a low-pass filter.

Chapter 4
Systems of Nonlinear Conservation Laws

THE PURPOSE of this chapter is to extend the exponential stability analysis to the case of systems of nonlinear conservation laws of the form

$$\mathbf{Y}_t + f(\mathbf{Y})_x = \mathbf{0}, \quad t \in [0, +\infty), \quad x \in [0, L], \tag{4.1}$$

where $\mathbf{Y} : [0, +\infty) \times [0, L] \to \mathbb{R}^n$ and $f \in C^2(\mathcal{Y}; \mathbb{R}^n)$.

Since the frequency domain is irrelevant for the nonlinear case, the main concern will be to examine how the Lyapunov approach is extended to nonlinear hyperbolic systems.

For a nonlinear differential system, the well-known *Lyapunov's indirect method* allows to deduce the local exponential stability of an equilibrium from the exponential stability of its linearization with the same Lyapunov function, see for instance (Khalil 1996, Chapter 3). In contrast, for hyperbolic systems, such an implication does not hold: the exponential stability of the linearization for a given norm does not directly induce the local exponential stability of the underlying nonlinear system for the same norm and with the same Lyapunov function. The analysis is more intricate as we shall see in the present chapter.

An additional difficulty comes from the fact that, even for smooth initial conditions that are close to the steady-state, the trajectories of nonlinear conservation laws may become discontinuous in finite time generating jump discontinuities which propagate on as *shocks*. Fortunately, the main result of this chapter will be to show that, if the boundary conditions are dissipative and if the initial conditions are smooth enough and sufficiently close to the steady state, the system trajectories are guaranteed to remain smooth (i.e., without shocks) for all time and that they exponentially converge locally to the steady state. Surprisingly enough, due to the nonlinearity of the system, even for smooth solutions, it will appear that the exponential stability strongly depends on the considered norm. In particular, using the Lyapunov approach, it is shown in Section 4.4 that the robust dissipativity test

© Springer International Publishing Switzerland 2016
G. Bastin, J.-M. Coron, *Stability and Boundary Stabilization of 1-D Hyperbolic Systems*, Progress in Nonlinear Differential Equations and Their Applications 88,
DOI 10.1007/978-3-319-32062-5_4

$\rho_2(\mathbf{K}) < 1$ of linear systems holds also in the nonlinear case for the H^2-norm, while, in the next section, we shall see that a more conservative test is needed for solutions in C^1.

For smooth solutions, the system of conservation laws (4.1) is equivalent to the quasi-linear system

$$\mathbf{Y}_t + F(\mathbf{Y})\mathbf{Y}_x = \mathbf{0}, \quad t \in [0, +\infty), \quad x \in [0, L], \tag{4.2}$$

where $F \in C^1(\mathcal{Y}; \mathcal{M}_{n,n}(\mathbb{R}))$ is the Jacobian matrix of f. Our main concern in this chapter is to address the exponential stability of the steady states of this quasi-linear system. For simplicity, we start the analysis with the special case of systems that can be written into characteristic form. The case of generic systems (4.2) that cannot be transformed into characteristic form is treated at the end of the chapter in Section 4.5.

The systems under consideration have the following form in Riemann coordinates:

$$\mathbf{R}_t + \Lambda(\mathbf{R})\mathbf{R}_x = \mathbf{0}, \quad t \in [0, +\infty), \quad x \in (0, L), \tag{4.3}$$

where $\mathbf{R} : [0, +\infty) \times [0, L] \to \mathbb{R}^n$. The map Λ is defined as:

$$\Lambda(\mathbf{R}) \triangleq \begin{pmatrix} \Lambda^+(\mathbf{R}) & 0 \\ 0 & -\Lambda^-(\mathbf{R}) \end{pmatrix}$$

with

$$\begin{cases} \Lambda^+(\mathbf{R}) = \mathrm{diag}\{\lambda_1(\mathbf{R}), \ldots, \lambda_m(\mathbf{R})\}, \\ \Lambda^-(\mathbf{R}) = \mathrm{diag}\{\lambda_{m+1}(\mathbf{R}), \ldots, \lambda_n(\mathbf{R})\}, \end{cases} \quad \lambda_i(\mathbf{R}) > 0, \ \forall i \in \{1, \ldots, n\} \text{ and } \forall \mathbf{R} \in \mathcal{Y}.$$

We assume that, for some $\sigma > 0$, the map $\Lambda : \mathcal{B}_\sigma \to \mathcal{D}^n$ is of class C^1, with $\mathcal{B}_\sigma$ an open ball of radius σ in $\mathbb{R}^n$. Hence there exist functions $\tilde{\lambda}_i \in C^0(\mathcal{B}_\sigma; \mathbb{R}^n)$ such that

$$\lambda_i(\mathbf{R}) \triangleq \lambda_i(\mathbf{0}) + \tilde{\lambda}_i(\mathbf{R})\mathbf{R}. \tag{4.4}$$

Our purpose is to analyze the exponential stability of the steady state $\mathbf{R}(t,x) \equiv 0$ under nonlinear boundary conditions in nominal form

$$\begin{pmatrix} \mathbf{R}^+(t,0) \\ \mathbf{R}^-(t,L) \end{pmatrix} = \mathcal{H}\begin{pmatrix} \mathbf{R}^+(t,L) \\ \mathbf{R}^-(t,0) \end{pmatrix}, \quad t \in [0, +\infty), \tag{4.5}$$

where $\mathcal{H} \in C^1(\mathcal{B}_\sigma; \mathbb{R}^n)$, $\mathcal{H}(\mathbf{0}) = \mathbf{0}$ and under an initial condition

$$\mathbf{R}(0, x) = \mathbf{R}_o(x), \quad x \in [0, L], \tag{4.6}$$

which satisfies the following compatibility conditions:

$$\begin{pmatrix} \mathbf{R}_o^+(0) \\ \mathbf{R}_o^-(L) \end{pmatrix} = \mathcal{H} \begin{pmatrix} \mathbf{R}_o^+(L) \\ \mathbf{R}_o^-(0) \end{pmatrix}, \tag{4.7}$$

$$\begin{pmatrix} \Lambda^+(\mathbf{R}_o(0))\partial_x\mathbf{R}_o^+(0) \\ -\Lambda^-(\mathbf{R}_o(L))\partial_x\mathbf{R}_o^-(L) \end{pmatrix} = \mathcal{H}' \begin{pmatrix} \mathbf{R}_o^+(L) \\ \mathbf{R}_o^-(0) \end{pmatrix} \begin{pmatrix} \Lambda^+(\mathbf{R}_o(L))\partial_x\mathbf{R}_o^+(L) \\ -\Lambda^-(\mathbf{R}_o(0))\partial_x\mathbf{R}_o^-(0) \end{pmatrix} \tag{4.8}$$

where $\mathcal{H}'$ denotes the Jacobian matrix of the map $\mathcal{H}$.

4.1   Dissipative Boundary Conditions for the C^1-Norm

In this section, our main result is to show how the exponential stability of systems of nonlinear conservation laws of the form (4.3), (4.5) can be established for the C^1-norm.

Let us first introduce some norm notations that will be useful to state and prove the result.

$$\forall \boldsymbol{\xi} \triangleq (\xi_1, \ldots, \xi_n)^\mathsf{T} \in \mathbb{R}^n, \ |\boldsymbol{\xi}|_0 \triangleq \max\{|\xi_j|; j \in \{1, \ldots, n\}\}.$$

For $f \in C^0([0, L]; \mathbb{R}^n)$ (resp. in $C^0([T_1, T_2] \times [0, L]; \mathbb{R}^n)$, we denote

$$|f|_0 \triangleq \max\{|f(x)|_0; x \in [0, L]\}, \tag{4.9}$$

$$(\text{resp. } |f|_0 \triangleq \max\{|f(t, x)|_0; t \in [T_1, T_2], x \in [0, L]\}). \tag{4.10}$$

For $f \in C^1([0, L]; \mathbb{R}^n)$ (resp. in $C^1([T_1, T_2] \times [0, L]; \mathbb{R}^n))$, we denote

$$|f|_1 \triangleq |f|_0 + |f'|_0, \tag{4.11}$$

$$(\text{resp. } |f|_1 \triangleq |f|_0 + |\partial_t f|_0 + |\partial_x f|_0). \tag{4.12}$$

The well-posedness of the Cauchy problem and the existence of a unique C^1-solution result from the following theorem.

Theorem 4.1. Let $T > 0$. There exist $C_1 > 0$ and $\varepsilon_1 > 0$ such that, for every $\mathbf{R}_o \in C^1([0, L]; \mathbb{R}^n)$ satisfying the compatibility conditions (4.7), (4.8) and such that

$$|\mathbf{R}_o|_1 \leqslant \varepsilon_1, \tag{4.13}$$

the Cauchy problem (4.3), (4.5), and (4.6) has one and only one solution

$$\mathbf{R} \in C^1([0, T] \times [0, L]; \mathbb{R}^n).$$

Moreover, this solution satisfies

$$|\mathbf{R}|_1 \leq C_1 |\mathbf{R}_0|_1. \tag{4.14}$$

Proof. A proof of this theorem was given by Li and Yu (1985) for the case of local boundary conditions of the form

$$\mathbf{R}^+(t,0) = \mathcal{H}^+(\mathbf{R}^-(t,0)), \qquad \mathbf{R}^-(t,L) = \mathcal{H}^-(\mathbf{R}^+(t,L)).$$

The general case can be reduced to this particular case by using the dummy doubling of the system size introduced in de Halleux et al. (2003): see Li et al. (2010) for further details. See also Wang (2006) for a generalization to the case of a nonautonomous coefficient matrix $\Lambda(\mathbf{R}, t, x)$ depending explicitly on x and t.   □

The definition of exponential stability is as follows.

Definition 4.2. The steady state $\mathbf{R}(t,x) \equiv \mathbf{0}$ of the system (4.3), (4.5) is exponentially stable for the C^1-norm if there exist $\varepsilon > 0$, $v > 0$, and $C > 0$ such that, for every $\mathbf{R}_0$ in the set $V \triangleq \{\mathbf{R} \in C^1([0,L]; \mathbb{R}^n) : |\mathbf{R}|_1 < \varepsilon\}$ and satisfying the compatibility conditions (4.7) and (4.8), the C^1-solution of the Cauchy problem (4.3), (4.5), (4.6) satisfies

$$|\mathbf{R}(t,.)|_1 \leq Ce^{-vt}|\mathbf{R}_0|_1, \quad \forall t \in [0, +\infty).$$ □

In order to state the next stability theorem, we define the matrix $\mathbf{K}$ as the linearization of the map $\mathcal{H}$ at the steady state:

$$\mathbf{K} \triangleq \mathcal{H}'(\mathbf{0}) \in \mathcal{M}_{n,n}(\mathbb{R})$$

and we recall the definition of the function ρ_∞ (see (3.6)) which can be stated as follows:

$$\rho_\infty(\mathbf{K}) \triangleq \inf\{\mathcal{R}_\infty(\Delta \mathbf{K} \Delta^{-1}); \ \Delta \in \mathcal{D}_n^+\}$$

where

$$\mathcal{R}_\infty(\mathbf{K}) \triangleq \max\{\sum_{j=1}^{n} |k_{ij}|; \ i \in \{1, \ldots, n\}\} \tag{4.15}$$

and k_{ij} denotes the (i,j)th entry of the matrix $\mathbf{K}$. Note that, by (Li 1994, Lemma 2.4, page 146), $\rho_\infty(\mathbf{K})$ is the spectral radius ρ of the matrix $|\mathbf{K}| \triangleq [|k_{ij}|]$:

$$\rho_\infty(\mathbf{K}) = \rho(|\mathbf{K}|). \tag{4.16}$$

We have the following theorem.

Theorem 4.3. If $\rho_\infty(\mathbf{K}) < 1$, the steady state $\mathbf{R}(t, x) \equiv 0$ of the system (4.3), (4.5) is exponentially stable for the C^1-norm.

From now on, in this section, $\mathbf{R} : [0, T] \times [0, L] \to \mathbb{R}^n$ denotes a C^1-solution of the system (4.3), (4.5). Let $\mathbf{W}_1$ and $\mathbf{W}_2$ be defined by

$$\mathbf{W}_1 \triangleq \left(\int_0^L \left[\sum_{i=1}^m p_i^p R_i^{2p} e^{-2p\mu x} + \sum_{i=m+1}^n p_i^p R_i^{2p} e^{2p\mu x} \right] dx \right)^{\frac{1}{2p}}, \tag{4.17}$$

$$\mathbf{W}_2 \triangleq \left(\int_0^L \left[\sum_{i=1}^m p_i^p (\partial_t R_i)^{2p} e^{-2p\mu x} + \sum_{i=m+1}^n p_i^p (\partial_t R_i)^{2p} e^{2p\mu x} \right] dx \right)^{\frac{1}{2p}}, \tag{4.18}$$

with $p \in \mathbb{N}_+$ and $p_i > 0\ \forall i \in \{1, \ldots, n\}$.

The proof of Theorem 4.3 will be based on two preliminary lemmas. The first lemma provides an estimate of $(d\mathbf{W}_1/dt)$ along the solutions of the system (4.3), (4.5).

Lemma 4.4. If $\rho_\infty(\mathbf{K}) < 1$, there exist $p_i > 0\ \forall i \in \{1, \ldots, n\}$, $\alpha > 0$, $\beta_1 > 0$ and $\mu_1 > 0$ such that, for every $\mu \in (0, \mu_1)$, for every $p \in (1/\mu_1, +\infty)$, for every solution of the system (4.3), (4.5) satisfying $|\mathbf{R}|_0 < \mu_1$,

$$\frac{d\mathbf{W}_1}{dt} \leq \left(-\mu\alpha_1 + \beta_1 |\mathbf{R}_t|_0 \right) \mathbf{W}_1. \tag{4.19}$$

Proof. Along the solutions the system (4.3), (4.5), the derivative of $\mathbf{W}_1$ is:

$$\frac{d\mathbf{W}_1}{dt} = \frac{1}{2p} \mathbf{W}_1^{1-2p} \int_0^L \left[\sum_{i=1}^m p_i^p 2p R_i^{2p-1} \partial_t R_i e^{-2p\mu x} \right.$$

$$\left. + \sum_{i=m+1}^n p_i^p 2p R_i^{2p-1} \partial_t R_i e^{2p\mu x} \right] dx$$

$$= \frac{1}{2p} \mathbf{W}_1^{1-2p} \int_0^L \left[-\sum_{i=1}^m p_i^p 2p R_i^{2p-1} (\lambda_i(\mathbf{R})\partial_x R_i) e^{-2p\mu x} \right.$$

$$\left. + \sum_{i=m+1}^n p_i^p 2p R_i^{2p-1} (\lambda_i(\mathbf{R})\partial_x R_i) e^{2p\mu x} \right] dx.$$

Hence, using (4.4),

$$
\frac{d\mathbf{W}_1}{dt} = \frac{1}{2p}\mathbf{W}_1^{1-2p}\int_0^L\Bigg[-\sum_{i=1}^m p_i^p\lambda_i(0)(\partial_x R_i^{2p})e^{-2p\mu x} + \sum_{i=m+1}^n p_i^p\lambda_i(0)(\partial_x R_i^{2p})e^{2p\mu x}
$$
$$
-\sum_{i=1}^m p_i^p 2pR_i^{2p-1}\Big(\tilde{\lambda}_i(\mathbf{R})\mathbf{R}\Big)\partial_x R_i e^{-2p\mu x}
$$
$$
+\sum_{i=m+1}^n p_i^p 2pR_i^{2p-1}\Big(\tilde{\lambda}_i(\mathbf{R})\mathbf{R}\Big)\partial_x R_i e^{2p\mu x}\Bigg]dx
$$

Then, using integration by parts, we get

$$
\frac{d\mathbf{W}_1}{dt} = \mathcal{T}_1 + \mathcal{T}_2 + \mathcal{T}_3
$$

with

$$
\mathcal{T}_1 \triangleq \frac{\mathbf{W}_1^{1-2p}}{2p}\Bigg[-\sum_{i=1}^m p_i^p\lambda_i(0)R_i^{2p}e^{-2p\mu x} + \sum_{i=m+1}^n p_i^p\lambda_i(0)R_i^{2p}e^{2p\mu x}\Bigg]_0^L,
$$
$$
\mathcal{T}_2 \triangleq -\mu\mathbf{W}_1^{1-2p}\int_0^L\Bigg[\sum_{i=1}^m p_i^p\lambda_i(0)R_i^{2p}e^{-2p\mu x} + \sum_{i=m+1}^n p_i^p\lambda_i(0)R_i^{2p}e^{2p\mu x}\Bigg]dx,
$$

$$
\mathcal{T}_3 \triangleq \mathbf{W}_1^{1-2p}\int_0^L\Bigg[\sum_{i=1}^m \frac{p_i^p}{\lambda_i(\mathbf{R})}2pR_i^{2p-1}\Big(\tilde{\lambda}_i(\mathbf{R})\mathbf{R}\Big)\partial_t R_i e^{-2p\mu x}
$$
$$
+\sum_{i=m+1}^n \frac{p_i^p}{\lambda_i(\mathbf{R})}2pR_i^{2p-1}\Big(\tilde{\lambda}_i(\mathbf{R})\mathbf{R}\Big)\partial_t R_i e^{2p\mu x}\Bigg]dx.
$$

Analysis of the First Term $\mathcal{T}_1$. Since $\rho_\infty(\mathbf{K}) < 1$, there exists $\Delta \triangleq$ diag$\{\delta_1,\dots,\delta_n\} \in \mathcal{D}_n^+$ such that

$$
\mathcal{R}_\infty(\Delta\mathbf{K}\Delta^{-1}) < 1. \tag{4.20}
$$

The parameters p_i are selected such that

$$
p_i^p\lambda_i(0) = \delta_i^{2p}, \quad i = 1,\dots,n. \tag{4.21}
$$

Then, using the boundary condition (4.5), $\mathcal{T}_1$ may be written

$$
\mathcal{T}_1 = -\frac{\mathbf{W}_1^{1-2p}}{2p}\Bigg[\sum_{i=1}^{m} \delta_i^{2p} R_i^{2p}(L) e^{-2p\mu L} + \sum_{i=m+1}^{n} \delta_i^{2p} R_i^{2p}(0)
$$

$$
- \sum_{i=1}^{m} \delta_i^{2p} \left(\sum_{j=1}^{m} k_{ij} R_j(L) + \sum_{j=m+1}^{n} k_{ij} R_j(0) \right)^{2p}
$$

$$
- \sum_{i=m+1}^{n} \delta_i^{2p} \left(\sum_{j=1}^{m} k_{ij} R_j(L) + \sum_{j=m+1}^{n} k_{ij} R_j(0) \right)^{2p} e^{2p\mu L} \Bigg].
$$

where we use the simplified notations $R_i(0) \triangleq R_i(t,0)$ and $R_i(L) \triangleq R_i(t,L)$. We define ξ_i, $i = 1, \ldots, n$ such that

$$
\xi_i \triangleq \delta_i R_i(L) \text{ for } i = 1, \ldots, m \quad \text{and} \quad \xi_i \triangleq \delta_i R_i(0) \text{ for } i = m+1, \ldots, n.
$$

Then, for $\mu = 0$, the term between brackets in $\mathcal{T}_1$ is written

$$
\sum_{i=1}^{n} \xi_i^{2p} - \sum_{i=1}^{n} \left(\sum_{j=1}^{n} k_{ij} \frac{\delta_i}{\delta_j} \xi_j \right)^{2p}.
$$

Let us now introduce $\xi_{\max}$ such that

$$
\xi_{\max}^2 = \max\{\xi_i^2, i = 1, \ldots, n\}
$$

and therefore

$$
\xi_{\max}^{2p} \leq \sum_{i=1}^{n} \xi_i^{2p} \leq n\xi_{\max}^{2p}.
$$

Then, we have

$$
\left(\sum_{j=1}^{n} k_{ij} \frac{\delta_i}{\delta_j} \xi_j \right)^{2p} \leq \left(\sum_{j=1}^{n} |k_{ij}| \frac{\delta_i}{\delta_j} |\xi_j| \right)^{2p} \leq \left(\sum_{j=1}^{n} |k_{ij}| \frac{\delta_i}{\delta_j} \right)^{2p} \xi_{\max}^{2p},
$$

and consequently,

$$
\sum_{i=1}^{n} \xi_i^{2p} - \sum_{i=1}^{n} \left(\sum_{j=1}^{n} k_{ij} \frac{\delta_i}{\delta_j} \xi_j \right)^{2p} \geq \xi_{\max}^{2p} \left(1 - n\big(\mathcal{R}_\infty(\Delta \mathbf{K} \Delta^{-1})\big)^{2p} \right).
$$

Now

$$\text{sign}\Big(1 - n\big(\mathcal{R}_\infty(\Delta\mathbf{K}\Delta^{-1})\big)^{2p}\Big) = \text{sign}\Big(1 - n^{1/(2p)}\mathcal{R}_\infty(\Delta\mathbf{K}\Delta^{-1})\Big).$$

Then, using (4.20), we easily check by continuity that there exists $\mu_{11} \in (0, \sigma]$ such that $\forall \mu \in (0, \mu_{11})$, $\forall p \in (1/\mu_{11}, +\infty)$ and $\forall \mathbf{R}$,

$$\mathcal{T}_1 \leqslant 0 \ \text{ if } |\mathbf{R}|_0 < \mu_{11}.$$

Analysis of the Second Term $\mathcal{T}_2$. Defining

$$\alpha \triangleq \tfrac{1}{2}\min(\lambda_1(0), \dots, \lambda_n(0)), \tag{4.22}$$

there is $\mu_{12} \in (0, \sigma]$ such that $\forall \mu \in (0, \mu_{12})$, $\forall p \in (1/\mu_{12}, +\infty)$ and $\forall \mathbf{R}$,

$$\mathcal{T}_2 \leqslant -\mu\alpha\mathbf{W}_1 \ \text{ if } |\mathbf{R}|_0 < \mu_{12}.$$

Analysis of the Third Term $\mathcal{T}_3$. The integrand of $\mathcal{T}_3$ is linear with respect to $\mathbf{R}_t$ and of order $2p$, at least, with respect to $\mathbf{R}$. It follows that there exist $\beta_1 > 0$ and $\mu_{13} \in (0, \sigma]$ such that $\forall \mu \in (0, \mu_{13})$, $\forall p \in (1/\mu_{13}, +\infty)$ and $\forall \mathbf{R}$

$$\mathcal{T}_3 \leqslant \beta_1 |\mathbf{R}_t|_0 \mathbf{W}_1 \ \text{ if } |\mathbf{R}|_0 < \mu_{13}.$$

Hence, with $\mu_1 \triangleq \min\{\mu_{11}, \mu_{12}, \mu_{13}\}$, we conclude that

$$\frac{d\mathbf{W}_1}{dt} \leqslant -\mu\alpha\mathbf{W}_1 + \beta_1 |\mathbf{R}_t|_0 \mathbf{W}_1$$

for all $\mu \in (0, \mu_1)$, for all $p \in (1/\mu_1, +\infty)$ and for all $\mathbf{R}$ such that $|\mathbf{R}|_0 < \mu_1$. This completes the proof of Lemma 4.4.   □

By time differentiation of the system equations (4.3), (4.5), $\mathbf{R}_t$ can be shown to satisfy the following hyperbolic dynamics:

$$\mathbf{R}_{tt} + \Lambda(\mathbf{R})\mathbf{R}_{tx} + \text{diag}[\Lambda'(\mathbf{R})\mathbf{R}_t]\mathbf{R}_x = \mathbf{0}, \tag{4.23}$$

$$\partial_t\begin{pmatrix}\mathbf{R}^+(t, 0)\\\mathbf{R}^-(t, L)\end{pmatrix} = \partial_t\left[\mathcal{H}\begin{pmatrix}\mathbf{R}^+(t, L)\\\mathbf{R}^-(t, 0)\end{pmatrix}\right], \tag{4.24}$$

where $\Lambda'(\mathbf{R})$ denotes the matrix with entries

$$[\Lambda'(\mathbf{R})]_{i,j} \triangleq \frac{\partial\lambda_i}{\partial R_j}$$

and

$$\mathrm{diag}[\Lambda'(\mathbf{R})\mathbf{R}_t] \triangleq \mathrm{diag}\{\lambda_1'(\mathbf{R})\mathbf{R}_t, \ldots, \lambda_m'(\mathbf{R})\mathbf{R}_t, -\lambda_{m+1}'(\mathbf{R})\mathbf{R}_t, \ldots, -\lambda_n'(\mathbf{R})\mathbf{R}_t\}.$$

The next lemma provides an estimate of the functional $(d\mathbf{W}_2/dt)$ along the solutions of the system (4.3), (4.5), (4.23), (4.24).

Lemma 4.5. If $\rho_\infty(\mathbf{K}) < 1$ and p_i $(i = 1, \ldots, n)$ are given by (4.21), there exist $\beta_2 > 0$ and $\mu_2 > 0$ such that, for every $\mu \in (0, \mu_2)$, for every $p \in (1/\mu_2, +\infty)$, for every solution of the system (4.3), (4.5), (4.23), (4.24), satisfying $|\mathbf{R}|_0 < \mu_2$,

$$\frac{d\mathbf{W}_2}{dt} \leq \left(-\mu\alpha + \beta_2 |\mathbf{R}_t|_0 \right)\mathbf{W}_2, \text{ in the distribution sense on } (0, T), \qquad (4.25)$$

with α defined by (4.22).

Since $\mathbf{R}$ is only of class C^1, we remark that $\mathbf{W}_2$ is only continuous and (4.25) has to be understood in the distribution sense on $(0, T)$, in contrast with inequality (4.19) that holds pointwise at every time $t \in [0, T]$.

Proof. In order to prove this lemma, we temporarily assume that $\mathbf{R}$ is of class C^2 on $[0, T] \times [0, L]$. The assumption will be relaxed at the end of the proof. Then along the solutions of the Cauchy problem (4.3) and (4.5), the derivative of $\mathbf{W}_2$ can be computed as

$$\frac{d\mathbf{W}_2}{dt} = \frac{1}{2p}\mathbf{W}_2^{1-2p} \int_0^L \left[\sum_{i=1}^m p_i^p 2p(\partial_t R_i)^{2p-1}\partial_{tt}R_i e^{-2p\mu x} \right.$$
$$\left. + \sum_{i=m+1}^n p_i^p 2p(\partial_t R_i)^{2p-1}\partial_{tt}R_i e^{2p\mu x} \right] dx.$$

Using (4.3) and (4.23), we have

$$\frac{d\mathbf{W}_2}{dt} = \frac{1}{2p}\mathbf{W}_2^{1-2p} \int_0^L \left[\sum_{i=1}^m p_i^p 2p(\partial_t R_i)^{2p-1}\left(-\lambda_i(\mathbf{R})(\partial_t R_i)_x + \frac{\lambda_i'(\mathbf{R})\mathbf{R}_t}{\lambda_i(\mathbf{R})}\partial_t R_i \right)e^{-2p\mu x} \right.$$
$$\left. + \sum_{i=m+1}^n p_i^p 2p(\partial_t R_i)^{2p-1}\left(-\lambda_i(\mathbf{R})(\partial_t R_i)_x + \frac{\lambda_i'(\mathbf{R})\mathbf{R}_t}{\lambda_i(\mathbf{R})}\partial_t R_i \right)e^{2p\mu x} \right] dx.$$

Integrating by parts, we obtain

$$\frac{d\mathbf{W}_2}{dt} = \mathcal{U}_1 + \mathcal{U}_2 + \mathcal{U}_3.$$

with

$$\mathcal{U}_1 \triangleq \frac{1}{2p}\mathbf{W}_2^{1-2p}\left[-\sum_{i=1}^n p_i^p \lambda_i(\mathbf{R})(\partial_t R_i)^{2p} e^{-2p\mu x} + \sum_{i=1}^n p_i^p \lambda_i(\mathbf{R})(\partial_t R_i)^{2p} e^{2p\mu x} \right]_0^L,$$

$$\mathcal{U}_2 \triangleq -\mu \mathbf{W}_2^{1-2p}\int_0^L \left[\sum_{i=1}^m p_i^p \lambda_i(0)(\partial_t R_i)^{2p} e^{-2p\mu x} + \sum_{i=m+1}^n p_i^p \lambda_i(0)(\partial_t R_i)^{2p} e^{2p\mu x} \right] dx,$$

$$\mathcal{U}_3 \triangleq \mathbf{W}_2^{1-2p}\int_0^L \left[\sum_{i=1}^m p_i^p (\partial_t R_i)^{2p} \lambda_i'(\mathbf{R}) L_i(\mathbf{R}) \mathbf{R}_t e^{-2p\mu x} \right.$$

$$\left. + \sum_{i=m+1}^n p_i^p (\partial_t R_i)^{2p} \lambda_i'(\mathbf{R}) L_i(\mathbf{R}) \mathbf{R}_t e^{2p\mu x} \right] dx,$$

with

$$L_i(\mathbf{R}) \triangleq \frac{1}{2p}\text{diag}\left\{ 2p\lambda_i^{-1}(\mathbf{R}) - \lambda_1^{-1}(\mathbf{R}), \quad \ldots \quad, 2p\lambda_i^{-1}(\mathbf{R}) - \lambda_n^{-1}(\mathbf{R}) \right\}.$$

Analysis of the First Term $\mathcal{U}_1$. In a way similar to the analysis of $\mathcal{T}_1$ in the proof of Lemma 4.4, it can be shown that, since $\rho_\infty(\mathbf{K}) < 1$, there exists $\mu_{21} \in (0, \eta]$ sufficiently small such that $\mathcal{U}_1 \leq 0$ for all $\mu \in (0, \mu_{21})$ and for all $p \in (1/\mu_{21}, +\infty)$ if $|\mathbf{R}|_0 < \mu_{21}$.

Analysis of the Second Term $\mathcal{U}_2$. There is $\mu_{22} \in (0, \eta]$ such that $\forall \mu \in (0, \mu_{22})$, $\forall p \in (1/\mu_{22}, +\infty)$,

$$\mathcal{U}_2 \leq -\mu\alpha \mathbf{W}_2.$$

Analysis of the Third Term $\mathcal{U}_3$. The integrand of $\mathcal{U}_3$ is of order $2p + 1$ with respect to $\mathbf{R}_t$ and it is easily checked that there exist $\beta_2 > 0$ and $\mu_{23} \in (0, \eta]$ such that $\forall \mu \in (0, \mu_{23})$, $\forall p \in (1/\mu_{23}, +\infty)$,

$$\mathcal{U}_3 \leq \beta_2 |\mathbf{R}_t|_0 \mathbf{W}_2.$$

if $|\mathbf{R}|_0 < \mu_{23}$.

Hence, with $\mu_2 \triangleq \min\{\mu_{21}, \mu_{22}, \mu_{23}\}$, we conclude that

$$\frac{d\mathbf{W}_2}{dt} \leq -\mu\alpha \mathbf{W}_2 + \beta_2 |\mathbf{R}_t|_0 \mathbf{W}_2 \tag{4.26}$$

for all $\mu \in (0, \mu_2)$, for all $p \in (1/\mu_2, +\infty)$ and for all $\mathbf{R}$ such that $|\mathbf{R}|_0 < \mu_2$. The above estimate of $d\mathbf{W}_2/dt$ has been obtained under the assumption that $\mathbf{R}$ is of class C^2. But the proof shows that the selection of μ, α, and β_2 does not depend on the C^2-norm of $\mathbf{R}$. Hence the estimate (4.26) remains valid, in the distribution sense, with $\mathbf{R}$ only of class C^1 as motivated in the following comment.

Comment 4.6. In this comment we explain why the estimate (4.26) is valid, in the distribution sense, when $\mathbf{R}$ is only of class C^1. Let $\Lambda^k : \mathbb{R}^n \to \mathcal{D}_n^+$ and $\mathcal{H}^k : \mathbb{R}^n \to \mathbb{R}^n$, $k \in \mathbb{N}$, of class C^2 be such that

$$\Lambda^k \underset{k \to \infty}{\longrightarrow} \Lambda \text{ in } C^1(\mathcal{B}_\sigma; \mathcal{D}_n^+), \tag{4.27}$$

$$\mathcal{H}^k \underset{k \to \infty}{\longrightarrow} \mathcal{H} \text{ in } C^1(\mathcal{B}_\sigma; \mathbb{R}^n), \qquad \mathcal{H}^k(\mathbf{0}) = \mathbf{0}. \tag{4.28}$$

Let $\mathbf{R}_0^k \in C^2([0, L]; \mathbb{R}^n)$ $k \in \mathbb{N}$ be a sequence of functions which satisfy the boundary conditions of order 2 as they are defined in page 154, such that $\mathbf{R}_0^k$ converges to $\mathbf{R}_0$ in the C^1-norm when $k \to \infty$. Let $\mathbf{R}^k : [0, T] \times [0, L] \to \mathbb{R}^n$ be the solutions of the Cauchy problem

$$\mathbf{R}_t^k + \Lambda^k(\mathbf{R}^k)\mathbf{R}_x^k = \mathbf{0}, \quad t \in [0, +\infty), \quad x \in (0, L), \tag{4.29}$$

$$\begin{pmatrix} \mathbf{R}^{k,+}(t, 0) \\ \mathbf{R}^{k,-}(t, L) \end{pmatrix} = \mathcal{H}^k \begin{pmatrix} \mathbf{R}^{k,+}(t, L) \\ \mathbf{R}^{k,-}(t, 0) \end{pmatrix}, \tag{4.30}$$

for initial data $\mathbf{R}_0^k$. We know that, for k large enough, $\mathbf{R}^k$ exists, is of class C^2 and that

$$\mathbf{R}^k \to \mathbf{R} \text{ in } C^1([0, T] \times [0, L]; \mathbb{R}^n) \text{ as } k \to \infty. \tag{4.31}$$

$\square$

Now, defining

$$\mathbf{W}_{2,k} \triangleq \left(\int_0^L \left[\sum_{i=1}^m p_i^p (\partial_t R_i^k)^{2p} e^{-2p\mu x} + \sum_{i=m+1}^n p_i^p (\partial_t R_i^k)^{2p} e^{2p\mu x} \right] dx \right)^{\frac{1}{2p}},$$

we have, if $|\mathbf{R}|_0 < \mu_2$ and for k sufficiently large, that

$$\frac{d\mathbf{W}_{2,k}}{dt} \leqslant -\mu\alpha\mathbf{W}_{2,k} + \beta_2|\mathbf{R}_t^k|_0\mathbf{W}_{2,k}. \tag{4.32}$$

Letting $k \longrightarrow \infty$ in (4.32) and using (4.31), we get

$$\frac{d\mathbf{W}_2}{dt} \leqslant -\mu\alpha\mathbf{W}_2 + \beta_2|\mathbf{R}_t|_0\mathbf{W}_2 \text{ in the sense of distributions.}$$

This completes the proof of Lemma 4.5.   □

Proof of Theorem 4.3. Let us choose $\mu \in \mathbb{R}$ such that

$$0 < \mu < \min\{\mu_1, \mu_2\}, \tag{4.33}$$

where μ_1 and μ_2 are as in Lemma 4.4 and Lemma 4.5 respectively.

We define the functionals $\mathbf{V}_1$ and $\mathbf{V}_2$ by

$$\mathbf{V}_1(\mathbf{R}) \triangleq \left| p_1 R_1 e^{-\mu x}, \dots, p_m R_m e^{-\mu x}, p_{m+1} R_{m+1} e^{\mu x}, \dots, p_n R_n e^{\mu x} \right|_0,$$

$$\mathbf{V}_2(\mathbf{R}_t) \triangleq \left| p_1 \partial_t R_1 e^{-\mu x}, \dots, p_m \partial_t R_m e^{-\mu x}, p_{m+1} \partial_t R_{m+1} e^{\mu x}, \dots, p_n \partial_t R_n e^{\mu x} \right|_0.$$

We consider the Lyapunov function candidate $\mathbf{V}$ defined by

$$\mathbf{V}(\mathbf{R}, \mathbf{R}_t) \triangleq \mathbf{V}_1(\mathbf{R}) + \mathbf{V}_2(\mathbf{R}_t). \tag{4.34}$$

There exists $\gamma \in (1, +\infty)$ such that

$$\frac{1}{\gamma} \mathbf{V}(\mathbf{R}, \mathbf{R}_t) \leqslant |\mathbf{R}|_0 + |\mathbf{R}_t|_0 \leqslant \gamma \mathbf{V}(\mathbf{R}, \mathbf{R}_t). \tag{4.35}$$

Let us select $T > 0$ large enough so that

$$\gamma^2 e^{-\mu \alpha T / 2} \leqslant \frac{1}{2} \tag{4.36}$$

with α defined by (4.22).

Let $\varepsilon_2 \in (0, +\infty)$ be such that

$$\varepsilon_2 < \min\left\{ \frac{\mu_1}{C_1}, \frac{\mu_2}{C_1}, \varepsilon_1 \right\}, \tag{4.37}$$

where ε_1 and C_1 are as in Theorem 4.1 while μ_1 and μ_2 are as in Lemma 4.4 and Lemma 4.5 respectively. Let us now assume that the initial condition $\mathbf{R}(0, .) = \mathbf{R}_0(.)$ be such that

$$|\mathbf{R}_0|_1 \leqslant \varepsilon_2. \tag{4.38}$$

Then, by Theorem 4.1, (4.37), and (4.38), the C^1-solutions of (4.3) and (4.5) satisfy

$$|\mathbf{R}|_0 \leqslant |\mathbf{R}|_1 < \min\{\mu_1, \mu_2\}. \tag{4.39}$$

By the definition of $\mathbf{W}_1$ and $\mathbf{W}_2$, we have

$$\mathbf{V}_1(\mathbf{R}) = \lim_{p \to \infty} \mathbf{W}_1(\mathbf{R}) \quad \text{and} \quad \mathbf{V}_2(\mathbf{R}_t) = \lim_{p \to \infty} \mathbf{W}_2(\mathbf{R}_t), \ \forall t \in [0, T], \qquad (4.40)$$

$$\exists M > 0 \text{ such that } \mathbf{W}_1(\mathbf{R}) + \mathbf{W}_2(\mathbf{R}_t) \leqslant M|\mathbf{R}|_1, \ \forall p \in [1, +\infty), \ \forall t \in [0, T]. \qquad (4.41)$$

Hence from Lemmas 4.4 and 4.5, (4.40) and (4.41), we have in the distribution sense in $(0, T)$,

$$\frac{d\mathbf{V}_1}{dt} \leqslant -\mu \alpha \mathbf{V}_1 + \beta_1 |\mathbf{R}_t|_0 \mathbf{V}_1, \qquad (4.42)$$

$$\frac{d\mathbf{V}_2}{dt} \leqslant -\mu \alpha \mathbf{V}_2 + \beta_2 |\mathbf{R}_t|_0 \mathbf{V}_2. \qquad (4.43)$$

Summing (4.42) and (4.43), we get, in the distribution sense in $(0, T)$,

$$\frac{d\mathbf{V}}{dt} \leqslant -\mu \alpha \mathbf{V} + \beta |\mathbf{R}|_1 \mathbf{V}, \qquad (4.44)$$

with $\beta \triangleq \max\{\beta_1, \beta_2\}$. Let us impose on ε_2, besides (4.37), that

$$\varepsilon_2 \leqslant \frac{\mu \alpha}{2 \beta C_1}. \qquad (4.45)$$

From (4.14), (4.38), and (4.45), we get that

$$\beta |\mathbf{R}|_1 \leqslant \frac{\mu \alpha}{2}. \qquad (4.46)$$

From (4.44) and (4.46), we have, in the distribution sense in $(0, T)$,

$$\frac{d\mathbf{V}}{dt} \leqslant -\frac{\mu}{2} \alpha \mathbf{V}, \qquad (4.47)$$

which implies that

$$\mathbf{V}(T) \leqslant e^{-\alpha \mu T/2} \mathbf{V}(0). \qquad (4.48)$$

From (4.35) and (4.48), we obtain that

$$|\mathbf{R}(T, .)|_1 \leqslant \gamma^2 e^{-\alpha RT/2} |\mathbf{R}_0|_1, \qquad (4.49)$$

which, together with (4.36), implies that

$$|\mathbf{R}(T, .)|_1 \leqslant \frac{1}{2} |\mathbf{R}_0|_1. \qquad (4.50)$$

Then, by repeating exactly the same argumentation, it can be iteratively shown that **R** is defined $[0, (j + 1)T] \times [0, L]$ and that

$$|R((j + 1)T, .)|_1 \leqslant \frac{1}{2}|R(jT, .)|_1, \quad j = 0, 1, 2, \dots.$$

This completes the proof of Theorem 4.3.   □

4.2   Control of Networks of Scalar Conservation Laws

The special case of nonlinear scalar conservation laws has been introduced in Section 1.14. Here we consider a network of scalar laws as illustrated in Fig. 4.1. The nodes of the network (i.e., the rectangular boxes) represent physical devices (called 'compartments') with dynamics expressed by scalar conservation laws of the form (1.63):

$$\partial_t \rho_j(t, x) + \partial_x q_j(t, x) = 0, \; t \geqslant 0, \; x \in (0, L), \; j = 1, \dots, n. \tag{4.51}$$

We assume that each flux q_j is a static monotone increasing function of the density ρ_j:

$$q_j = \varphi_j(\rho_j).$$

This relation is supposed to be invertible as

$$\rho_j = \varphi_j^{-1}(q_j),$$

in such a way that the system is also written in the quasi-linear form

$$\partial_t q_j + \lambda_j(q_j)\partial_x q_j = 0, \quad j = 1, \dots, n, \tag{4.52}$$

with

$$\lambda_j(q_j) \triangleq \frac{1}{\left[\dfrac{\partial\varphi_j^{-1}(q)}{\partial q}(q_j)\right]} > 0. \tag{4.53}$$

Fig. 4.1 Network of scalar
conservation laws

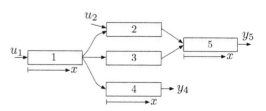

The directed edges $\boxed{i} \longrightarrow \boxed{j}$ of the network represent instantaneous transfer flows between the compartments. The flow from the output of a compartment i to the input of a compartment j is denoted $f_{ij}(t)$. Additional input and output arcs represent inflows $u_j(t)$ injected from the outside into some compartments or outflows $y_j(t)$ from some compartments to the outside. Hence, the set of PDEs (4.52) is subject to boundary conditions of the form:

$$q_j(t, 0) = \sum_{i \neq j} f_{ij}(t) + u_j(t), \tag{4.54a}$$

$$q_j(t, L) = \sum_{k \neq j} f_{jk}(t) + y_j(t), \quad j = 1, \dots, n. \tag{4.54b}$$

In equations (4.54), only the terms corresponding to actual edges of the network are explicitly written. Otherwise stated, all the u_j, y_j, and f_{ij} for non-existing edges do not appear in the equations.

It is assumed here that the flows f_{ij} and y_i are fractions of the outgoing flow $q_i(t, L)$ from compartment i:

$$f_{ij}(t) \triangleq a_{ij} q_i(t, L), \ 0 < a_{ij} \leqslant 1 \quad \text{and} \quad y_i(t) \triangleq a_{i0} q_i(t, L), \ 0 < a_{i0} \leqslant 1.$$

The conservation of flows then imposes the following obvious constraints:

$$\sum_{j=0}^{n} a_{ij} = 1, \ i = 1, \dots, n. \tag{4.55}$$

We introduce the following vector and matrix notations:

$$\mathbf{q} \triangleq (q_1, q_2, \dots, q_n)^{\mathsf{T}},$$

$$\Lambda(\mathbf{q}) \triangleq \operatorname{diag} \{\lambda_1(q_1), \dots, \lambda_n(q_n)\},$$

$$\mathbf{u} = \text{vector including only the actual input flows } u_j,$$

$$\mathbf{y} = \text{vector including only the actual output flows } y_j,$$

$$A = \text{matrix with entries } a_{ji} \ (j = 1, \dots, n; i = 1, \dots, n).$$

With these notations, the system (4.52), (4.54) may be written in the following compact form:

$$\partial_t \mathbf{q} + \Lambda(\mathbf{q}) \partial_x \mathbf{q} = \mathbf{0}, \tag{4.56a}$$

$$\mathbf{q}(t, 0) = A\mathbf{q}(t, L) + B\mathbf{u}(t), \tag{4.56b}$$

$$\mathbf{y}(t) = C\mathbf{q}(t, L). \tag{4.56c}$$

where the definition of B and C is obvious. The first equation (4.56a) is a system of hyperbolic quasi-linear PDEs that defines the system state dynamics. The second equation defines the boundary conditions of the system, some of them being assignable by the system input $\mathbf{u}(t)$. The third equation can be interpreted as an output equation with system output $\mathbf{y}(t)$ being the set of outflows.

For any constant input $\mathbf{u}^*$, a steady state (or equilibrium state) of the system is defined as a constant state $\mathbf{q}^*$ which satisfies the state equation (4.56a) and the boundary condition (4.56b):

$$(A - I)\mathbf{q}^* + B\mathbf{u}^* = 0.$$

Under the constraints (4.55) it can be verified that the matrix $A - I$ is a full-rank compartmental matrix (see, e.g., Bastin and Guffens (2006) for more details on compartmental systems). It follows that, for any positive $\mathbf{u}^*$ there exists a unique positive steady state:

$$\mathbf{q}^* = -(A - I)^{-1} B\mathbf{u}^*.$$

Our goal is to analyze the exponential stability of the steady state $\mathbf{q}^*$ of the control system (4.56) when the system is under a linear state feedback control of the form

$$\mathbf{u}(t) = \mathbf{u}^* + G\left(\mathbf{q}(t, L) - \mathbf{q}^*\right) \tag{4.57}$$

where the matrix G is the control gain.

Defining the Riemann coordinates $\mathbf{R} = (R_1, \ldots, R_n)^\mathsf{T} \triangleq \mathbf{q} - \mathbf{q}^*$, the Cauchy problem associated with the closed-loop control system (4.56), (4.57) is equivalently written as:

$$\mathbf{R}_t + \tilde{\Lambda}(\mathbf{R})\mathbf{R}_x = 0, \tag{4.58a}$$

$$\mathbf{R}(t, 0) = K\mathbf{R}(t, L), \tag{4.58b}$$

$$\mathbf{R}(0, x) = \mathbf{R}_0(x), \tag{4.58c}$$

with $\tilde{\Lambda}(\mathbf{R}) \triangleq \Lambda(\mathbf{q}^* + \mathbf{R})$ and $K \triangleq A + BG$. According to Theorem 4.3, the steady state $\mathbf{q}^*$ is exponentially stable for the C^1-norm if the control gain G is selected such that $\rho_\infty(A + BG) < 1$.

4.2.1 Example: Ramp-Metering Control in Road Traffic Networks

In the fluid paradigm for road traffic modeling, the traffic state is represented by a macroscopic variable $\varrho(t, x)$ which represents the density of the vehicles (# veh/km) at time t and at position x along the road. The traffic dynamics are represented by a conservation law

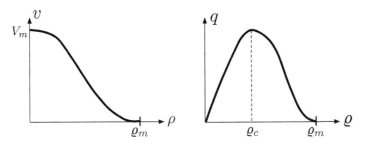

Fig. 4.2 Velocity v and flux q viz. density ϱ

$$\partial_t \varrho(t, x) + \partial_x \left(q(t, x) \right) = 0.$$

which expresses the conservation of the number of vehicles on a road segment without entries nor exits. In this equation, $q(t, x)$ is the traffic *flux* representing the flow rate of the vehicles at (t, x). By definition, we have $q(t, x) \triangleq \varrho(t, x)v(t, x)$ where $v(t, x)$ is the velocity of the vehicles at (t, x). As we have seen in Section 1.14, the basic assumption of the so-called LWR model is that the drivers instantaneously adapt their speed to the local traffic density, which is expressed by a function $v(t, x) = V(\varrho(t, x))$. The LWR traffic model is therefore written as

$$\partial_t \varrho(t, x) + \partial_x \left(\varrho(t, x)V(\varrho(t, x)) \right) = 0. \tag{4.59}$$

In accordance with the physical observations, the velocity-density relation is a monotonic decreasing function $(dV/d\varrho < 0)$ on the interval $[0, \varrho_m]$ (see Figs. 4.2 and 1.12, in Chapter 1, for an illustration of this function with experimental data) such that:

1. $V(0) = V_m$ the maximal vehicle velocity when the road is (almost) empty;
2. $V(\varrho_m) = 0$: the velocity is zero when the density is maximal, the vehicles are stopped and the traffic is totally congested.

Then the flux $q(\varrho) = \rho V(\varrho)$ is a non-monotonic function with $q(0) = 0$ and $q(\varrho_m) = 0$ which is maximal at some critical value ϱ_c which separates free-flow and traffic-congestion: the traffic is flowing freely when $\varrho < \varrho_c$ while the traffic is congested when $\varrho > \varrho_c$ (see Fig. 4.2).

As a matter of example, let us now consider the network of interconnected one-way road segments as depicted in Fig. 4.3. The network is made up of nine road segments with four entries and three exits. The densities and flows on the road segments are denoted ϱ_j and $q_j, j = 1, 9$. The flow rate v_1 is a disturbance input and the flow rates u_2, u_3, u_4 at the three other entries are control inputs.

Our objective is to analyze the stability of this network under a feedback ramp-metering strategy which consists in using traffic lights for modulating the entry flows u_i. The motivation behind such control strategy is that a temporary limitation of the

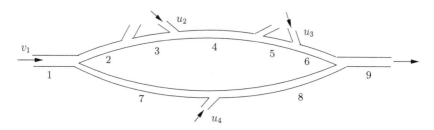

Fig. 4.3 A road network

flow entering a highway can prevent the appearance of traffic jams and improve the network efficiency (possibly at the price of temporary queue formation at the ramps). The reader can refer to Reilly et al. (2015) and the references therein for more details on ramp-metering. The traffic dynamics are described by a set of LWR models (4.59):

$$\partial_t \varrho_j(t, x) + \partial_x \big(\varrho_j(t, x) V(\varrho_j(t, x)) \big) = 0, \ j = 1, \dots, 9. \tag{4.60}$$

Under free-flow conditions, the flows $q_j(\varrho_j) = \varrho_j V(\varrho_j)$ are monotonic increasing functions and the model for the network of Fig. 4.3 is written as a set of kinematic wave equations

$$\partial_t q_j(t, x) + c(q_j(t, x))\partial_x q_j(t, x) = 0, \ c(q_j) > 0, \ j = 1, \dots, 9 \tag{4.61}$$

with the boundary conditions

$$
\begin{aligned}
&q_1(t, 0) = v_1(t), && q_4(t, 0) = q_3(t, L) + u_2(t), \ q_7(t, 0) = (1 - \alpha) q_1(t, L), \\
&q_2(t, 0) = \alpha q_1(t, L), \ q_5(t, 0) = \gamma q_4(t, L), && q_8(t, 0) = q_7(t, L) + u_4(t), \\
&q_3(t, 0) = \beta q_2(t, L), \ q_6(t, 0) = q_5(t, L) + u_3(t), \ q_9(t, 0) = q_6(t, L) + q_8(t, L)
\end{aligned}
$$

where α, β, γ are traffic splitting factors at the diverging junction and the two exits of the network. Obviously the set-point for the feedback traffic regulation is selected as a free-flow steady state $(\bar{q}_1, \bar{q}_2, \dots, \bar{q}_9)^\mathsf{T}$. A linear feedback is then defined for the ramp-metering of the three input flows:

$$u_2(t) = \bar{u}_2 + k_2 \big(q_6(t, L) - \bar{q}_6 \big),$$
$$u_3(t) = \bar{u}_3 + k_3 \big(q_6(t, L) - \bar{q}_6 \big),$$
$$u_4(t) = \bar{u}_4 + k_4 \big(q_8(t, L) - \bar{q}_8 \big),$$

where k_2, k_3, k_4 are tuning control parameters. Defining the Riemann coordinates $R_i = q_i - \bar{q}_i$, the boundary conditions of the system under the ramp-metering control are:

$$
\begin{pmatrix} R_1(t,0) \\ R_2(t,0) \\ R_3(t,0) \\ R_4(t,0) \\ R_5(t,0) \\ R_6(t,0) \\ R_7(t,0) \\ R_8(t,0) \\ R_9(t,0) \end{pmatrix} = \underbrace{\begin{pmatrix} 0 & 0 & 0 & 0 & 0 & 0 & 0 & 0 & 0 \\ \alpha & 0 & 0 & 0 & 0 & 0 & 0 & 0 & 0 \\ 0 & \beta & 0 & 0 & 0 & 0 & 0 & 0 & 0 \\ 0 & 0 & 1 & 0 & 0 & k_2 & 0 & 0 & 0 \\ 0 & 0 & 0 & \gamma & 0 & 0 & 0 & 0 & 0 \\ 0 & 0 & 0 & 0 & 1 & k_3 & 0 & 0 & 0 \\ 1-\alpha & 0 & 0 & 0 & 0 & 0 & 0 & 0 & 0 \\ 0 & 0 & 0 & 0 & 0 & 0 & 1 & k_4 & 0 \\ 0 & 0 & 0 & 0 & 0 & 1 & 0 & 1 & 0 \end{pmatrix}}_{\mathbf{K}} \begin{pmatrix} R_1(t,L) \\ R_2(t,L) \\ R_3(t,L) \\ R_4(t,L) \\ R_5(t,L) \\ R_6(t,L) \\ R_7(t,L) \\ R_8(t,L) \\ R_9(t,L) \end{pmatrix}.
$$

In this example, with parameter values $\alpha = 0.8$, $\beta = 0.9$, $\gamma = 0.8$, the stability condition $\rho_\infty(\mathbf{K}) < 1$ can be shown to be satisfied if and only if the control parameters are selected such that:

$$0.8|k_2| + |k_3| < 1 \qquad |k_4| < 1.$$

These bounds on the values of the k_i parameters are computed with an algorithm for the computation of the maximal stability region of nonnegative matrices which is described in Haut et al. (2009) (see also Haut (2007) for further details on this example).

4.3   Interlude: Solutions Without Shocks

For systems of *nonlinear* conservation laws (4.1) which may be written in quasi-linear characteristic form (4.3), it is well known that, even for small smooth initial conditions, the trajectories of the system may become discontinuous in finite time generating jump discontinuities which propagate on as *shocks*.

An important consequence of the analysis of Section 4.1 is that if the initial condition is of class C^1 and *small enough*, then the dissipativity condition $\rho_\infty(\mathbf{K}) < 1$ guarantees that shocks will not develop and that the solutions remain smooth (i.e., of class C^1) for all $t \in [0, +\infty)$, while exponentially converging to zero for the C^1-norm.

Moreover, in Chapter 3, we remember that, for *linear* systems of conservation laws, the steady state is exponentially stable for the L^2-norm under another dissipativity condition $\rho_2(\mathbf{K}) < 1$ which appears to be weaker according to the following proposition.

Proposition 4.7. For every $\mathbf{K} \in \mathcal{M}_{n,n}(\mathbb{R})$,

$$\rho_2(\mathbf{K}) \leqslant \rho_\infty(\mathbf{K}). \tag{4.62}$$

□

Proof. The proof is given in Appendix C.

It is also worth to point out that there are matrices $\mathbf{K}$ for which the inequality (4.62) is strict. For example, for $a > 0$, we have

$$\mathbf{K} \triangleq \begin{pmatrix} a & a \\ -a & a \end{pmatrix} \quad \text{and} \quad \rho_2(\mathbf{K}) = \sqrt{2}a < 2a = \rho_\infty(\mathbf{K}).$$

So we have the following natural question: *does the condition $\rho_2(\mathbf{K}) < 1$ imply also the exponential convergence of smooth solutions for nonlinear conservation laws ?* Surprisingly enough, the answer is negative for the C^1-norm as stated in the next proposition.

Proposition 4.8. For the C^1-norm the condition $\rho_2(\mathbf{K}) < 1$ is not sufficient for the exponential stability of the steady state of systems of nonlinear conservation laws.

Proof. This proposition is a special case of Theorem 2 in Coron and Nguyen (2015). $\qquad\square$

Hence, in contrast with the linear case, it appears that the considered norm matters when looking at dissipative boundary conditions for systems of nonlinear conservation laws. In the next section, we prove that the condition $\rho_2(\mathbf{K}) < 1$ is in fact a dissipativity condition for the H^2-norm.

4.4 Dissipative Boundary Conditions for the H^2-Norm

In this section, we consider again the system (4.3), (4.5), i.e.

$$\mathbf{R}_t + \Lambda(\mathbf{R})\mathbf{R}_x = \mathbf{0}, \quad t \in [0, +\infty), \quad x \in (0, L), \tag{4.63}$$

$$\begin{pmatrix} \mathbf{R}^+(t, 0) \\ \mathbf{R}^-(t, L) \end{pmatrix} = \mathcal{H} \begin{pmatrix} \mathbf{R}^+(t, L) \\ \mathbf{R}^-(t, 0) \end{pmatrix}, \tag{4.64}$$

under the initial condition (4.6), i.e.

$$\mathbf{R}(0, x) = \mathbf{R}_o(x), \quad x \in [0, L], \tag{4.65}$$

which satisfies the compatibility conditions (4.7) and (4.8).

However, in contrast with the previous section, we now assume that Λ and $\mathcal{H}$ are of class C^2 on the ball B_η. Our main result is to show that the exponential stability of the steady state can be established for the H^2-norm under the condition $\rho_2(\mathbf{K}) < 1$. For a function $\phi : x \to \phi(x)$, $\phi \in H^2((a, b); \mathbb{R}^n)$, the definition of the H^2-norm is

$$\|\phi\|_{H^2((a,b);\mathbb{R}^n)} \triangleq \left(\int_a^b (\|\phi\|^2 + \|\phi_x\|^2 + \|\phi_{xx}\|^2) dx \right)^{1/2}. \tag{4.66}$$

The well-posedness of the Cauchy problem and the existence of a unique solution result from the following theorem.

Theorem 4.9. There exists $\delta_0 > 0$ such that, for every $\mathbf{R}_o \in H^2((0,L), \mathbb{R}^n)$ satisfying

$$\|\mathbf{R}_o\|_{H^2((0,L),\mathbb{R}^n)} \leqslant \delta_0$$

and the compatibility conditions (4.7), (4.8), the Cauchy problem (4.63), (4.64), and (4.65) has a unique maximal H^2-solution

$$\mathbf{R} \in C^0([0,T); H^2((0,L); \mathbb{R}^n)) \cap C^1([0,T) \times [0,L]; \mathbb{R}^n)$$

with $T \in (0, +\infty]$. Moreover, if

$$\|\mathbf{R}(t,\cdot)\|_{H^2((0,L);\mathbb{R}^n)} \leqslant \delta_0, \ \forall t \in [0,T),$$

then $T = +\infty$.

Proof. See Appendix B. □

The definition of the exponential stability is as follows.

Definition 4.10. The steady state $\mathbf{R}(t,x) \equiv 0$ of the system (4.63), (4.64) is exponentially stable for the H^2-norm if there exist $\delta > 0$, $\nu > 0$, and $C > 0$ such that, for every $\mathbf{R}_o \in H^2((0,L); \mathbb{R}^n)$ satisfying $\|\mathbf{R}_o\|_{H^2((0,L);\mathbb{R}^n)} \leqslant \delta$ and the compatibility conditions (4.7), (4.8), the H^2-solution of the Cauchy problem (4.63), (4.64), (4.65) is defined on $[0, +\infty) \times [0,L]$ and satisfies

$$\|\mathbf{R}(t,.)\|_{H^2((0,L);\mathbb{R}^n)} \leqslant Ce^{-\nu t}\|\mathbf{R}_o\|_{H^2((0,L);\mathbb{R}^n)}, \ \ \forall t \in [0, +\infty). \quad □$$

We then have the following stability theorem.

Theorem 4.11. If $\rho_2(\mathbf{K}) < 1$, the steady state $\mathbf{R}(t,x) \equiv 0$ of the system (4.63), (4.64) is exponentially stable for the H^2-norm. □

From now on in this section, $\mathbf{R} : [0,T] \times [0,L] \to \mathbb{R}^n$ denotes a solution in $C^0([0,T]; H^2((0,L); \mathbb{R}^n))$ to the system (4.63), (4.64).

In this case, as we shall see in the proof of the theorem, we need to expand the analysis up to the dynamics of $\mathbf{R}_{tt}$. Therefore, we introduce the candidate Lyapunov function defined by:

$$\mathbf{V} \triangleq \underbrace{\int_0^L \mathbf{R}^\mathsf{T} P(\mu x)\mathbf{R}dx}_{\mathbf{V}_1} + \underbrace{\int_0^L \mathbf{R}_t^\mathsf{T} P(\mu x)\mathbf{R}_t dx}_{\mathbf{V}_2} + \underbrace{\int_0^L \mathbf{R}_{tt}^\mathsf{T} P(\mu x)\mathbf{R}_{tt} dx}_{\mathbf{V}_3}. \tag{4.67}$$

with $P(\mu x) \triangleq \operatorname{diag}\{P^+ e^{-\mu x}, P^- e^{\mu x}\}$, $P^+ \triangleq \operatorname{diag}\{p_1, \ldots, p_m\}$, $P^- \triangleq \operatorname{diag}\{p_{m+1}, \ldots, p_n\}$. Let us remark that, if $\mathbf{R} \in C^0([0, T]; H^2((0, L); \mathbb{R}^n))$, $\mathbf{V}$ is a continuous function of t.

4.4.1 Proof of Theorem 4.11

In order to prove the theorem, we temporarily assume that $\mathbf{R}$ is of class C^3 on $[0, T] \times [0, L]$ and therefore that $\mathbf{V}$ is of class C^1 in $[0, T]$. This assumption (that will be relaxed later on) allows, by time differentiation of the equations (4.23), (4.24), to express the dynamics of $\mathbf{R}_{tt}$ as follows:

$$\mathbf{R}_{ttt} + \Lambda(\mathbf{R})\mathbf{R}_{ttx} + 2\operatorname{diag}[\Lambda'(\mathbf{R})\mathbf{R}_t]\mathbf{R}_{tx} + \operatorname{diag}[\Lambda'(\mathbf{R})\mathbf{R}_t]_t\mathbf{R}_x = 0, \tag{4.68}$$

$$\partial_{tt}\begin{pmatrix} \mathbf{R}^+(t, 0) \\ \mathbf{R}^-(t, L) \end{pmatrix} = \partial_{tt}\left[\mathcal{H}\begin{pmatrix} \mathbf{R}^+(t, L) \\ \mathbf{R}^-(t, 0) \end{pmatrix}\right], \tag{4.69}$$

and, consequently, to compute the time derivatives of $\mathbf{V}_1$, $\mathbf{V}_2$ and $\mathbf{V}_3$ along the solutions of the system (4.3), (4.5), (4.23), (4.24), (4.68), (4.69) in the following way:

$$\frac{d\mathbf{V}_1}{dt} = \int_0^L 2\mathbf{R}^\mathsf{T} P(\mu x)\mathbf{R}_t dx = -\int_0^L 2\mathbf{R}^\mathsf{T} P(\mu x)\Lambda(\mathbf{R})\mathbf{R}_x dx. \tag{4.70}$$

$$\frac{d\mathbf{V}_2}{dt} = \int_0^L 2\mathbf{R}_t^\mathsf{T} P(\mu x)\mathbf{R}_{tt} dx$$

$$= -\int_0^L 2\mathbf{R}_t^\mathsf{T} P(\mu x)\left(\Lambda(\mathbf{R})\mathbf{R}_{tx} - \operatorname{diag}\Lambda^{-1}(\mathbf{R})\mathbf{R}_t\right)dx. \tag{4.71}$$

$$\frac{d\mathbf{V}_3}{dt} = \int_0^L 2\mathbf{R}_{tt}^\mathsf{T} P(\mu x)\mathbf{R}_{ttt} dx = -\int_0^L 2\mathbf{R}_{tt}^\mathsf{T} P(\mu x)\Big(\Lambda(\mathbf{R})\mathbf{R}_{ttx}$$

$$+ 2\operatorname{diag}[\Lambda'(\mathbf{R})\mathbf{R}_t]\mathbf{R}_{tx} - \operatorname{diag}[\Lambda'(\mathbf{R})\mathbf{R}_t]_t\Lambda^{-1}(\mathbf{R})\mathbf{R}_t\Big)dx. \tag{4.72}$$

Since $\rho_2(\mathbf{K}) < 1$ by assumption, there exist $D_0 \in \mathcal{D}_m^+$, $D_1 \in \mathcal{D}_{n-m}^+$ and $\Delta \triangleq$ diag$\{D_0, D_1\}$ such that

$$\|\Delta \mathbf{K} \Delta^{-1}\| < 1.$$

The parameters p_i are selected such that $P^+ \Lambda^+(0) = D_0^2$ and $P^- \Lambda^-(0) = D_1^2$.

The proof of the theorem is then based on the following lemma which provides estimates of the functionals $\mathbf{V}_i$ and $(d\mathbf{V}_i/dt)$ $(i=1,2,3)$ along the system solutions.

Lemma 4.12.

(a) There exists $\mu_1 > 0$ such that , $\forall \mu \in (0, \mu_1)$, there exist positive real constants $\alpha_1, \beta_1, \delta_1$ independent of $\mathbf{R}$, such that, if $|\mathbf{R}|_0 < \delta_1$,

$$\frac{1}{\beta_1} \int_0^L \|\mathbf{R}(t,x)\|^2 dx \leqslant \mathbf{V}_1(t) \leqslant \beta_1 \int_0^L \|\mathbf{R}(t,x)\|^2 dx, \tag{4.73}$$

$$\frac{d\mathbf{V}_1}{dt}(t) \leqslant -\alpha_1 \mathbf{V}_1(t) + \beta_1 \int_0^L \|\mathbf{R}(t,x)\|^2 \|\mathbf{R}_t(t,x)\| dx. \tag{4.74}$$

(b) There exists $\mu_2 > 0$ such that , $\forall \mu \in (0, \mu_2)$, there exist positive real constants $\alpha_2, \beta_2, \delta_2$ independent of $\mathbf{R}$, such that, if $|\mathbf{R}|_0 < \delta_2$,

$$\frac{1}{\beta_2} \int_0^L \|\mathbf{R}_t(t,x)\|^2 dx \leqslant \mathbf{V}_2(t) \leqslant \beta_2 \int_0^L \|\mathbf{R}_t(t,x)\|^2 dx, \tag{4.75}$$

$$\frac{d\mathbf{V}_2}{dt}(t) \leqslant -\alpha_2 \mathbf{V}_2(t) + \beta_2 \int_0^L \|\mathbf{R}_t(t,x)\|^3 dx. \tag{4.76}$$

(c) There exists $\mu_3 > 0$ such that, $\forall \mu \in (0, \mu_3)$, there exist positive real constants $\alpha_3, \beta_3, \delta_3$ independent of $\mathbf{R}$, such that, if $|\mathbf{R}|_0 + |\mathbf{R}_t|_0 < \delta_3$,

$$\frac{1}{\beta_3} \int_0^L \|\mathbf{R}_{tt}(t,x)\|^2 dx \leqslant \mathbf{V}_3(t) \leqslant \beta_3 \int_0^L \|\mathbf{R}_{tt}(t,x)\|^2 dx, \tag{4.77}$$

$$\frac{d\mathbf{V}_3}{dt}(t) \leqslant -\alpha_3 \mathbf{V}_3(t) + \beta_3 \int_0^L (\|\mathbf{R}_t(t,x)\|^2 \|\mathbf{R}_{tt}(t,x)\| + \|\mathbf{R}_t(t,x)\| \|\mathbf{R}_{tt}(t,x)\|^2) dx. \tag{4.78}$$

Proof. Here, we give only the proof of item (a) of Lemma 4.12. The proofs of items (b) and (c) are given in Appendix D and are constructed in a similar way.

Using integration by parts, from (4.70), we can decompose $d\mathbf{V}_1/dt$ as follows:

$$\frac{d\mathbf{V}_1}{dt} = \mathcal{T}_1 + \mathcal{T}_2 + \mathcal{T}_3,$$

with

$$\mathcal{T}_1 \triangleq -\int_0^L (\mathbf{R}^\mathsf{T} P(\mu x) \Lambda (\mathbf{R}) \mathbf{R})_x dx, \tag{4.79}$$

$$\mathcal{T}_2 \triangleq \int_0^L \left(\mathbf{R}^\mathsf{T} P(\mu x) [\Lambda'(\mathbf{R}) \mathbf{R}_x] \mathbf{R} \right) dx, \tag{4.80}$$

$$\mathcal{T}_3 \triangleq -\mu \int_0^L \left(\mathbf{R}^\mathsf{T} P(\mu x) |\Lambda (\mathbf{R})| \mathbf{R} \right) dx. \tag{4.81}$$

Analysis of the First Term $\mathcal{T}_1$. We have

$$\mathcal{T}_1 = -\left[\mathbf{R}^\mathsf{T} P(\mu x) \Lambda (\mathbf{R}) \mathbf{R} \right]_0^L = -\left[\mathbf{R}^\mathsf{T} P(\mu x) \Lambda (0) \mathbf{R} \right]_0^L + \text{h.o.t.} \tag{4.82}$$

where "h.o.t." stands for "higher order terms."

Let us introduce a notation in order to deal with estimates of these "higher order terms." We denote by $\mathcal{O}(X; Y)$, with $X \geqslant 0$ and $Y \geqslant 0$, quantities such that there exist $C > 0$ and $\varepsilon > 0$, independent of $\mathbf{R}$ and $\mathbf{R}_t$ such that

$$(Y \leqslant \varepsilon) \Rightarrow (|\mathcal{O}(X; Y)| \leqslant CX).$$

Hence, using $P^+ \Lambda^+ (0) = D_0^2$ and $P^- \Lambda^- (0) = D_1^2$, we have

$$\mathcal{T}_1 = -\left(\mathbf{R}^{+\mathsf{T}}(t, L) D_0 \ \mathbf{R}^{-\mathsf{T}}(t, 0) D_1 \right) \Omega(\mu) \begin{pmatrix} D_0 \mathbf{R}^+(t, L) \\ \\ D_1 \mathbf{R}^-(t, 0) \end{pmatrix}$$

$$+ \mathcal{O} \left(\left\| \begin{matrix} \mathbf{R}^+(t, L) \\ \mathbf{R}^-(t, 0) \end{matrix} \right\|^3 ; \left\| \begin{matrix} \mathbf{R}^+(t, L) \\ \mathbf{R}^-(t, 0) \end{matrix} \right\| \right),$$

with

$$\Omega(\mu) \triangleq \begin{pmatrix} I_n e^{-\mu L} & 0 \\ 0 & I \end{pmatrix} -$$

$$\begin{pmatrix} D_0 K_{00} D_0^{-1} & D_0 K_{01} D_1^{-1} \\ D_1 K_{10} D_0^{-1} & D_1 K_{11} D_1^{-1} \end{pmatrix}^\mathsf{T} \begin{pmatrix} D_0 K_{00} D_0^{-1} & D_0 K_{01} D_1^{-1} \\ D_1 K_{10} D_0^{-1} e^{\mu L} & D_1 K_{11} D_1^{-1} e^{\mu L} \end{pmatrix}.$$

Following the same argumentation as in the proof of Theorem 3.2, we know that there exists $\mu_1 > 0$ such that $\Omega(\mu)$ is positive definite for all $\mu \in (0, \mu_1)$. Consequently, there exists $\delta_{11} > 0$ such that $\mathcal{T}_{11} \leqslant 0$ for all $\mu \in (0, \mu_1)$ if $|\mathbf{R}^{+\mathsf{T}}(t, L), \mathbf{R}^{-\mathsf{T}}(t, 0)|_0 \leqslant \delta_{11}$.

Analysis of the Second Term $\mathcal{T}_2$. The absolute value of the integrand of $\mathcal{T}_2$ is bounded by $\|\mathbf{R}\|^2\|\mathbf{R}_t\|\|P(\mu x)\Lambda'(\mathbf{R})\Lambda^{-1}(\mathbf{R})\|$. It follows that $\beta_1 > 0$ can be taken sufficiently large such that

- inequalities (4.73) hold for every $\mu \in (0, \mu_1)$,
- there exists δ_{12} such that

$$\mathcal{T}_2 \leqslant \beta_1 \int_0^L \|\mathbf{R}\|^2\|\mathbf{R}_t\|dx$$

if $|\mathbf{R}|_0 \leqslant \delta_{12}$.

Analysis of the Third Term $\mathcal{T}_3$. By (4.73) there exist $\alpha_1 > 0$ and $\delta_{13} > 0$ such that

$$\mathcal{T}_3 \leqslant -\alpha_1 \mathbf{V}_1.$$

for every $\mu \in (0, \mu_1)$ if $|\mathbf{R}|_0 \leqslant \delta_{13}$.

Then we conclude that

$$\frac{d\mathbf{V}_1}{dt} = \mathcal{T}_1 + \mathcal{T}_2 + \mathcal{T}_3 \leqslant -\alpha_1 \mathbf{V}_1 + \beta_1 \int_0^L \|\mathbf{R}\|^2\|\mathbf{R}_t\|dx$$

if $|\mathbf{R}|_0 \leqslant \delta_1 \triangleq \min(\delta_{11}, \delta_{12}, \delta_{13})$. $\square$

In the next lemma, we now show how it follows from the previous estimates that $\mathbf{V} = \mathbf{V}_1 + \mathbf{V}_2 + \mathbf{V}_3$ exponentially decreases along the system solutions.

Lemma 4.13. For every $\mu \in (0, \min\{\mu_1, \mu_2, \mu_3\})$, there exist positive real constants α, β and δ such that, for every $\mathbf{R}$ such that $|\mathbf{R}|_0 + |\mathbf{R}_t|_0 < \delta$,

$$\frac{1}{\beta} \int_0^L (\|\mathbf{R}\|^2 + \|\mathbf{R}_x\|^2 + \|\mathbf{R}_{xx}\|^2)dx \leqslant \mathbf{V} \leqslant \beta \int_0^L (\|\mathbf{R}\|^2 + \|\mathbf{R}_x\|^2 + \|\mathbf{R}_{xx}\|^2)dx,$$
$$(4.83)$$

$$\frac{d\mathbf{V}}{dt} \leq -\alpha \mathbf{V}.   \tag{4.84}$$

Proof. Let β, δ be selected such that

$$\beta \geqslant \max\{\beta_1, \beta_2, \beta_3\}, \quad 0 < \delta \leqslant \min\{\delta_1, \delta_2, \delta_3\}.$$

From (4.3) and (4.23), we know that

$$\mathbf{R}_t = -\Lambda(\mathbf{R})\mathbf{R}_x,   \tag{4.85}$$

$$\mathbf{R}_{tt} = \Lambda(\mathbf{R})(\Lambda(\mathbf{R})\mathbf{R}_x)_x - [\Lambda'(\mathbf{R})\Lambda(\mathbf{R})\mathbf{R}_x]\mathbf{R}_x.   \tag{4.86}$$

Using these expressions, it readily follows from (4.73), (4.75), and (4.77) that, if $|\mathbf{R}|_0 + |\mathbf{R}_t|_0 < \delta$, then (4.83) holds if β is large enough. For every $\eta > 0$, we have

$$
\int_0^L \|\mathbf{R}_t\|^2 \|\mathbf{R}_{tt}\| dx \leq \int_0^L \left(\frac{n}{4\eta} \|\mathbf{R}_t\|^4 + \eta \|\mathbf{R}_{tt}\|^2 \right) dx
$$

$$
\leq \frac{1}{4\eta} |\mathbf{R}_t|_0^2 \int_0^L \|\mathbf{R}_t\|^2 dx + \eta \int_0^L \|\mathbf{R}_{tt}\|^2 dx. \tag{4.87}
$$

In order to get (4.84), it is sufficient to combine inequalities (4.74), (4.76), (4.78), and (4.87) with $\eta \triangleq \alpha_3/(2\beta_3)^2$, and to point out that

$$
\int_0^L \|\mathbf{R}\|^2 \|\mathbf{R}_t\| dx \leq n |\mathbf{R}_t|_0 \int_0^L \|\mathbf{R}\|^2 dx,
$$

$$
\int_0^L \|\mathbf{R}_t\|^3 dx \leq n |\mathbf{R}_t|_0 \int_0^L \|\mathbf{R}_t\|^2 dx,
$$

$$
\int_0^L \|\mathbf{R}_t\| \|\mathbf{R}_{tt}\|^2 dx \leq n |\mathbf{R}_t|_0 \int_0^L \|\mathbf{R}_{tt}\|^2 dx.
$$

$\square$

Proof of Theorem 4.11. The estimates (4.83) and (4.84) have been obtained under the assumption that Λ and $\mathcal{H}$ of class C^3 and $\mathbf{R}$ is of class C^3. But the selection of α and β_2 does depend neither on the C^3-norms of Λ and $\mathcal{H}$ nor on the C^3-norm of $\mathbf{R}$: they depend only on the C^2-norm of Λ and $\mathcal{H}$ and the $C^0([0, T]; H^2((0, L); \mathbb{R}^n))$-norm of $\mathbf{R}$. Hence, using a density argument similar to Comment 4.6, the estimates (4.83) and (4.84) remain valid, in the distribution sense, with Λ, $\mathcal{H}$, $\mathbf{R}$ only of class C^2.

By the Sobolev inequality (see, for instance, (Brezis 1983, Theorem VII, page 129)), there exists $C_0 > 0$ such that, for every φ in the Sobolev space $H^2((0, L); \mathbb{R}^n)$,

$$
|\varphi|_0 + |\varphi'|_0 \leq C_0 \|\varphi\|_{H^2((0,L);\mathbb{R}^n)}. \tag{4.88}
$$

We choose $\mu \in (0, \min\{\mu_1, \mu_2, \mu_3\}]$. Let

$$
\varepsilon \triangleq \min \left\{ \frac{\delta}{2 C_0 \beta}, \frac{\delta_0}{\beta} \right\}. \tag{4.89}
$$

Note that $\beta \geq 1$ and therefore that $\delta \leq \delta_0$. Using Lemma 4.13, (4.88), and (4.89), for every $t \in [0, T]$:

$$\left(\|\mathbf{R}(t,.)\|_{H^2((0,L);\mathbb{R}^n)} \leqslant \varepsilon\right) \Longrightarrow \left(|\mathbf{R}(t,.)|_0 + |\mathbf{R}_x(t,.)|_0 \leqslant \frac{\delta}{2} \text{ and } \mathbf{V}(t) \leqslant \beta\varepsilon^2\right),$$
$$(4.90)$$

$$\left(|\mathbf{R}|_0 + |\mathbf{R}_x|_0 \leq \delta \text{ and } \mathbf{V} \leqslant \beta\varepsilon^2\right)$$
$$\Longrightarrow \left(|\mathbf{R}(t,.)|_0 + |\mathbf{R}(t,.)_t|_0 \leqslant \frac{\delta}{2} \text{ and } \|\mathbf{R}(t,.)\|_{H^2((0,L);\mathbb{R}^n)} \leqslant \delta_0\right), \qquad (4.91)$$

$$\text{and} \quad (|\mathbf{R}|_0 + |\mathbf{R}_x|_0 \leq \delta) \Longrightarrow \left(\frac{d\mathbf{V}}{dt} \leqslant 0\right) \text{ in the distribution sense.} \qquad (4.92)$$

Let $\mathbf{R}_o \in H^2((0,L);\mathbb{R}^n)$ satisfy the compatibility conditions (4.7),(4.8) and

$$\|\mathbf{R}_o\|_{H^2((0,L);\mathbb{R}^n)} < \varepsilon.$$

Let $\mathbf{R} \in C^0([0,T^*), H^2((0,L);\mathbb{R}^n))$ be the maximal classical solution of the Cauchy problem (4.3), (4.5), (4.6). Using implications (4.90) to (4.92) for $T \in [0,T^*)$, we get that

$$|\mathbf{R}(t,\cdot)|_{H^2((0,L);\mathbb{R}^n)} \leqslant \delta_0, \ \forall t \in [0,T^*), \qquad (4.93)$$
$$|\mathbf{R}(t,\cdot)|_0 + |\mathbf{R}_t(t,\cdot)|_0 \leqslant \delta, \ \forall t \in [0,T^*). \qquad (4.94)$$

Using (4.93) and Theorem 4.9, we have that $T = +\infty$. Using Lemma 4.13 and (4.94), we finally obtain that

$$\|\mathbf{R}(t,\cdot)\|^2_{H^2((0,L);\mathbb{R}^n)} \leqslant \beta\mathbf{V}(t) \leqslant \beta\mathbf{V}(0)e^{-\alpha t} \leqslant \beta^2\|\mathbf{R}_o\|^2_{H^2((0,L);\mathbb{R}^n)}e^{-\alpha t}.$$

This concludes the proof of Theorem 4.11. □

4.5   Stability of General Systems of Nonlinear Conservation Laws in Quasi-Linear Form

In this section, we now consider systems having the general quasi-linear form (4.2):

$$\mathbf{Y}_t + F(\mathbf{Y})\mathbf{Y}_x = 0, \quad t \in [0,+\infty), \ x \in [0,L], \qquad (4.95)$$

where $\mathbf{Y} : [0,+\infty) \times [0,L] \to \mathbb{R}^n$ and $F : \mathcal{Y} \to \mathcal{M}_{n,n}(\mathbb{R})$ is of class C^1. We do not assume that the system can be transformed into characteristic form with Riemann coordinates. We shall nevertheless be able to analyze the exponential stability of a steady state $\mathbf{Y}(t,x) \equiv 0$ when the matrix $F(0)$ has n real nonzero eigenvalues denoted λ_i, $i = 1,\ldots,n$ and is diagonalizable. In this section, for simplicity, we treat only the special case where all eigenvalues are positive: $\lambda_i > 0$, $i = 1, \ldots, n$. The generalization to the case where $F(\mathbf{0})$ has both positive and negative eigenvalues can be found in Coron and Bastin (2015).

The key point is that, by a linear change of coordinates, it is always possible to rewrite the system (4.95) in the equivalent form

$$\mathbf{Z}_t + A(\mathbf{Z})\mathbf{Z}_x = \mathbf{0}, \quad t \in [0, +\infty), \quad x \in [0, L], \tag{4.96}$$

where $A \in C^1(\mathcal{Y}; \mathcal{M}_{n,n}(\mathbb{R}))$ and $A(\mathbf{0}) \in \mathcal{D}_n^+$ is the positive diagonal matrix $\mathrm{diag}\{\lambda_1, \ldots, \lambda_n\}$.

Our concern is to analyze the exponential stability of the steady state $\mathbf{Z}(t, x) \equiv \mathbf{0}$ under nonlinear boundary conditions in nominal form

$$\mathbf{Z}(t, 0) = \mathcal{H}\big(\mathbf{Z}(t, L)\big), \quad t \in [0, +\infty), \tag{4.97}$$

where $\mathcal{H} \in C^1(\mathcal{B}_\sigma, \mathbb{R}^n)$, $\mathcal{H}(\mathbf{0}) = 0$ and under an initial condition

$$\mathbf{Z}(0, x) = \mathbf{Z}_o(x), \quad x \in [0, L], \tag{4.98}$$

which satisfies the compatibility conditions

$$\mathbf{Z}_o(0) = \mathcal{H}\big(\mathbf{Z}_o(L)\big), \tag{4.99}$$

$$A\big(\mathbf{Z}_o(0)\big)\partial_x\mathbf{Z}_o(0) = \mathcal{H}'\big(\mathbf{Z}_o(L)\big)A\big(\mathbf{Z}_o(L)\big)\partial_x\mathbf{Z}_o(0), \tag{4.100}$$

where $\mathcal{H}'$ denotes the Jacobian matrix of the map $\mathcal{H}$.

The matrix $\mathbf{K}$ is defined as the linearization of the map $\mathcal{H}$ at the steady state:

$$\mathbf{K} \triangleq \mathcal{H}'(\mathbf{0}).$$

In this section, we shall show that $\rho_\infty(\mathbf{K}) < 1$ and $\rho_2(\mathbf{K}) < 1$ are dissipativity conditions for the C^p-norm and the H^p-norm respectively.

In order to define appropriate Lyapunov functions for this analysis, we introduce the following assumption.

Assumption 4.14. Let $D(\mathbf{Z})$ be the diagonal matrix whose diagonal entries are the eigenvalues $\lambda_i(\mathbf{Z})$, $i = 1, \ldots, n$, of the matrix $A(\mathbf{Z})$. There exist a positive real number η and a map $Q : \mathcal{B}_\eta \to \mathcal{M}_{n,n}(\mathbb{R})$ of class C^2 such that

$$Q(\mathbf{Z})A(\mathbf{Z}) = D(\mathbf{Z})Q(\mathbf{Z}), \quad \forall \mathbf{Z} \in \mathcal{B}_\eta, \tag{4.101}$$

$$Q(\mathbf{0}) = \mathrm{Id}_n, \tag{4.102}$$

where Id_n is the identity matrix of $\mathcal{M}_{n,n}(\mathbb{R})$.

Remark 4.15. This assumption is the expression of the fact that the matrix $A(\mathbf{Z})$ can be diagonalized in a neighborhood of the origin. It is however important to realize that this does not imply that the quasi-linear hyperbolic system (4.96) itself can be written in characteristic form. Otherwise the present section would obviously be irrelevant!

Remark 4.16. It is worth noting that Assumption 4.14 is generic. Indeed it is satisfied as soon as the eigenvalues λ_i of the matrix $F(0)$ are distinct. But, there are interesting cases where there is a matrix Q satisfying equality (4.101) even with non-distinct eigenvalues. For example, in the case where the matrix $A(\mathbf{Z})$ is block diagonal, it may be sufficient to have distinct eigenvalues in each block, but the different blocks may share identical eigenvalues. This situation will appear in particular in the proofs of Theorems 4.22 and 4.24 which deal with stability for C^p-norms with $p > 1$ and for H^p-norms with $p > 2$.

4.5.1 Stability Condition for the C^1-Norm

Our purpose is to show that the dissipativity condition $\rho_\infty(\mathbf{K}) < 1$ of Section 4.1 for systems in Riemann coordinates can be extended to general systems (4.96), (4.97) in a similar fashion. The definition of exponential stability is as follows.

Definition 4.17. The steady state $\mathbf{Z}(t,x) \equiv 0$ of the system (4.96), (4.97) is exponentially stable for the C^1-norm if there exist $\varepsilon > 0$, $\nu > 0$ and $C > 0$ such that, for every $\mathbf{Z}_o$ in the set $\mathcal{V} \triangleq \{\mathbf{Z} \in C^1([0,L]; \mathbb{R}^n) : |\mathbf{Z}|_1 < \varepsilon\}$ and satisfying the compatibility conditions (4.99) and (4.100), the C^1-solution of the Cauchy problem (4.96), (4.97), and (4.98) satisfies

$$|\mathbf{Z}(t,.)|_1 \leq Ce^{-\nu t}|\mathbf{Z}_o|_1, \quad \forall t \in [0, +\infty).$$

$\square$

We have the following stability theorem.

Theorem 4.18. If $\rho_\infty(\mathbf{K}) < 1$, the steady state $\mathbf{Z}(t,x) \equiv 0$ of the system (4.96), (4.97) is exponentially stable for the C^1 norm.

In order to prove this theorem, we introduce the following generalizations of the functionals $\mathbf{W}_1$ and $\mathbf{W}_2$ that were used in Section 4.1:

$$\mathbf{W}_1 \triangleq \left(\int_0^L \sum_{i=1}^n p_i^p \left(\sum_{j=1}^n q_{ij}(\mathbf{Z})Z_j \right)^{2p} e^{-2p\mu x} dx \right)^{\frac{1}{2p}}, \tag{4.103}$$

$$\mathbf{W}_2 \triangleq \left(\int_0^L \sum_{i=1}^n p_i^p \left(\sum_{j=1}^n q_{ij}(\mathbf{Z})\partial_t Z_j \right)^{2p} e^{-2p\mu x} dx \right)^{\frac{1}{2p}}, \tag{4.104}$$

with $p \in \mathbb{N}_+$ and $p_i > 0$ $\forall i$. In (4.103) and (4.104), $q_{ij}(\mathbf{Z})$ denotes the (i,j)th entry of the matrix $Q(\mathbf{Z})$.

The key point for the analysis is that Lemma 4.4 and Lemma 4.5 can be extended to these generalized forms of $\mathbf{W}_1$ and $\mathbf{W}_2$.

Lemma 4.19. If $\rho_\infty(K) < 1$, there exist $p_i > 0$ $\forall i \in \{1,\dots,n\}$, positive real constants α, β_1 and δ_1 such that, for every $\mu \in (0, \delta_1)$, for every $p \in (1/\delta_1, +\infty)$, for every $T > 0$, for every C^1-solution $\mathbf{Z} : [0,T] \times [0,L] \to \mathbb{R}^n$ of (4.96) and (4.97) satisfying $|\mathbf{Z}|_0 < \delta_1$, we have

$$\frac{d\mathbf{W}_1}{dt} \leq \left(-\mu\alpha + \beta_1 |\mathbf{Z}_x(t)|_0 \right) \mathbf{W}_1.$$

Proof. Let $\mathbf{Z} : [0,T] \times [0,L] \to \mathbb{R}^n$ be a C^1-solution of (4.96) and (4.97). The time derivative of $\mathbf{W}_1$ is:

$$\frac{d\mathbf{W}_1}{dt} = \frac{1}{2p} \mathbf{W}_1^{1-2p} \int_0^L \sum_{i=1}^n 2p\, p_i^p \left(\sum_{j=1}^n q_{ij}(\mathbf{Z}) Z_j \right)^{2p-1}$$
$$\left(\left[\sum_{j=1}^n q_{ij}(\mathbf{Z}) \partial_t Z_j \right] + \sum_{j=1}^n (\partial_t q_{ij}(\mathbf{Z})) Z_j \right) e^{-2p\mu x} dx. \qquad (4.105)$$

Using (4.96), the term between brackets can be written as

$$\sum_{j=1}^n q_{ij}(\mathbf{Z}) \partial_t Z_j = -\sum_{j=1}^n q_{ij}(\mathbf{Z}) \left(\sum_{k=1}^n a_{jk}(\mathbf{Z}) \partial_x Z_k \right) \qquad (4.106)$$
$$= -\sum_{k=1}^n \left(\sum_{j=1}^n q_{ij}(\mathbf{Z}) a_{jk}(\mathbf{Z}) \partial_x Z_k \right),$$

where $a_{jk}(\mathbf{Z})$ is the (j,k)th entry of the matrix $A(\mathbf{Z})$. Now, from (4.101), we have

$$\sum_{j=1}^n q_{ij}(\mathbf{Z}) a_{jk}(\mathbf{Z}) \partial_x Z_k = \sum_{j=1}^n d_{ij}(\mathbf{Z}) q_{jk}(\mathbf{Z}) \partial_x Z_k = \lambda_i(\mathbf{Z}) q_{ik}(\mathbf{Z}) \partial_x Z_k, \qquad (4.107)$$

where $d_{ij}(\mathbf{Z})$ is the (i,j)th entry of the matrix $D(\mathbf{Z})$. From (4.106) and (4.107), we have

$$\sum_{j=1}^n q_{ij}(\mathbf{Z}) \partial_t Z_j = -\lambda_i(\mathbf{Z}) \sum_{k=1}^n q_{ik}(\mathbf{Z}) \partial_x Z_k = -\lambda_i(\mathbf{Z}) \sum_{j=1}^n q_{ij}(\mathbf{Z}) \partial_x Z_j. \qquad (4.108)$$

By substituting this expression for the term between brackets in (4.105), we get

$$\frac{d\mathbf{W}_1}{dt} = \frac{1}{2p}\mathbf{W}_1^{1-2p} \int_0^L \sum_{i=1}^n 2p\, p_i^p \left(\sum_{j=1}^n q_{ij}(\mathbf{Z})Z_j \right)^{2p-1}$$

$$\left(\left(-\lambda_i(\mathbf{Z}) \sum_{j=1}^n q_{ij}(\mathbf{Z})\partial_x Z_j \right) + \left(\sum_{j=1}^n (\partial_t q_{ij}(\mathbf{Z}))Z_j \right) \right) e^{-2p\mu x} dx,$$

which leads to

$$\frac{d\mathbf{W}_1}{dt} = \frac{1}{2p}\mathbf{W}_1^{1-2p} \left[\int_0^L -\sum_{i=1}^n p_i^p \lambda_i(\mathbf{Z}) \left(\left(\sum_{j=1}^n q_{ij}(\mathbf{Z})Z_j \right)^{2p} \right)_x e^{-2p\mu x} dx \right.$$

$$+ \int_0^L 2p \sum_{i=1}^n p_i^p \left(\sum_{j=1}^n q_{ij}(\mathbf{Z})Z_j \right)^{2p-1} \left(\lambda_i(\mathbf{Z}) \left(\sum_{j=1}^n (\partial_x q_{ij}(\mathbf{Z}))Z_j \right) \right)$$

$$\left. + \left(\sum_{j=1}^n (\partial_t q_{ij}(\mathbf{Z}))Z_j \right) \right) e^{-2p\mu x} dx \right]. \qquad (4.109)$$

Using integrations by parts, we now get

$$\frac{d\mathbf{W}_1}{dt} = \mathcal{T}_1 + \mathcal{T}_2 + \mathcal{T}_3, \qquad (4.110)$$

with

$$\mathcal{T}_1 \triangleq \frac{\mathbf{W}_1^{1-2p}}{2p} \left[-\sum_{i=1}^n p_i^p \lambda_i(\mathbf{Z}) \left(\sum_{j=1}^n q_{ij}(\mathbf{Z})Z_j e^{-\mu x} \right)^{2p} \right]_0^1, \qquad (4.111)$$

$$\mathcal{T}_2 \triangleq -\mu \mathbf{W}_1^{1-2p} \int_0^L \sum_{i=1}^n p_i^p \lambda_i(\mathbf{Z}) \left(\sum_{j=1}^n q_{ij}(\mathbf{Z})Z_j \right)^{2p} e^{-2p\mu x} dx, \qquad (4.112)$$

$$\mathcal{T}_3 \triangleq \mathbf{W}_1^{1-2p} \int_0^L \sum_{i=1}^n p_i^p \left(\sum_{j=1}^n q_{ij}(\mathbf{Z})Z_j \right)^{2p-1} \left(\left(\lambda_i(\mathbf{Z}) \sum_{j=1}^n (\partial_x q_{ij}(\mathbf{Z}))Z_j \right) \right. \qquad (4.113)$$

$$\left. + \left(\sum_{j=1}^n (\partial_t q_{ij}(\mathbf{Z}))Z_j \right) + \frac{1}{2p}\left(\sum_{j=1}^n q_{ij}(\mathbf{Z})Z_j \right) \frac{\partial \lambda_i}{\partial \mathbf{Z}}(\mathbf{Z})\partial_x \mathbf{Z} \right) e^{-2p\mu x} dx.$$

Analysis of the First Term $\mathcal{T}_1$. From (4.111), we have

$$
\mathcal{T}_1 = -\frac{\mathbf{W}_1^{1-2p}}{2p} \left[\sum_{i=1}^{n} p_i^p \lambda_i(\mathbf{Z}(t,L)) \left(\sum_{j=1}^{n} q_{ij}(\mathbf{Z}(t,L)) Z_j(t,L) e^{-\mu L} \right)^{2p} \right.
$$

$$
\left. - \sum_{i=1}^{n} p_i^p \lambda_i(\mathbf{Z}(t,0)) \left(\sum_{j=1}^{n} q_{ij}(\mathbf{Z}(t,0)) Z_j(t,0) \right)^{2p} \right]. \qquad (4.114)
$$

According to (Li 1994, Lemma 2.4, page 146),

$$
\rho_\infty(\mathbf{K}) = \sup \left\{ \sum_{j=1}^{n} |K_{ij}| \frac{\delta_i}{\delta_j} ; i \in \{1,\ldots,n\}, \delta_j > 0, \forall j \in \{1,\ldots,n\} \right\}. \qquad (4.115)
$$

Since $\rho_\infty(\mathbf{K}) < 1$ and by (4.115), there exist $\delta_i > 0$, $i \in \{1,\ldots,n\}$, such that

$$
\theta \triangleq \sum_{j=1}^{n} |K_{ij}| \frac{\delta_i}{\delta_j} < 1. \qquad (4.116)
$$

The parameters p_i are selected such that

$$
p_i^p \lambda_i = \delta_i^{2p}, \qquad i = 1,\ldots,n. \qquad (4.117)
$$

We define $\xi_i : [0,T] \to \mathbb{R}$, $i = 1,\ldots,n$, by

$$
\xi_i(t) \triangleq \delta_i Z_i(t,L), \forall t \in [0,T]. \qquad (4.118)
$$

From (4.114), (4.117), and (4.118), we have

$$
\mathcal{T}_1 = -\frac{\mathbf{W}_1^{1-2p}}{2p} \left[\sum_{i=1}^{n} \frac{\lambda_i(\mathbf{Z}(t,L))}{\Lambda_i} \left(\sum_{j=1}^{n} q_{ij}(\mathbf{Z}(t,L)) \frac{\delta_i}{\delta_j} \xi_j(t) e^{-\mu L} \right)^{2p} \right.
$$

$$
\left. - \sum_{i=1}^{n} \frac{\lambda_i(\mathbf{Z}(t,0))}{\Lambda_i} \left(\sum_{j=1}^{n} q_{ij}(\mathbf{Z}(t,0)) \delta_i Z_j(t,0) \right)^{2p} \right]. \qquad (4.119)
$$

Let $t \in [0,T]$. Without loss of generality, we may assume that

$$
\xi_1^2(t) = \max\{\xi_i^2(t), i = 1,\ldots,n\}. \qquad (4.120)
$$

Let us denote by δ and C various positive constants which may vary from place to place but are independent of $t \in [0, T]$, $\mathbf{Z}$, and $p \in \mathbb{N}_+$. From (4.102) and (4.120), we have, for $|\mathbf{Z}(t, L)|_0 \leq \delta$,

$$
\sum_{i=1}^{n} \frac{\lambda_i(\mathbf{Z}(t, L))}{\Lambda_i} \left(\sum_{j=1}^{n} q_{ij}(\mathbf{Z}(t, L)) \frac{\delta_i}{\delta_j} \xi_j(t) e^{-\mu L} \right)^{2p}
$$

$$
\geq \frac{\lambda_1(\mathbf{Z}(t, L))}{\Lambda_1} \left(\sum_{j=1}^{n} q_{1j}(\mathbf{Z}(t, L)) \frac{\delta_1}{\delta_j} \xi_j(t) e^{-\mu L} \right)^{2p} \tag{4.121}
$$

$$
\geq e^{-2p\mu} \left(1 - C|\xi_1(t)| \right) \left(|\xi_1(t)| - C|\xi_1(t)|^2 \right)^{2p}
$$

$$
= e^{-2p\mu} \left(1 - C|\xi_1(t)| \right)^{2p+1} (\xi_1(t))^{2p} .
$$

From (4.97), (4.116), (4.118), and (4.120), we have, for $|\mathbf{Z}(t, 0)|_0 \leq \delta$,

$$
\sum_{i=1}^{n} \frac{\lambda_i(\mathbf{Z}(t, 0))}{\Lambda_i} \left(\sum_{j=1}^{n} q_{ij}(\mathbf{Z}(t, 0)) \delta_i Z_j(t, 0) \right)^{2p}
$$

$$
\leq (1 + C|\xi_1(t)|) \left(\sum_{i=1}^{n} \left(C|\xi_1(t)|^2 + \sum_{j=1}^{n} |K_{ij}| \frac{\delta_i}{\delta_j} |\xi_j(t)| \right)^{2p} \right)
$$

$$
\leq n \left(1 + C|\xi_1(t)| \right) \left(\theta |\xi_1(t)| + C|\xi_1(t)|^2 \right)^{2p} . \tag{4.122}
$$

From (4.116), (4.119), (4.120), and (4.122), there exists $\delta_{11} \in (0, L)$, independent of $\mathbf{Z}$, such that, for every $\mu \in (0, \delta_{11})$, for every $p \in (1/\delta_{11}, +\infty) \cap \mathbb{N}_+$ and for every $\mathbf{Z}$, we have

$$
\mathcal{T}_1(t) \leq 0 \text{ if } |\mathbf{Z}(t)|_0 < \delta_{11}. \tag{4.123}
$$

Analysis of the Second Term $\mathcal{T}_2$. Let

$$
\alpha \triangleq \min(\Lambda_1, \dots, \Lambda_n)/2. \tag{4.124}
$$

From (4.112), (4.117) and (4.124) there is a $\delta_{12} \in (0, \eta]$ such that, for every $\mu \in (0, +\infty)$, for every $p \in \mathbb{N}_+$ and for every $\mathbf{Z}$,

$$
\mathcal{T}_2 \leq -\mu \alpha \mathbf{W}_1 \text{ if } |\mathbf{Z}|_0 < \delta_{12}. \tag{4.125}
$$

Analysis of the Third Term T_3. Using (4.96) and (4.113), we have

$$
T_3 = \mathbf{W}_1^{1-2p} \int_0^L \sum_{i=1}^n p_i^p \left(\sum_{j=1}^n q_{ij}(\mathbf{Z})Z_j \right)^{2p-1}
$$

$$
\left(\frac{1}{2p} \frac{\partial \lambda_i}{\partial \mathbf{Z}} \mathbf{Z}_x \left(\sum_{j=1}^n q_{ij}(\mathbf{Z})Z_j \right) + \sum_{j=1}^n \left(\frac{\partial q_{ij}}{\partial \mathbf{Z}} (-A(\mathbf{Z}) + \lambda_i(\mathbf{Z}))\mathbf{Z}_x \right) Z_j \right) e^{-2p\mu x} dx.
$$

$$(4.126)$$

From (4.102) and (4.126) we get the existence of $\beta_1 > 0$ and $\delta_{13} > 0$ such that, for every $\mu \in (0, +\infty)$, for every $p \in \mathbb{N}_+$ and for every $\mathbf{Z}$,

$$
T_3 \le \beta_1 |\mathbf{Z}_x|_0 \mathbf{W}_1 \text{ if } |\mathbf{Z}|_0 < \delta_{13}. \tag{4.127}
$$

Let $\delta_1 \triangleq \min\{\delta_{11}, \delta_{12}, \delta_{13}\}$. From (4.110), (4.123), (4.125), and (4.127), we conclude that

$$
\frac{d\mathbf{W}_1}{dt} \le -\mu\alpha_1 \mathbf{W}_1 + \beta_1 |\mathbf{Z}_x|_0 \mathbf{W}_1
$$

provided that $\mathbf{Z}$ is such that $|\mathbf{Z}|_0 < \delta_1$, that $p \in (1/\delta_1, +\infty) \cap \mathbb{N}_+$ and that $\mu \in (0, \delta_1)$.   □

Lemma 4.20. Let p_i $(i = 1, \ldots, n)$ be given by (4.117). If $\rho(\mathbf{K}) < 1$, there exist β_2 and δ_2 such that, for every $\mu \in (0, \delta_2)$, for every $p \in (1/\delta_2, +\infty) \cap \mathbb{N}_+$ and for every C^1-solution $\mathbf{Z} : [0, T] \times [0, L] \to \mathbb{R}^n$ of (4.96) and (4.97) such that $|\mathbf{Z}|_0 < \delta_2$, we have, in the sense of distributions in $(0, T)$,

$$
\frac{d\mathbf{W}_2}{dt} \le \left(-\mu\alpha + \beta_2 |\mathbf{Z}_x|_0 \right) \mathbf{W}_2
$$

with α defined by (4.124).

Let us remark that, since $\mathbf{Z}$ is only of class C^1, $\mathbf{W}_2$ is only continuous and $d\mathbf{W}_2/dt$ has to be understood in the distribution sense.

Proof. In order to prove the lemma, we temporarily assume that $\mathbf{Z}$ is of class C^2 on $[0, T] \times [0, L]$. The assumption will be relaxed at the end of the proof. By time differentiation of (4.96) and (4.97), we see that $\mathbf{Z}_t$ satisfy the following hyperbolic dynamics for $t \in [0, T]$ and $x \in [0, L]$:

$$
(\mathbf{Z}_t)_t + A(\mathbf{Z})(\mathbf{Z}_t)_x + A'(\mathbf{Z}, \mathbf{Z}_t)\mathbf{Z}_x = 0, \tag{4.128}
$$

$$
\partial_t \mathbf{Z}(t, 0) = \frac{\partial \mathcal{H}(\mathbf{Z}(t, L))}{\partial \mathbf{Z}(t, L)} \partial_t \mathbf{Z}(t, L), \tag{4.129}
$$

where $A'(\mathbf{Z}, \mathbf{Z}_t)$ is a compact notation for the matrix whose entries are

$$A'(\mathbf{Z}, \mathbf{Z}_t)_{i,j} \triangleq \frac{\partial a_{ij}(\mathbf{Z})}{\partial \mathbf{Z}} \mathbf{Z}_t, \ i \in \{1, \ldots, n\}, j \in \{1, \ldots, n\}.$$

Using (4.128)–(4.129), we see that the time derivative of $\mathbf{W}_2$ is:

$$\frac{d\mathbf{W}_2}{dt} = \frac{1}{2p} \mathbf{W}_2^{1-2p} \int_0^L \sum_{i=1}^n 2p \, p_i^p \left(\sum_{j=1}^n q_{ij}(\mathbf{Z}) \partial_t Z_j \right)^{2p-1}$$

$$\left(\left[\sum_{j=1}^n q_{ij}(\mathbf{Z}) \partial_{tt} Z_j \right] + \sum_{j=1}^n \left(\partial_t q_{ij}(\mathbf{Z}) \right) \partial_t Z_j \right) e^{-2p\mu x} dx. \qquad (4.130)$$

From (4.101), similarly as for (4.108), it can be shown that

$$\sum_{j=1}^n q_{ij}(\mathbf{Z}) \partial_{tt} Z_j = -\lambda_i(\mathbf{Z}) \sum_{j=1}^n q_{ij}(\mathbf{Z}) \partial_x \left(\partial_t Z_j \right) + \sum_{j=1}^n q_{ij}(\mathbf{Z}) \left(\sum_{k=1}^n \tilde{a}_{jk}(\mathbf{Z}, \mathbf{Z}_t) \partial_t Z_k \right),$$

where $\tilde{a}_{ij}(\mathbf{Z}, \mathbf{Z}_t)$ is the (i, j)th entry of the matrix $\tilde{A}(\mathbf{Z}, \mathbf{Z}_t) \triangleq A'(\mathbf{Z}, \mathbf{Z}_t) A^{-1}(\mathbf{Z})$. Then, by substituting this expression for the term between brackets in (4.130), we get

$$\frac{d\mathbf{W}_2}{dt} = \frac{1}{2p} \mathbf{W}_2^{1-2p} \int_0^L \sum_{i=1}^n 2p \, p_i^p \left(\sum_{j=1}^n q_{ij}(\mathbf{Z}) \partial_t Z_j \right)^{2p-1}$$

$$\left[-\lambda_i(\mathbf{Z}) \sum_{j=1}^n q_{ij}(\mathbf{Z}) \partial_x \left(\partial_t Z_j \right) \right.$$

$$\left. + \sum_{j=1}^n \left(\left(\sum_{k=1}^n q_{ik}(\mathbf{Z}) \tilde{a}_{kj}(\mathbf{Z}, \mathbf{Z}_t) \right) + \partial_t q_{ij}(\mathbf{Z}) \right) \partial_t Z_j \right] e^{-2p\mu x} dx.$$

Using integration by parts as in the proof of Lemma 4.19, we get

$$\frac{d\mathbf{W}_2}{dt} = \mathcal{U}_1 + \mathcal{U}_2 + \mathcal{U}_3,$$

with

$$\mathcal{U}_1 \triangleq \frac{\mathbf{W}_2^{1-2p}}{2p} \left[-\sum_{i=1}^n p_i^p \lambda_i(\mathbf{Z}) \left(\sum_{j=1}^n q_{ij}(\mathbf{Z}) (\partial_t Z_j) e^{-\mu x} \right)^{2p} \right]_0^1,$$

$$\mathcal{U}_2 \triangleq -\mu \mathbf{W}_2^{1-2p} \int_0^L \sum_{i=1}^n p_i^p \lambda_i(\mathbf{Z}) \left(\sum_{j=1}^n q_{ij}(\mathbf{Z})(\partial_t Z_j) \right)^{2p} e^{-2p\mu x} dx,$$

$$\mathcal{U}_3 \triangleq \mathbf{W}_2^{1-2p} \int_0^L \sum_{i=1}^n p_i^p \left(\sum_{j=1}^n q_{ij}(\mathbf{Z})(\partial_t Z_j) \right)^{2p-1} \left[\sum_{j=1}^n \left(\lambda_i(\mathbf{Z})(\partial_x q_{ij}(\mathbf{Z})) \right. \right.$$

$$\left. + \left(\sum_{k=1}^n q_{ik}(\mathbf{Z})\tilde{a}_{kj}(\mathbf{Z}, \mathbf{Z}_t) \right) + \partial_t q_{ij}(\mathbf{Z}) \right)(\partial_t Z_j)$$

$$\left. + \frac{1}{2p} \left(\sum_{j=1}^n q_{ij}(\mathbf{Z}) \partial_t Z_j \right) \frac{\partial \lambda_i}{\partial \mathbf{Z}}(\mathbf{Z}) \mathbf{Z}_x \right] e^{-2p\mu x} dx.$$

Analysis of the First Term $\mathcal{U}_1$. Using the boundary conditions (4.97) and (4.129), we have

$$\mathcal{U}_1 = -\frac{\mathbf{W}_2^{1-2p}}{2p} \left[\sum_{i=1}^n p_i^p \lambda_i(\mathbf{Z}(t, L)) \left(\sum_{j=1}^n q_{ij}(\mathbf{Z}(t, L))(\partial_t Z_j(t, L)) \right)^{2p} e^{-2p\mu} \right.$$

$$\left. - \sum_{i=1}^n p_i^p \lambda_i(\mathcal{H}(\mathbf{Z}(t, L))) \left(\sum_{j=1}^n q_{ij}(\mathcal{H}(\mathbf{Z}(t, L))) \frac{\partial \mathcal{H}_j(\mathbf{Z}(t, L))}{\partial \mathbf{Z}(t, L)} \partial_t \mathbf{Z}(t, L) \right)^{2p} \right]$$

Then, in a way similar to the analysis of $\mathcal{T}_1$ in the proof of Lemma 4.19, we can show that, if $\rho_\infty(\mathbf{K}) < 1$, there exists $\delta_{21} \in (0, 1)$, such that, for every $\mu \in (0, \delta_{21})$, for every $p \in (1/\delta_{21}, +\infty) \cap \mathbb{N}_+$ and for every $\mathbf{Z}$, we have $\mathcal{U}_1 \leqslant 0$ provided that $|\mathbf{Z}|_0 < \delta_{21}$.

Analysis of the Second Term $\mathcal{U}_2$. Proceeding as in the proof of (4.125), we get the existence of $\delta_{22} \in (0, \eta]$ such that, for every $\mu \in (0, +\infty)$, for every $p \in \mathbb{N}_+$ and for every $\mathbf{Z}$,

$$\mathcal{U}_2 \leq -\mu \alpha \mathbf{W}_2 \text{ if } |\mathbf{Z}|_0 < \delta_{22}, \tag{4.131}$$

with α defined as in (4.124).

Analysis of the Third Term $\mathcal{U}_3$. Proceeding as in the proof of (4.127), we get the existence of $\beta_2 > 0$ and $\delta_{23} > 0$ such that, for every $\mu \in (0, +\infty)$, for every $p \in \mathbb{N}_+$, and for every $\mathbf{Z}$,

$$\mathcal{U}_3 \leq \beta_2 |\mathbf{Z}_x|_0 \mathbf{W}_2 \text{ if } |\mathbf{Z}|_0 < \delta_{23}.$$

From the analysis of $\mathcal{U}_1$, $\mathcal{U}_2$, and $\mathcal{U}_3$, we conclude that, with $\delta_2 \triangleq$ $\min\{\delta_{21}, \delta_{22}, \delta_{23}\}$,

$$\frac{d\mathbf{W}_2}{dt} \leq -\mu\alpha\mathbf{W}_2 + \beta_2|\mathbf{Z}_x|_0\mathbf{W}_2 \tag{4.132}$$

for all $\mathbf{Z}$ such that $|\mathbf{Z}|_0 < \delta_2$ provided that $\mu \in (0, \delta_2)$ and that $p \in (1/\delta_2, +\infty) \cap \mathbb{N}_+$.

The above estimate of $d\mathbf{W}_2/dt$ has been obtained under the assumption that A, $\mathcal{H}$, and $\mathbf{Z}$ are of class C^2. But the selection of $\mu\alpha$ and β_2 does not depend on the C^2-norm of A, $\mathcal{H}$ and $\mathbf{Z}$. Hence, using a density argument similar to Comment 4.6, the estimate (4.132) remains valid with A, $\mathcal{H}$, and $\mathbf{Z}$ only of class C^1. This completes the proof of Lemma 4.20.   □

The proof of Theorem 4.18 then follows from Lemma 4.19, 'word-for-word' as the proof of Theorem 4.3 follows from Lemmas 4.4 and 4.5.

4.5.2 Stability Condition for the C^p-Norm for Any $p \in \mathbb{N} \smallsetminus \{0\}$

In the previous subsection, we have seen that $\rho_\infty(\mathbf{K}) < 1$ is a dissipativity condition with respect to the C^1-norm for general quasi-linear systems

$$\mathbf{Z}_t + A(\mathbf{Z})\mathbf{Z}_x = 0 \tag{4.133}$$

$$\mathbf{Z}(t, 0) = \mathcal{H}\big(\mathbf{Z}(t, L)\big) \quad t \in [0, +\infty), \tag{4.134}$$

with initial conditions

$$\mathbf{Z}(0, x) = \mathbf{Z}_o(x), \quad x \in [0, L] \tag{4.135}$$

satisfying the compatibility conditions (4.99), (4.100).

Our purpose is now to emphasize that the exponential stability with respect to the C^p-norm holds in fact for any $p \in \mathbb{N} \smallsetminus \{0\}$ under the same dissipativity condition $\rho_\infty(\mathbf{K}) < 1$ when the maps A and $\mathcal{H}$ are of class C^p. To establish this property it is first needed to generalize the definition of the compatibility of the initial conditions. For that, for a map $\mathcal{G} : C^0([0, L]; \mathbb{R}^n) \to C^0([0, L]; \mathbb{R}^n)$ of class C^p, we introduce the sequence $D^j\mathcal{G} : C^j([0, L]; \mathbb{R}^n) \to C^0([0, L]; \mathbb{R}^n)$ defined by induction as follows:

$$(D^0\mathcal{G})(\mathbf{Z}) \triangleq \mathcal{G} \quad \forall \mathbf{Z} \in C^0([0, L]; \mathbb{R}^n),$$

$$(D^j\mathcal{G})(\mathbf{Z}) \triangleq \Big((D^{j-1}\mathcal{G})'(\mathbf{Z})\Big)A(\mathbf{Z})\mathbf{Z}_x \quad \forall \mathbf{Z} \in C^j([0, L]; \mathbb{R}^n) \quad \forall j \in \{1, \ldots, p\}.$$

Let $\mathcal{I}$ be the identity map from $C^0([0, L]; \mathbb{R}^n)$ into $C^0([0, L]; \mathbb{R}^n)$ and let us define $\mathcal{H} : C^0([0, L]; \mathbb{R}^n) \to C^0([0, L]; \mathbb{R}^n)$ by

$$\left(\mathcal{H}(\varphi)\right)(x) = \mathcal{H}\left(\varphi(x)\right) \quad \forall \varphi \in C^0([0, L]; \mathbb{R}^n) \text{ and } \forall x \in [0, L].$$

Then, we say that the function $\mathbf{Z}_0 \in C^p([0, L]; \mathbb{R}^n)$ satisfies the compatibility conditions of order p if

$$\left((D^j \mathcal{I})(\mathbf{Z}_0)\right)(0) = \left((D^j \mathcal{H})(\mathbf{Z}_0)\right)(L) \text{ for every } j \in \{0, 1, \ldots, p\}. \qquad (4.136)$$

Less formally, let us explain how the explicit expression of the compatibility conditions of order 2 can be obtained. The *first compatibility condition* for $j = 0$ is obtained by evaluating the boundary condition (4.134) at $t = 0$ which, using the initial condition (4.135), gives:

$$\mathbf{Z}_0(0) = \mathcal{H}\left(\mathbf{Z}_0(L)\right). \qquad (4.137)$$

Let us now differentiate the boundary condition (4.134) with respect to t. We get:

$$\mathbf{Z}_t(t, 0) = \mathcal{H}'(\mathbf{Z}(t, L))\mathbf{Z}_t(t, L).$$

Using the system equation (4.133), this latter relation is rewritten as

$$A(\mathbf{Z}(t, 0))\mathbf{Z}_x(t, 0) = \mathcal{H}'(\mathbf{Z}(t, L))A(\mathbf{Z}(t, L))\mathbf{Z}_x(t, L). \qquad (4.138)$$

Then the *second compatibility condition* for $j = 1$ is obtained by evaluating (4.138) at $t = 0$ which, using the initial condition (4.135), gives:

$$A(\mathbf{Z}_0(0))\partial_x \mathbf{Z}_0(0) = \mathcal{H}'(\mathbf{Z}_0(L))A(\mathbf{Z}_0(L))\partial_x \mathbf{Z}_0(L). \qquad (4.139)$$

Let us now differentiate (4.138) with respect to t. We get:

$$A(\mathbf{Z}(t, 0))\mathbf{Z}_{xt}(t, 0) + [A'(\mathbf{Z}(t, 0))\mathbf{Z}_t(t, 0)]\mathbf{Z}_x(t, 0) = \mathcal{H}'(\mathbf{Z}(t, L))A(\mathbf{Z}(t, L))\mathbf{Z}_{xt}(t, L) \qquad (4.140)$$

$$+ \left\{\mathcal{H}'(\mathbf{Z}(t, L))\left[A'(\mathbf{Z}(t, L))\mathbf{Z}_t(t, L)\right] + \left[\mathcal{H}''(\mathbf{Z}(t, L))\mathbf{Z}_t(t, L)\right]A(\mathbf{Z}(t, L))\right\}\mathbf{Z}_x(t, L).$$

Differentiating the system equation (4.133) with respect to x, we have

$$\mathbf{Z}_{tx} + A(\mathbf{Z})\mathbf{Z}_{xx} + \left[A'(\mathbf{Z})\mathbf{Z}_x\right]\mathbf{Z}_x = \mathbf{0}. \qquad (4.141)$$

Then, using (4.133) and (4.141), the relation (4.140) is rewritten as

$$
A(\mathbf{Z}(t,0))\Big[A(\mathbf{Z}(t,0))\mathbf{Z}_{xx}(t,0) + \big[A'(\mathbf{Z}(t,0))\mathbf{Z}_x(t,0)\big]\mathbf{Z}_x(t,0)\Big]\mathbf{Z}_x(t,0)
$$
$$
+ [A'(\mathbf{Z}(t,0))A(\mathbf{Z}(t,0))\mathbf{Z}_x(t,0)]\mathbf{Z}_x(t,0)
$$
$$
= \mathcal{H}'(\mathbf{Z}(t,L))A(\mathbf{Z}(t,L))\Big[A(\mathbf{Z}(t,L))\mathbf{Z}_{xx}(t,L) + \big[A'(\mathbf{Z}(t,L))\mathbf{Z}_x(t,L)\big]\mathbf{Z}_x(t,L)\Big]
$$
$$
+ \Big\{\mathcal{H}'(\mathbf{Z}(t,L))\big[A'(\mathbf{Z}(t,L))A(\mathbf{Z}(t,L))\mathbf{Z}_x(t,L)\big]
$$
$$
+ \big[\mathcal{H}''(\mathbf{Z}(t,L))A(\mathbf{Z}(t,L))\mathbf{Z}_x(t,L)\big]A(\mathbf{Z}(t,L))\Big\}\mathbf{Z}_x(t,L).
$$

Finally, the *third compatibility condition* for $j = 2$ is obtained by evaluating this latter relation at $t = 0$ which, using the initial condition (4.135), gives:

$$
A(\mathbf{Z}_0(0))\Big[A(\mathbf{Z}_0(0))\partial_{xx}\mathbf{Z}_0(0) + \big[A'(\mathbf{Z}_0(0))\partial_x\mathbf{Z}_0(0)\big]\partial_x\mathbf{Z}_0(0)\Big] \tag{4.142}
$$
$$
+ [A'(\mathbf{Z}_0(0))A(\mathbf{Z}_0(0))\partial_x\mathbf{Z}_0(0)]\partial_x\mathbf{Z}_0(0)
$$
$$
= \mathcal{H}'(\mathbf{Z}_0(L))A(\mathbf{Z}_0(L))\Big[A(\mathbf{Z}_0(L))\partial_{xx}\mathbf{Z}_0(L) + \big[A'(\mathbf{Z}_0(L))\partial_x\mathbf{Z}_0(L)\big]\partial_x\mathbf{Z}_0(L)\Big]
$$
$$
+ \Big\{\mathcal{H}'(\mathbf{Z}_0(L))\big[A'(\mathbf{Z}_0(L))A(\mathbf{Z}_0(L))\partial_x\mathbf{Z}_0(L)\big]
$$
$$
+ \big[\mathcal{H}''(\mathbf{Z}_0(L))A(\mathbf{Z}_0(L))\partial_x\mathbf{Z}_0(L)\big]A(\mathbf{Z}_0(L))\Big\}\partial_x\mathbf{Z}_0(L).
$$

So, the set of compatibility conditions of order $p = 2$ is given by equations (4.137), (4.139), (4.142).

We then have the two following theorems, respectively for the well-posedness of the Cauchy problem associated with the system (4.133), (4.134) and for the exponential stability of the steady state.

Theorem 4.21. Let $p \in \mathbb{N}_+$. Let $T > 0$. There exist $\varepsilon > 0$ and $C > 0$ such that, for every $\mathbf{Z}_0 \in C^p([0,L];\mathbb{R}^n)$ satisfying the compatibility conditions of order p (4.136) and such that $\|\mathbf{Z}_0\|_{C^p([0,L];\mathbb{R}^n)} \leqslant \varepsilon$, the Cauchy problem (4.133), (4.134), (4.135) has one and only one solution $\mathbf{Z} \in C^p([0,T] \times [0,L];\mathbb{R}^n)$. Furthermore,

$$
\|\mathbf{Z}\|_{C^p([0,T]\times[0,L];\mathbb{R}^n)} \leqslant C\|\mathbf{Z}_0\|_{C^p([0,L];\mathbb{R}^n)}.
$$

□

Theorem 4.22. If $\rho_\infty(\mathbf{K}) < 1$, there exist $\varepsilon > 0$, $\nu > 0$, and $C > 0$ such that, for every $\mathbf{Z}_0$ in the set $\mathcal{V} \triangleq \{\mathbf{Z} \in C^p([0,L];\mathbb{R}^n) : \|\mathbf{Z}\|_{C^p([0,L];\mathbb{R}^n)} < \varepsilon\}$ and satisfying the compatibility conditions of order p (4.136), the C^p-solution of the Cauchy problem (4.133), (4.134), (4.135) satisfies

$$
\|\mathbf{Z}(t,.)\|_{C^p([0,L];\mathbb{R}^n)} \leq Ce^{-\nu t}\|\mathbf{Z}_0\|_{C^p([0,L];\mathbb{R}^n)}, \quad \forall t \in [0,+\infty).
$$

□

Here, for the sake of conciseness, we will not give the proofs of these theorems which are, indeed, straightforward but rather tedious extensions of the proofs of Theorems 4.1 and Theorem 4.18 respectively obtained, roughly speaking, by considering the augmented hyperbolic system of balance laws with state variables $\mathbf{Z}, \partial_t \mathbf{Z}, \ldots, \partial_t^{p-1} \mathbf{Z}$. These systems are systems of balance laws with a uniform zero steady state and a quadratic source term which is analyzed in Section 6.1, see Corollary 6.3. In this case, the augmented matrix $A(\mathbf{Z}, \partial_t \mathbf{Z}, \ldots, \partial_t^{p-1} \mathbf{Z})$ is block diagonal and Remark 4.16 applies.

4.5.3   Stability Condition for the H^p-Norm for Any $p \in \mathbb{N} \smallsetminus \{0, 1\}$

Let $p \in \mathbb{N} \smallsetminus \{0, 1\}$. The definition of exponential stability for the H^p-norm is as follows.

Definition 4.23. The steady state $\mathbf{Z}(t, x) \equiv 0$ of the system (4.133), (4.134) is exponentially stable for the H^p-norm if there exist $\delta > 0$, $\nu > 0$ and $C > 0$ such that, for every $\mathbf{Z}_o \in H^p((0, L); \mathbb{R}^n)$ satisfying $\|\mathbf{Z}_o\|_{H^p((0,L);\mathbb{R}^n)} \leqslant \delta$ and the compatibility conditions of order $p - 1$ (4.136), the solution $\mathbf{Z}$ of the Cauchy problem (4.133), (4.134), (4.135) is defined on $[0, +\infty) \times [0, L]$ and satisfies

$$\|\mathbf{Z}(t, .)\|_{H^p((0,L);\mathbb{R}^n)} \leq Ce^{-\nu t} \|\mathbf{Z}_o\|_{H^p((0,L);\mathbb{R}^n)}, \quad \forall t \in [0, +\infty). \qquad \square$$

We then have the following stability theorem.

Theorem 4.24. If $\rho_2(\mathbf{K}) < 1$, the steady state $\mathbf{Z}(t, x) \equiv 0$ of the system (4.133), (4.134) is exponentially stable for the H^p-norm. $\qquad \square$

The proof of this theorem is an extension of Theorem 4.11 which deals with the case $p = 2$. Here also, similar to the case of the C^p-norm (see page 153), the proof is carried out by considering an augmented hyperbolic system of balance laws with state variables $\mathbf{Z}, \partial_t \mathbf{Z}, \ldots, \partial_t^{p-2} \mathbf{Z}$.

4.6   References and Further Reading

For nonlinear systems of conservation laws, the issue of finding sufficient dissipative boundary conditions was addressed in the literature for more than thirty years. To our knowledge, first results were published by Slemrod (1983) and by Greenberg and Li (1984) for the special case of systems of size $n = 2$. A generalization to systems of size n was then progressively elaborated by the Li Ta-Tsien group, in particular by Qin (1985) and by Zhao (1986). All these contributions were dealing with the particular case of 'local' boundary conditions having the specific form

$$\mathbf{R}^+(t, 0) = \mathcal{H}^+(\mathbf{R}^-(t, 0)), \qquad \mathbf{R}^-(t, L) = \mathcal{H}^-(\mathbf{R}^+(t, L)). \tag{4.143}$$

With these boundary conditions, indeed, the analysis can be based on the method of characteristics which allows to exploit an explicit computation of the 'reflection' of the solutions at the boundaries along the characteristic curves. This has given rise to the sufficient condition

$$\rho_\infty \left(\begin{pmatrix} \mathbf{0} & (\mathcal{H}^+)'(\mathbf{0}) \\ (\mathcal{H}^-)'(\mathbf{0}) & \mathbf{0} \end{pmatrix} \right) < 1, \tag{4.144}$$

for the dissipativity of the boundary conditions (4.143) for the C^1-norm. This result was first proved by Qin (1985) and Zhao (1986) using the method of characteristics and was also given by (Li 1994, Theorem 1.3, page 173) in his seminal book of 1994 on the stability of the classical solutions of quasi-linear hyperbolic systems. By using an appropriate dummy doubling of the system size, (de Halleux et al. 2003, Theorem 4) showed how the method of characteristics can be used to prove Theorem 4.3 for systems with the general 'nonlocal' boundary condition (4.5). This dummy doubling of the size of the system was also used by Li et al. (2010) to prove the well-posedness of the Cauchy problem associated with (4.96) and (4.97) still in the framework of C^1-solutions. In contrast the proof of Theorem 4.3 given in Section 4.1 uses the Lyapunov method and was published much later in Coron and Bastin (2015).

Historically, the first approach for a Lyapunov stability analysis of 1-D hyperbolic systems was to use entropies as Lyapunov functions expressed in the physical coordinates. This was done for instance by Coron et al. (1999) or by Leugering and Schmidt (2002). The drawback of this approach was however that the time derivatives of such entropy-based Lyapunov functions are necessarily only *semidefinite negative*. In such a case it is well known that the exponential convergence of the solutions may be proved with the LaSalle invariant set principle, see, e.g., Khalil (1996, Section 3.2) for nonlinear ODEs and Coron (2007, Chapter 13) for linear PDEs. Unfortunately, the LaSalle invariance principle requires the precompactness of the trajectories, a property which is quite difficult to get in the case of nonlinear partial differential equations. This difficulty is alleviated by using the *strict* Lyapunov functions with exponential weights that are repeatedly utilized in this book. Such functions were initially introduced by Coron (1999) for 2-D Euler equations of incompressible fluids, then by Xu and Sallet (2002) for a class of symmetric *linear* hyperbolic systems, and finally by Coron et al. (2007) and Coron et al. (2008) for general systems of nonlinear conservation laws.

Various interesting generalizations of the results presented in this chapter are worth mentioning.

- The stabilization of systems of two nonlinear conservation laws with integral actions is analyzed in Drici (2011).
- The design of dead-beat controllers for the stabilization of systems of two nonlinear conservation laws is discussed in Perollaz and Rosier (2013).

- The design of boundary observers for linear and quasi-linear hyperbolic systems of conservation laws is addressed by Castillo et al. (2013). The analysis is illustrated with an application to flow control.
- The boundary feedback stabilization of gas pipelines modeled by isentropic Euler equations is studied by Dick et al. (2010) and Gugat and Herty (2011).
- The output feedback stabilization of a scalar conservation law with a nonlocal characteristic velocity is addressed in Coron and Wang (2013).
- The stabilization of extrusion processes modeled by coupled conservation laws with moving boundaries is studied in Diagne et al. (2016a).

It is also interesting to notice that these boundary stabilization results of hyperbolic systems all assume that the characteristic velocities do not vanish. To our knowledge, the characterization of boundary conditions ensuring the exponential stability of systems with vanishing characteristic velocities remains an open question. Indeed in this case, the state components corresponding to vanishing characteristic velocities cannot be directly controlled from the boundary, but only indirectly through their coupling to other states. Let us however mention that *controllability* results for hyperbolic systems with vanishing characteristic velocities are already available and can be found for instance in Glass (2007), Coron et al. (2009), Glass (2014), Hu and Wang (2015).

Chapter 5
Systems of Linear Balance Laws

T HE THREE PREVIOUS CHAPTERS have dealt with the stability and the boundary stabilization of systems of *conservation laws*. From now on, we shall move to the analysis of systems of *balance laws*.

In this chapter, we begin with the case of *linear* balance laws. We consider the class of systems of linear hyperbolic balance laws with nonuniform coefficients of the form:

$$\mathbf{Y}_t + \big(F(x)\mathbf{Y}\big)_x + G(x)\mathbf{Y} = \mathbf{0} \quad t \in [0, +\infty), \quad x \in (0, L), \tag{5.1}$$

where $\mathbf{Y} : [0, +\infty) \times [0, L] \rightarrow \mathcal{Y}$, $F : [0, L] \rightarrow \mathcal{M}_{n,n}(\mathbb{R})$, $G : [0, L] \rightarrow \mathcal{M}_{n,n}(\mathbb{R})$. The maps F and G are of class C^1. The matrix $F(x)$ has real nonzero eigenvalues and is diagonalizable for all $x \in [0, L]$ through a change of variables which is of class C^1 with respect to x.

Let us remark that any system of balance laws of the form (5.1) can also be written as the following linear hyperbolic system with nonuniform coefficients:

$$\mathbf{Y}_t + A(x)\mathbf{Y}_x + B(x)\mathbf{Y} = 0, \tag{5.2}$$

with $A(x) \triangleq F(x)$ and $B(x) \triangleq G(x) + F'(x)$. Conversely, any linear system of the form (5.2) can be viewed as a system of balance laws of the form (5.1) with $F(x) \triangleq A(x)$ and $G(x) = B(x) - A'(x)$.

With respect to conservation laws, the presence of the *source term* $G(x)\mathbf{Y}$ brings a big additional difficulty for the stability analysis. In fact the tests for dissipative boundary conditions of conservation laws are directly extendable to balance laws only if the source terms themselves have appropriate dissipativity properties that may be expressed in terms of matrix inequalities. This is the topic of the next section. Otherwise, as shown in Sections 5.3 and 5.6, it is only known (through the special

© Springer International Publishing Switzerland 2016   159
G. Bastin, J.-M. Coron, *Stability and Boundary Stabilization of 1-D Hyperbolic Systems*, Progress in Nonlinear Differential Equations and Their Applications 88, DOI 10.1007/978-3-319-32062-5_5

case of systems of two balance laws) that there are intrinsic limitations to the system stabilizability with local controls. The other sections of the chapter are devoted to special cases and illustrating examples.

5.1   Lyapunov Exponential Stability

As we have seen in Chapter 1, the system (5.1) can always be represented by an equivalent linear system having the following form in Riemann coordinates:

$$\mathbf{R}_t + \Lambda(x)\mathbf{R}_x + M(x)\mathbf{R} = \mathbf{0} \quad t \in [0, +\infty), \quad x \in (0, L). \tag{5.3}$$

where $\mathbf{R} : [0, +\infty) \times [0, L] \to \mathbb{R}^n$, $\Lambda : [0, L] \to \mathcal{D}_n$, $M : [0, L] \to \mathcal{M}_{n,n}(\mathbb{R})$. The maps Λ and M are of class C^1. The matrix $\Lambda(x)$ is diagonal and defined as

$$\Lambda(x) = \begin{pmatrix} \Lambda^+(x) & 0 \\ 0 & -\Lambda^-(x) \end{pmatrix}$$

with

$$\Lambda^+(x) = \mathrm{diag}\{\lambda_1(x), \dots, \lambda_m(x)\},$$
$$\Lambda^-(x) = \mathrm{diag}\{\lambda_{m+1}(x), \dots, \lambda_n(x)\}, \qquad \lambda_i(x) > 0 \ \forall i, \quad \forall x \in [0, L],$$

The diagonal entries of $\Lambda(x)$ are the eigenvalues of the matrix $F(x)$.

As in the previous chapters, our concern is to analyze the exponential stability of this system under linear boundary conditions in canonical form

$$\begin{pmatrix} \mathbf{R}^+(t, 0) \\ \mathbf{R}^-(t, L) \end{pmatrix} = \mathbf{K} \begin{pmatrix} \mathbf{R}^+(t, L) \\ \mathbf{R}^-(t, 0) \end{pmatrix}, \quad t \in [0, +\infty), \tag{5.4}$$

and an initial condition

$$\mathbf{R}(0, x) = \mathbf{R}_0(x), \quad x \in (0, L). \tag{5.5}$$

The well-posedness of the Cauchy problem (5.3), (5.4), (5.5) in L^2 is addressed in Appendix A, see Theorem A.4.

For the stability analysis, we adopt a L^2 Lyapunov function candidate:

$$\mathbf{V} = \int_0^L \mathbf{R}^\mathsf{T} Q(x) \mathbf{R} \, dx,$$

with

$$Q(x) \triangleq \text{diag}\{Q^+(x), Q^-(x)\}, \quad Q^+ \in C^1([0, L]; \mathcal{D}_m^+), \quad Q^- \in C^1([0, L]; \mathcal{D}_{n-m}^+).$$
(5.6)

The time derivative of this function along the solutions of the system (5.3), (5.4), (5.5) is

$$\frac{d\mathbf{V}}{dt} = \mathbf{W}_1 + \mathbf{W}_2$$
(5.7)

with

$$\mathbf{W}_1 \triangleq -\left[\mathbf{R}^\mathsf{T} Q(x) \Lambda(x) \mathbf{R}\right]_0^L$$

$$= \begin{pmatrix} \mathbf{R}^+(t, L) \\ \mathbf{R}^-(t, 0) \end{pmatrix}^\mathsf{T} \begin{pmatrix} Q^+(L)\Lambda^+(L) & 0 \\ 0 & Q^-(0)\Lambda^-(0) \end{pmatrix} \begin{pmatrix} \mathbf{R}^+(t, L) \\ \mathbf{R}^-(t, 0) \end{pmatrix}$$

$$- \begin{pmatrix} \mathbf{R}^+(t, L) \\ \mathbf{R}^-(t, 0) \end{pmatrix}^\mathsf{T} \mathbf{K}^\mathsf{T} \begin{pmatrix} Q^+(0)\Lambda^+(0) & 0 \\ 0 & Q^-(L)\Lambda^-(L) \end{pmatrix} \mathbf{K} \begin{pmatrix} \mathbf{R}^+(t, L) \\ \mathbf{R}^-(t, 0) \end{pmatrix},$$

$$\mathbf{W}_2 \triangleq \int_0^L \mathbf{R}^\mathsf{T} \left((Q(x)\Lambda(x))_x - Q(x)M(x) - M^\mathsf{T}(x)Q(x) \right) \mathbf{R}\, dx.$$

From (5.7), we have the following straightforward stability result (which was already sketched in the early papers by Rauch and Taylor (1974) and Russell (1978)).

Proposition 5.1. The solution $\mathbf{R}(t, x)$ of the Cauchy problem (5.3), (5.4), (5.5) exponentially converges to $\mathbf{0}$ for the L^2–norm if there exists a map Q satisfying (5.6) such that the following *Matrix Inequalities* hold:

(i) the matrix

$$\begin{pmatrix} Q^+(L)\Lambda^+(L) & 0 \\ 0 & Q^-(0)\Lambda^-(0) \end{pmatrix} - \mathbf{K}^\mathsf{T} \begin{pmatrix} Q^+(0)\Lambda^+(0) & 0 \\ 0 & Q^-(L)\Lambda^-(L) \end{pmatrix} \mathbf{K}$$

is positive semi-definite;

(ii) the matrix

$$-(Q(x)\Lambda(x))_x + Q(x)M(x) + M^\mathsf{T}(x)Q(x)$$

is positive definite $\forall x \in [0, L]$.

$\square$

In Chapter 3, for linear conservation laws, we have considered the special case where

$$Q(x) \triangleq P(\mu x) = \text{diag}\{P^+ e^{-\mu x}, P^- e^{\mu x}\},$$

with

$$P^+ = \text{diag}\{p_1,\dots,p_m\}, \quad P^- \triangleq \text{diag}\{p_{m+1},\dots,p_n\}, \quad p_i > 0 \text{ for } i = 1,\dots,n.$$

In that case, Proposition 5.1 is specialized as follows.

Proposition 5.2. The solution $\mathbf{R}(t,x)$ of the Cauchy problem (5.3), (5.4), (5.5) exponentially converges to $\mathbf{0}$ for the L^2-norm if there exists real $\mu \neq 0$ and $p_i > 0$, $i = 1,\dots,n$, such that the following *Matrix Inequalities* hold:

(i) the matrix

$$\begin{pmatrix} P^+\Lambda^+(L)e^{-\mu L} & 0 \\ 0 & P^-\Lambda^-(0) \end{pmatrix} - \mathbf{K}^{\mathsf{T}} \begin{pmatrix} P^+\Lambda^+(0) & 0 \\ 0 & P^-\Lambda^-(L)e^{\mu L} \end{pmatrix} \mathbf{K}$$

is positive semi-definite;

(ii) the matrix

$$\mu P(\mu x)|\Lambda(x)| - P(\mu x)\Lambda'(x) + M^{\mathsf{T}}(x)P(\mu x) + P(\mu x)M(x)$$

is positive definite $\forall x \in (0,L)$.

$\square$

For general linear systems of the form (5.3), it is rather clear that more explicit convergence conditions can be derived only when the internal structure and the numerical values of the involved matrices $\Lambda(x)$, $M(x)$ and $\mathbf{K}$ are at least partially specified. The special case of balance laws with uniform (i.e., independent of x) Λ and M matrices will be addressed in the next section. Another interesting special case is when $\mathbf{K} = 0$ and the characteristic velocities are all constant and positive, i.e., $m = n$ and $\Lambda = \Lambda^+ \in \mathcal{D}_n^+$. In that case, the following corollary of Proposition 5.2 states that the convergence is unconditionally guaranteed, in accordance with the physical intuition.

Corollary 5.3. If $\mathbf{K} = 0$ and $\Lambda \in \mathcal{D}_n^+$ is constant, the solution $\mathbf{R}(t,x)$ of the Cauchy problem (5.3), (5.4), (5.5) exponentially converges to zero in L^2-norm.

Proof. If the matrix $\Lambda \in \mathcal{D}_n^+$ is constant and $\mathbf{K} = 0$, the first stability condition of Proposition 5.2 reduces to

(i) $P\Lambda e^{-\mu L}$ is positive semi-definite,

which is trivially satisfied for any $P \triangleq \text{diag}\{p_1,\dots,p_n\} \in \mathcal{D}_n^+$. The second stability condition of Proposition 5.2 reduces to:

(ii) $e^{-\mu x}\left(\mu P\Lambda + M^{\mathsf{T}}(x)P + PM(x)\right)$ is positive definite,

which is satisfied if $\mu > 0$ is taken sufficiently large. $\square$

In practical applications, we may have $\mathbf{K} = 0$ either by the nature of the physical boundary conditions or by the choice of the boundary control laws as we illustrate with the example of a chemical plug flow reactor hereafter.

5.1.1 Example: Feedback Control of an Exothermic Plug Flow Reactor

According to the description given in Section 1.7, the dynamics of a parallel plug flow reactor (PFR) are represented by the following system of balance laws:

$$
\begin{aligned}
&\partial_t T_r + V_r \partial_x T_r + k_o(T_c - T_r) - k_1 r(T_r, C_A, C_B) = 0, \\
&\partial_t C_A + V_r \partial_x C_A + r(T_r, C_A, C_B) = 0, \\
&\partial_t C_B + V_r \partial_x C_B - r(T_r, C_A, C_B) = 0, \\
&\partial_t T_c + V_c \partial_x T_c - k_o(T_c - T_r) = 0,
\end{aligned}
\tag{5.8}
$$

where $V_r(t)$ is the reactive fluid velocity in the reactor, $V_c(t)$ is the coolant velocity in the jacket, $T_r(t, x)$ is the reactor temperature, $T_c(t, x)$ is the coolant temperature. The variables $C_A(t, x)$ and $C_B(t, x)$ denote the concentrations of the chemicals in the reaction mixture. The function $r(T_r, C_A, C_B)$ representing the reaction rate is defined as follows:

$$
r(T_r, C_A, C_B) = (aC_A - bC_B) \exp\left(-\frac{E}{RT_r}\right),
$$

where a and b are rate constants, E is the activation energy, and R is the Boltzmann constant. Remark that here we consider a plug flow reactor where the chemical flow and the coolant flow are parallel while the model in Section 1.7 was for a countercurrent system.

The characteristic velocities V_r and V_c are supposed to be constant and the system is subject to the following constant boundary conditions:

$$
T_r(t, 0) = T_r^{in}, \quad C_A(t, 0) = C_A^{in}, \quad C_B(t, 0) = 0, \quad T_c(t, 0) = T_c^{in}.
\tag{5.9}
$$

From (5.8), by summing the second and third equations, it follows that the dynamics of the total concentration $C_T \triangleq C_A + C_B$ are simply described by the delay equation

$$
\partial_t C_T + V_r \partial_x C_T = 0.
$$

Therefore, for the stability analysis, there is no loss of generality if we assume that $C_T = C_A^{in}$ is constant $\forall x$ and $\forall t$. Then the reaction rate function may be redefined as

$$r(T_r, C_A) = \left((a + b)C_A - bC_A^{in}\right) \exp\left(-\frac{E}{RT_r}\right)$$

and the third equation of the system (5.8) may be ignored.

A steady state $T_r^*(x)$, $C_A^*(x)$, $T_c^*(x)$ is a solution, over the interval $[0, L]$, of the differential system

$$V_r \partial_x T_r^* = -k_o(T_c^* - T_r^*) + k_1 r(T_r^*, C_A^*),$$
$$V_r \partial_x C_A^* = -r(T_r^*, C_A^*),$$
$$V_c \partial_x T_c^* = k_o(T_c^* - T_r^*),$$

under the boundary conditions

$$T_r^*(0) = T_r^{in}, \qquad C_A^*(0) = C_A^{in}, \qquad T_c^*(0) = T_c^{in}.$$

In order to linearize the model, we define the deviations of the states T_c, T_r, and C_A with respect to the steady states:

$$R_1 = T_r - T_r^*, \qquad R_2 = C_A - C_A^*, \qquad R_3 = T_c - T_c^*.$$

Then the linearized system expressed in Riemann coordinates around the steady state is

$$\partial_t \begin{pmatrix} R_1 \\ R_2 \\ R_3 \end{pmatrix} + \underbrace{\begin{pmatrix} V_r & 0 & 0 \\ 0 & V_r & 0 \\ 0 & 0 & V_c \end{pmatrix}}_{\Lambda} \partial_x \begin{pmatrix} R_1 \\ R_2 \\ R_3 \end{pmatrix} + \underbrace{\begin{pmatrix} -k_0 - k_1\phi_1(x) & -k_1\phi_0(x) & k_0 \\ \phi_1(x) & \phi_0(x) & 0 \\ k_0 & 0 & -k_0 \end{pmatrix}}_{M(x)} \begin{pmatrix} R_1 \\ R_2 \\ R_3 \end{pmatrix},$$

with

$$\phi_0(x) \triangleq (a + b) \exp\left(-\frac{E}{RT_r^*(x)}\right),$$

$$\phi_1(x) \triangleq \left(C_A^*(x) - \frac{b}{a + b} C_A^{in}\right) \frac{E}{R(T_r^*(x))^2} \phi_0(x).$$

Moreover, the boundary conditions (5.9) expressed in Riemann coordinates are:

$$R_1(t, 0) = 0, \qquad R_2(t, 0) = 0, \qquad R_3(t, L) = 0.$$

This implies trivially that we are in the special case where Λ is constant and belongs to $\mathcal{D}_n^+$ while $\mathbf{K} = \mathbf{0}$. Therefore the stability is guaranteed according to Corollary 5.3.

Although the reactor is stable, the use of feedback control is nevertheless required in practice because of the risk of peaks in the temperature profile (hot spots) and the possibility of thermal runaway. Indeed, by quoting Karafyllis and Daoutidis (2002), "the occurrence of excessive temperatures can have detrimental consequences on the operation of the reactor, such as catalyst deactivation, undesired side reactions, and thermal decomposition of the products."

For simplicity, we consider here the case of a simple proportional control. The command signal is the inlet cooling temperature $U(t) = T_c^{in}(t)$ and the regulated output variable is the exit reactor temperature $T_r(t, L)$. The control law is defined as:

$$U(t) = T_{ref} + k_P(T_{sp} - T_r(t, L)).$$

where T_{sp} is the temperature set-point, k_P is the tuning parameter of the controller, and T_{ref} is an arbitrary reference temperature freely chosen by the user. For the stability analysis, without loss of generality, we may assume that

$$T_{ref} = T_c^*(0)$$

Then, in Riemann coordinates, the boundary conditions are:

$$\begin{pmatrix} R_1(t,0) \\ R_2(t,0) \\ R_3(t,0) \end{pmatrix} = \underbrace{\begin{pmatrix} 0 & 0 & 0 \\ 0 & 0 & 0 \\ -k_P & 0 & 0 \end{pmatrix}}_{K} \begin{pmatrix} R_1(t,L) \\ R_2(t,L) \\ R_3(t,L) \end{pmatrix}.$$

The two stability conditions of Proposition 5.2 may be expressed as follows:

$$\text{(i)} \quad k_p^2 < e^{-\mu L}$$

$$\text{(ii)} \quad \begin{pmatrix} p_1(\mu V_r - 2(k_0 + k_1\phi_1(x))) & -p_1 k_1\phi_0(x) + p_2\phi_1(x) & (p_1 + p_3)k_0 \\ -p_1 k_1\phi_0(x) + p_2\phi_1(x) & p_2(\mu V_r + 2\phi_0(x)) & 0 \\ (p_1 + p_3)k_0 & 0 & p_3(\mu V_c - 2k_0) \end{pmatrix}$$

is positive semi-definite.

It is then clear that, for any positive p_1, p_2, p_3, μ can be taken sufficiently large to satisfy condition (ii), and that the stability is therefore guaranteed as long as the control tuning parameter k_p satisfies condition (i).

5.2 Linear Systems with Uniform Coefficients

Let us now consider the special case of linear systems with uniform coefficients:

$$\mathbf{R}_t + \Lambda \mathbf{R}_x + M\mathbf{R} = \mathbf{0}, \quad t \in [0, +\infty), \quad x \in (0, L), \tag{5.10}$$

where the matrices Λ and M are constant. For μ sufficiently small, it is clear that the condition (ii) of Proposition 5.2 is satisfied if there exist $p_i > 0$ such that $M^\mathsf{T} P(0) + P(0)M$ is positive definite. In such a case, the matrix $-M$ is said to be *diagonally stable* because it is stable and the associated Lyapunov equation is satisfied with a diagonal weighting matrix (see, e.g., Barker et al. (1978) and Shorten et al. (2009) for more information on diagonally stable matrices).

In the next theorem, we shall show, with an appropriate choice of the weighting matrix $Q(x)$ of the Lyapunov function, that condition (ii) of Proposition 5.1 is satisfied even if there exists $P \in \mathcal{D}_n^+$ such that $M^\mathsf{T} P + PM$ is only positive *semi-definite*.

Theorem 5.4. If there exists $P \in \mathcal{D}_n^+$ such that

$$M^\mathsf{T} P + PM \text{ is positive semi-definite} \tag{5.11}$$

and

$$\| \Delta K \Delta^{-1} \| < 1 \tag{5.12}$$

with $\Delta \triangleq \sqrt{P|\Lambda|}$, then the system (5.4), (5.10) is exponentially stable for the L^2-norm (in the sense of Definition 3.3).

Proof. We use the Lyapunov function

$$\mathbf{V} = \int_0^L \mathbf{R}^\mathsf{T} Q(x) \mathbf{R} \, dx,$$

with $Q(x)$ defined as follows:

$$Q(x) \triangleq \begin{pmatrix} P^+ e^{-\phi(x)} & 0 \\ 0 & P^- e^{\phi(x)} \end{pmatrix},$$

$$\phi(x) \triangleq \mu(x + 1)^c, \quad \mu > 0, \quad c > 0,$$

where $P^+ \in \mathcal{D}_m^+$, $P^- \in \mathcal{D}_{n-m}^+$ are defined by $P \triangleq \text{diag}\{P^+, P^-\}$.

With this definition, the matrix involved in the second stability condition (ii) of Proposition 5.1 is

$$P_2(\mu) \triangleq \mu c(x+1)^{c-1}|\Lambda|Q(x) + M^\mathsf{T}Q(x) + Q(x)M.$$

The Taylor expansion of $P_2(\mu)$ is as follows:

$$P_2(\mu) = (M^\mathsf{T}P + PM) + \mathcal{S}(c)\mu + \text{ higher order terms in } \mu,$$

$$\text{with } \mathcal{S}(c) \triangleq (x+1)^{c-1}\Big[c|\Lambda|P - (x+1)(M^\mathsf{T}P + PM)\Big].$$

Then, $c > 0$ is selected sufficiently large such that the matrix $\mathcal{S}(c)$ is positive definite for all $x \in [0, L]$. Hence, by (5.11), we know that for sufficiently small $\mu > 0$, $P_2(\mu)$ is positive definite for all $x \in [0, L]$. Moreover, the matrix of the first stability condition (i) of Proposition 5.1 is

$$P_1(\mu) \triangleq \begin{pmatrix} P^+\Lambda^+e^{-\mu\phi(L)} & 0 \\ 0 & P^-\Lambda^-e^{\mu\phi(0)} \end{pmatrix} - \mathbf{K}^\mathsf{T}\begin{pmatrix} P^+\Lambda^+e^{-\mu\phi(0)} & 0 \\ 0 & P^-\Lambda^-e^{\mu\phi(L)} \end{pmatrix}\mathbf{K}.$$

Then, using (5.12), we know that $P_1(0)$ is positive definite (same argument as for $\mathbf{W}(0)$ in the proof of Theorem 3.2) and therefore that $\mu > 0$ can be taken sufficiently small such that condition (i) of Proposition 5.1 holds.   □

In the next section, we illustrate this theorem with the example of a linearized Saint-Venant-Exner model.

5.2.1   Application to a Linearized Saint-Venant-Exner Model

In this section adapted from Diagne et al. (2012), we consider a pool of a prismatic sloping open channel with a rectangular cross-section, a unit width, and a moving bathymetry (because of sediment transportation), as described in Section 1.5 and represented in Fig. 5.1.

The state variables of the model are: the water depth $H(t, x)$, the water velocity $V(t, x)$, and the bathymetry $B(t, x)$ which is the depth of the sediment layer above the channel bottom. The dynamics of the system are described by the coupling of Saint-Venant and Exner equations:

$$\partial_t H + V\partial_x H + H\partial_x V = 0,$$

$$\partial_t V + V\partial_x V + g\partial_x H + g\partial_x B - gS_b + C\frac{V^2}{H} = 0, \tag{5.13}$$

$$\partial_t B + aV^2\partial_x V = 0.$$

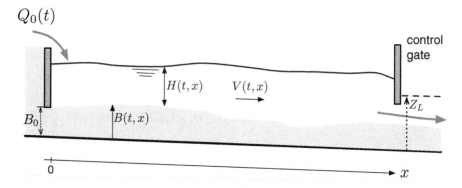

Fig. 5.1 A pool of a sloping channel with a moving bathymetry and a boundary control gate.

In these equations, g is the gravity constant, S_b is the bottom slope of the channel, C is a friction coefficient, and a is a parameter that encompasses porosity and viscosity effects on the sediment dynamics.

5.2.1.1 Steady State and Linearization

Here we assume a uniform *steady state* H^*, V^*, B^* which satisfies the relation

$$gS_bH^* = CV^{*2}.$$

In order to linearize the model, we define the deviations of the state $H(t, x), V(t, x), B(t, x)$ with respect to the steady state:

$$h(x, t) = H(x, t) - H^*,$$
$$v(x, t) = V(x, t) - V^*,$$
$$b(x, t) = B(x, t) - B^*.$$

Then the Saint-Venant-Exner model (5.13) linearized around the constant steady state is

$$\partial_t h + V^* \partial_x h + H^* \partial_x v = 0,$$
$$\partial_t v + V^* \partial_x v + g\partial_x h + g\partial_x b - C\frac{V^{*2}}{H^{*2}}h + 2C\frac{V^*}{H^*}v = 0, \qquad (5.14)$$
$$\partial_t b + aV^{*2}\partial_x v = 0.$$

5.2.1.2 Riemann Coordinates

In matrix form, the linearized model (5.14) can be written as

$$\mathbf{Y}_t + A\mathbf{Y}_x + B\mathbf{Y} = \mathbf{0} \tag{5.15}$$

where

$$\mathbf{Y} \triangleq \begin{pmatrix} h \\ v \\ b \end{pmatrix}, \quad A \triangleq \begin{pmatrix} V^* & H^* & 0 \\ g & V^* & g \\ 0 & aV^{*2} & 0 \end{pmatrix}, \quad B \triangleq \begin{pmatrix} 0 & 0 & 0 \\ -C\dfrac{V^{*2}}{H^{*2}} & 2C\dfrac{V^*}{H^*} & 0 \\ 0 & 0 & 0 \end{pmatrix}.$$

As we have seen in Section 1.5 the eigenvalues of the matrix A can be approximated as

$$\lambda_1 \approx V^* + \sqrt{gH^*}, \quad \lambda_2 \approx \frac{agV^{*3}}{gH^* - V^{*2}}, \quad \lambda_3 \approx V^* - \sqrt{gH^*}. \tag{5.16}$$

Obviously exact, but rather more complicated, expressions of the eigenvalues of A could also be obtained by using the *Cardano-Vieta* method, see, e.g., Hudson and Sweby (2003). Remark also that here, exceptionally, we use the notation λ_3 (instead of $-\lambda_3$) for the negative eigenvalue, in order to keep the symmetry of the ongoing mathematical developments.

Once the eigenvalues λ_i of the matrix A are obtained, the corresponding left eigenvectors can be computed as

$$L_k = \frac{1}{(\lambda_k - \lambda_i)(\lambda_k - \lambda_j)} \begin{pmatrix} (V^* - \lambda_i)(V^* - \lambda_j) + gH^* \\ H^*\lambda_k \\ gH^* \end{pmatrix}^T,$$

$$k \neq i \neq j \in \{1, 2, 3\}.$$

We multiply (5.15) by L_k in order to write the model in terms of the Riemann coordinates R_k ($k = 1, 2, 3$). Then we obtain

$$\partial_t R_k + \lambda_k \partial_x R_k + L_k B\mathbf{Y} = 0, \quad k = 1, 2, 3, \tag{5.17}$$

where

$$R_k = \frac{1}{(\lambda_k - \lambda_i)(\lambda_k - \lambda_j)} \left[\left((V^* - \lambda_i)(V^* - \lambda_j) + gH^* \right)h + H^*\lambda_k v + gH^* b \right].$$

Conversely, we can express h, u, and b in terms of the Riemann coordinates:

$$h = R_1 + R_2 + R_3,$$

$$v = \frac{1}{H^*}\left[(\lambda_1 - V^*)R_1 + (\lambda_2 - V^*)R_2 + (\lambda_3 - V^*)R_3\right],$$

$$b = \frac{1}{gH^*}\left[\left((\lambda_1 - V^*)^2 - gH^*\right)R_1\right.$$

$$\left.+\left((\lambda_2 - V^*)^2 - gH^*\right)R_2 + \left((\lambda_3 - V^*)^2 - gH^*\right)R_3\right].$$

Using the coordinates R_k, the last term of (5.17) writes:

$$L_k BY = \gamma_1 l_2^k h + \gamma_2 l_2^k v$$

$$= \sum_{s=1}^{3} \left(\gamma_1 + \gamma_2 \frac{\lambda_s - V^*}{H^*}\right) l_2^k R_s, \tag{5.18}$$

where

$$\gamma_1 = -C\frac{V^{*2}}{H^{*2}}, \qquad \gamma_2 = 2C\frac{V^*}{H^*},$$

and l_2^k is the second component of L_k. Equation (5.18) can be rewritten as:

$$L_k BY = -C\frac{V^*}{H^*}\frac{\lambda_k}{(\lambda_k - \lambda_i)(\lambda_k - \lambda_j)}\sum_{\ell=1}^{3}\left(3V^* - 2\lambda_\ell\right)R_\ell,$$

$$k \neq i \neq j \in \{1, 2, 3\}.$$

For the sake of simplicity, we introduce the following notation θ_k:

$$\theta_k = -C\frac{V^*}{H^*}\frac{\lambda_k}{(\lambda_k - \lambda_i)(\lambda_k - \lambda_j)}.$$

Then equation (5.17) writes:

$$\partial_t S_k + \lambda_k \partial_x S_k + \sum_{\ell=1}^{3}(2\lambda_\ell - 3V^*)\theta_\ell S_\ell = 0, \qquad (k = 1, 2, 3), \tag{5.19}$$

where the characteristic coordinates are now redefined as

$$S_k = \frac{1}{\theta_k}R_k.$$

From (5.19), the linearized model (5.17) in characteristic form may now be written as

$$\mathbf{S}_t + \Lambda \mathbf{S}_x + M\mathbf{S} = 0$$

where

$$\mathbf{S} = (S_1, S_2, S_3)^\mathsf{T}, \quad \Lambda = \mathrm{diag}\{\lambda_1, \lambda_2, \lambda_3\},$$

$$M = \begin{pmatrix} \alpha_1 \ \alpha_2 \ \alpha_3 \\ \alpha_1 \ \alpha_2 \ \alpha_3 \\ \alpha_1 \ \alpha_2 \ \alpha_3 \end{pmatrix}, \quad \text{with } \alpha_k = \left(2\lambda_k - 3V^*\right)\theta_k.$$

From Section 1.5, we know that the three eigenvalues of the matrix A are such that

$$\lambda_1 \gg \lambda_2 > 0 > \lambda_3 \tag{5.20}$$

with λ_1 and λ_3 the characteristic velocities of the water flow and λ_2 the characteristic velocity of the sediment motion. On the basis of (5.20), we are now going to determine the sign of the coefficients α_k in M.

For α_1, we have

$$\alpha_1 = C\frac{V^*}{H^*}\left(2\lambda_1 - 3V^*\right)\frac{\lambda_1}{(\lambda_1 - \lambda_2)(\lambda_1 - \lambda_3)}.$$

From (5.20), we have

$$\lambda_1 > 0, \quad \lambda_1 - \lambda_3 > 0 \quad \text{and} \quad \lambda_1 - \lambda_2 > 0.$$

Using the trace of A, we have also

$$3V^* - 2\lambda_1 = 2\lambda_3 + 2\lambda_2 - V^*.$$

Since λ_2 is small, $3V^* - 2\lambda_1$ has the same sign as $2\lambda_3 - V^*$. Since $\lambda_3 < 0$ is negative, we obtain: $3V^* - 2\lambda_1 < 0$ and consequently $\alpha_1 > 0$.

For α_2, we have

$$\alpha_2 = C\frac{V^*}{H^*}\left(2\lambda_2 - 3V^*\right)\frac{\lambda_2}{(\lambda_2 - \lambda_1)(\lambda_2 - \lambda_3)}.$$

Since the sediment motion is much slower than the water flow, we may assume that $3V^* - 2\lambda_2 > 0$. Moreover from (5.20), we have also

$$\lambda_2 > 0, \quad \lambda_2 - \lambda_3 > 0 \quad \text{and} \quad \lambda_2 - \lambda_1 < 0.$$

From these inequalities, we conclude that $\alpha_2 > 0$.

Finally, for α_3, we have:

$$\alpha_3 = C\frac{V^*}{H^*}\left(2\lambda_3 - 3V^*\right)\frac{\lambda_3}{(\lambda_3 - \lambda_2)(\lambda_3 - \lambda_1)}.$$

Since $\lambda_3 < 0$, we have $3V^* - 2\lambda_3 > 0$. Using (5.20), we infer that:

$$\lambda_3 - \lambda_2 < 0 \quad \text{and} \quad \lambda_3 - \lambda_1 < 0.$$

From the above inequalities, we conclude that $\alpha_3 > 0$.

Hence all the coefficients α_k in matrix M are strictly positive.

5.2.1.3 Lyapunov Stability

We are now going to show how Theorem 5.4 may be applied to analyze the stability of an open channel represented by Saint-Venant-Exner equations (5.13). As seen in Fig. 5.1 the channel is provided at the upstream boundary with a hydraulic device allowing to assign the value of the flow-rate $Q_0(t)$. At the downstream boundary there is an underflow control gate. Therefore we have the following boundary conditions in physical coordinates:

$$Q_0(t) = H(t,0)V(t,0),$$

$$Q_L(t) = H(t,L)V(t,L) = \left(k_G\sqrt{2g}\right)(U(t) - B(t,L))\sqrt{H(t,L) + B(t,L) - Z_L},$$

$$B(t,0) = B_0, \tag{5.21}$$

where k_G is a constant positive parameter, Z_L is the external water level, and $U(t)$ denotes the aperture of the control gate which is the command signal of the control system (See Section 1.4 for more details). Remark that the aperture may be partially blocked by the sediment.

For a given constant inflow rate Q_0, a uniform steady state of the system is a quadruple H^*, V^*, B^*, U^* which satisfies the four equations

$$Q_0 = H^*V^*, \quad gS_bH^* = CV^{*2}, \quad B^* = B_0,$$

$$Q_0 = \left(k_G\sqrt{2g}\right)(U^* - B_0)\sqrt{H^* + B_0 - Z_L}.$$

It is evident that physical equilibria exist only for $H^* > Z_L - B_0 > 0$. For any Q_0, there is a unique steady state with the value of H^* such that

$$Q_0 = \theta H^*\sqrt{gH^*}, \quad \text{with} \quad \theta \triangleq \sqrt{\frac{S_b}{C}} \ll 1,$$

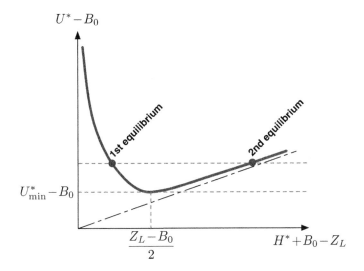

Fig. 5.2 Locus of steady states for the Saint-Venant-Exner system.

and the control gate position U^* given by

$$U^* = B_0 + \frac{\theta}{k_G \sqrt{2g}} \frac{H^* \sqrt{gH^*}}{\sqrt{H^* + B_0 - Z_L}}.$$

This function is represented in Fig. 5.2. It is interesting to observe that there is a minimal aperture U^* of the control gate which is given by

$$U^*_{min} = B_0 + \left(\frac{3}{2}\right)^{\frac{3}{2}} \frac{\theta}{k_G} (Z_L - B_0),$$

Under this value, there is no uniform steady state (but a nonuniform steady state may exist obviously). On the other hand, it can be seen that there are two distinct equilibria for each admissible position of the control gate. We shall now show how the stability of those equilibria can be analyzed using Theorem 5.4.

Expressed in the **S** Riemann coordinates, linearized around a steady state, the boundary conditions (5.21) are as follows:

$$\begin{pmatrix} S_1(t, 0) \\ S_2(t, 0) \\ S_3(t, L) \end{pmatrix} = \underbrace{\begin{pmatrix} 0 & 0 & \pi_1 \\ 0 & 0 & \pi_2 \\ \chi_1 & \chi_2 & 0 \end{pmatrix}}_{\mathbf{K}} \begin{pmatrix} S_1(t, L) \\ S_2(t, L) \\ S_3(t, 0) \end{pmatrix},$$

$$\pi_1 \triangleq \left(\frac{\lambda_3}{\lambda_1}\right) \left(\frac{gH^* - V^{*2} + \lambda_2 \lambda_3}{gH^* - V^{*2} + \lambda_1 \lambda_2}\right),$$

$$\pi_2 \triangleq \left(\frac{\lambda_3}{\lambda_2}\right)\left(\frac{gH^* - V^{*2} + \lambda_1\lambda_3}{gH^* - V^{*2} + \lambda_1\lambda_2}\right),$$

$$\chi_1 \triangleq \left(\frac{\lambda_1}{\lambda_3}\right)\left(\frac{\lambda_2 - \lambda_3}{\lambda_1 - \lambda_2}\right)\left(\frac{\lambda_1 - \phi(H^*) + \psi(H^*)((\lambda_1 - V^*)^2 - gH^*)}{\lambda_3 - \phi(H^*) + \psi(H^*)((\lambda_3 - V^*)^2 - gH^*)}\right),$$

$$\chi_2 \triangleq \left(\frac{\lambda_2}{\lambda_3}\right)\left(\frac{\lambda_1 - \lambda_3}{\lambda_2 - \lambda_1}\right)\left(\frac{\lambda_2 - \phi(H^*) + \psi(H^*)((\lambda_2 - V^*)^2 - gH^*)}{\lambda_3 - \phi(H^*) + \psi(H^*)((\lambda_3 - V^*)^2 - gH^*)}\right),$$

with

$$\phi(H^*) \triangleq \frac{\theta\sqrt{gH^*}}{2}\frac{H^*}{H^* + B_0 - Z_L}, \qquad \psi(H^*) \triangleq k_G\sqrt{2}\frac{\sqrt{g(H^* + B_0 - Z_L)}}{gH^*}.$$

In order to analyze the stability of the equilibria by invoking Theorem 5.4, we have

1) to find a matrix $\mathbf{P} = \mathrm{diag}\{p_1, p_2, p_3\}$ such that $\mathbf{S}^\mathsf{T}(\mathbf{M}^\mathsf{T}\mathbf{P} + \mathbf{PM})\mathbf{S}$ is a positive semi-definite quadratic form,
2) to check that the dissipative boundary condition $\|\Delta\mathbf{K}\Delta^{-1}\| < 1$ is satisfied.

For the matrix $\mathbf{P}$, a natural choice is $p_i = \alpha_i$ ($i = 1,2,3$) since then the quadratic form is

$$\mathbf{S}^\mathsf{T}(\mathbf{M}^\mathsf{T}\mathbf{P} + \mathbf{PM})\mathbf{S} = 2\left(\sum_{i=1}^{3}\alpha_i S_i\right)^2.$$

In order to check the dissipativity condition $\|\Delta\mathbf{K}\Delta^{-1}\| < 1$, we have to compute the matrix $\Delta \triangleq \sqrt{P|\Lambda|}$. We know that $\mathbf{P} = \mathrm{diag}\{\alpha_1, \alpha_2, \alpha_3\}$ and $|\Lambda| = \mathrm{diag}\{\lambda_1, \lambda_2, |\lambda_3|\}$ by definition. Consequently:

$$\Delta = \mathrm{diag}\left\{\sqrt{\lambda_1\alpha_1}, \sqrt{\lambda_2\alpha_2}, \sqrt{|\lambda_3|\alpha_3}\right\}$$

and

$$\Delta\mathbf{K}\Delta^{-1} = \begin{pmatrix} 0 & 0 & \pi_1\sqrt{\dfrac{\lambda_1\alpha_1}{|\lambda_3|\alpha_3}} \\[2ex] 0 & 0 & \pi_2\sqrt{\dfrac{\lambda_2\alpha_2}{|\lambda_3|\alpha_3}} \\[2ex] \chi_1\sqrt{\dfrac{|\lambda_3|\alpha_3}{\lambda_1\alpha_1}} & \chi_2\sqrt{\dfrac{|\lambda_3|\alpha_3}{\lambda_2\alpha_2}} & 0 \end{pmatrix}.$$

The inequality

$$\|\Delta \mathbf{K} \mathbf{\Delta}^{-1}\| < 1$$

holds if and only if

$$\pi_1^2 \frac{\lambda_1 \alpha_1}{|\lambda_3| \alpha_3} + \pi_2^2 \frac{\lambda_2 \alpha_2}{|\lambda_3| \alpha_3} < 1 \tag{5.22}$$

and

$$\chi_1^2 \frac{|\lambda_3| \alpha_3}{\lambda_1 \alpha_1} + \chi_2^2 \frac{|\lambda_3| \alpha_3}{\lambda_2 \alpha_2} < 1. \tag{5.23}$$

Using the definitions of α_i, we get

$$\pi_1^2 \frac{\lambda_1 \alpha_1}{|\lambda_3| \alpha_3} = \left(\frac{\lambda_3 - \lambda_2}{\lambda_1 - \lambda_2} \right) \left(\frac{2\lambda_1 - 3V^*}{2\lambda_3 - 3V^*} \right) \left(\frac{gH^* - V^{*2} + \lambda_2 \lambda_3}{gH^* - V^{*2} + \lambda_1 \lambda_2} \right)^2,$$

$$\pi_2^2 \frac{\lambda_2 \alpha_2}{|\lambda_3| \alpha_3} = \left(\frac{\lambda_3 - \lambda_1}{\lambda_2 - \lambda_1} \right) \left(\frac{2\lambda_2 - 3V^*}{2\lambda_3 - 3V^*} \right) \left(\frac{gH^* - V^{*2} + \lambda_1 \lambda_3}{gH^* - V^{*2} + \lambda_1 \lambda_2} \right)^2,$$

$$\chi_1^2 \frac{|\lambda_3| \alpha_1}{\lambda_1 \alpha_1} = \left(\frac{\lambda_3 - \lambda_2}{\lambda_1 - \lambda_2} \right) \left(\frac{2\lambda_3 - 3V^*}{2\lambda_1 - 3V^*} \right) \left(\frac{\lambda_1 - \phi(H^*) + \psi(H^*)((\lambda_1 - V^*)^2 - gH^*)}{\lambda_3 - \phi(H^*) + \psi(H^*)((\lambda_3 - V^*)^2 - gH^*)} \right)^2,$$

$$\chi_2^2 \frac{|\lambda_3| \alpha_1}{\lambda_2 \alpha_2} = \left(\frac{\lambda_3 - \lambda_1}{\lambda_2 - \lambda_1} \right) \left(\frac{2\lambda_3 - 3V^*}{2\lambda_2 - 3V^*} \right) \left(\frac{\lambda_2 - \phi(H^*) + \psi(H^*)((\lambda_2 - V^*)^2 - gH^*)}{\lambda_3 - \phi(H^*) + \psi(H^*)((\lambda_3 - V^*)^2 - gH^*)} \right)^2.$$

Let us now compute the values of these four expressions corresponding to the approximate values of the characteristic velocities given by

$$\lambda_1 \approx V^* + \sqrt{gH^*}, \qquad \lambda_2 \approx \frac{ag\theta^3}{1 - \theta^2} \sqrt{gH^*}, \qquad \lambda_3 \approx V^* - \sqrt{gH^*}. \tag{5.24}$$

We get

$$\pi_1^2 \frac{\lambda_1 \alpha_1}{|\lambda_3| \alpha_3} \approx \left(\frac{1 - \theta}{1 + \theta} \right) \left(\frac{2 - \theta}{2 + \theta} \right),$$

$$\pi_2^2 \frac{\lambda_2 \alpha_2}{|\lambda_3| \alpha_3} = \mathcal{O}(\theta^3) \text{ as } \theta \to 0,$$

$$\chi_1^2 \frac{|\lambda_3| \alpha_1}{\lambda_1 \alpha_1} \approx \left(\frac{1 - \theta}{1 + \theta} \right) \left(\frac{2 + \theta}{2 - \theta} \right) \left(\frac{2(1 - \theta)(H^* + B_0 - Z_L) - \theta H^*}{2(1 + \theta)(H^* + B_0 - Z_L) - \theta H^*} \right)^2,$$

$$\chi_2^2 \frac{|\lambda_3| \alpha_1}{\lambda_2 \alpha_2} = \mathcal{O}(\theta^3) \text{ as } \theta \to 0.$$

Since $0 < \theta \ll 1$ and $H^* > Z_L - B_0 > 0$, it follows immediately that inequalities (5.22) and (5.23) are satisfied and that all uniform steady states are exponentially stable.

5.3 Existence of a Basic Quadratic Control Lyapunov Function for a System of Two Linear Balance Laws

In this section, we shall now focus on the special case of systems of two linear balance laws in characteristic form

$$\partial_t R_1 + \lambda_1(x)\partial_x R_1 + \gamma_1(x)R_1 + \delta_1(x)R_2 = 0,$$

$$\partial_t R_2 - \lambda_2(x)\partial_x R_2 + \gamma_2(x)R_1 + \delta_2(x)R_2 = 0, \tag{5.25}$$

where the functions λ_1, λ_2 are in $C^1([0, L]; (0, +\infty))$ and the functions γ_i, δ_i are in $C^1([0, L]; \mathbb{R})$. For this particular class of systems, the purpose is to generalize the previous results by giving explicit necessary and sufficient conditions for the existence of stabilizing boundary controls that can be derived from quadratic Lyapunov functions. This analysis is made possible because, for the system (5.25), there always exists a useful coordinate transformation which was emphasized in (Krstic and Smyshlyaev 2008b, Chapter 9). In order to state this coordinate transformation, we introduce the notations

$$\varphi_1(x) \triangleq \exp\left(\int_0^x \frac{\gamma_1(s)}{\lambda_1(s)}ds\right), \quad \varphi_2(x) \triangleq \exp\left(-\int_0^x \frac{\delta_2(s)}{\lambda_2(s)}ds\right), \quad \varphi(x) \triangleq \frac{\varphi_1(x)}{\varphi_2(x)}, \tag{5.26}$$

and the new variables

$$S_1(t, x) = \varphi_1(x)R_1(t, x), \quad S_2(t, x) = \varphi_2(x)R_2(t, x). \tag{5.27}$$

Then the system (5.25) is transformed into the following system expressed in these new coordinates:

$$\partial_t S_1 + \lambda_1(x)\partial_x S_1 + a(x)S_2 = 0,$$

$$\partial_t S_2 - \lambda_2(x)\partial_x S_2 + b(x)S_1 = 0, \tag{5.28}$$

with

$$a(x) = \varphi(x)\delta_1(x), \qquad b(x) = \frac{\gamma_2(x)}{\varphi(x)}.$$

Let us consider this system under boundary conditions of the form

$$S_1(t,0) = k_1 S_2(t,0), \qquad S_2(t,L) = k_2 S_1(t,L). \qquad (5.29)$$

For this system, we have the stability condition given in the following corollary of Theorem 5.1.

Corollary 5.5. The steady state solution $S_1(t,x) \equiv 0, S_2(t,x) \equiv 0$ of the system (5.28), (5.29) is exponentially stable for the L^2-norm if there exist positive parameters $p_1 > 0, p_2 > 0$ and $\mu > 0$ such that

(i) $k_1^2 \leq \dfrac{p_2 \lambda_2(0)}{p_1 \lambda_1(0)}, \qquad k_2^2 \leq \dfrac{p_1 \lambda_1(L)}{p_2 \lambda_2(L)} e^{-2\mu L},$

(ii) The matrix $\begin{pmatrix} (\mu\lambda_1(x) - \partial_x\lambda_1(x))p_1 e^{-\mu x} & a(x)p_1 e^{-\mu x} + b(x)p_2 e^{\mu x} \\ a(x)p_1 e^{-\mu x} + b(x)p_2 e^{\mu x} & (\mu\lambda_2(x) + \partial_x\lambda_2(x))p_2 e^{\mu x} \end{pmatrix}$ is positive

definite $\forall x \in [0, L]$.

□

Based on a Lyapunov function of the form

$$\mathbf{V} = \int_0^L \left(p_1 S_1^2(t,x) e^{-\mu x} + p_2 S_2^2(t,x) e^{\mu x} \right) dx,$$

the stability conditions of Corollary 5.5 are only sufficient, possibly conservative, and given in a rather implicit way which is not easily checked.

Our purpose, in the sequel of this section, is to show that explicit necessary and sufficient conditions can be given for the existence of basic quadratic control Lyapunov functions, as defined in Definition 5.6 below, that can be used to guarantee the stabilizability of system (5.28) with decentralized boundary feedback control laws.

We now consider the system (5.28) under the boundary conditions

$$S_1(t,0) = u_1(t), \qquad S_2(t,L) = u_2(t). \qquad (5.30)$$

Equations (5.28) and (5.30) form a control system where, at time t, the state is $\mathbf{S}(t,\cdot) = (S_1(t,\cdot), S_2(t,\cdot))^{\mathsf{T}} \in L^2(0,L)^2$ and the control is $\mathbf{U}(t) = (u_1(t), u_2(t))^{\mathsf{T}} \in \mathbb{R}^2$.

We introduce the following control Lyapunov function candidate (see, e.g., Coron (2007, Section 12.1) for the classical concept of control Lyapunov function):

$$\mathbf{V(S)} \triangleq \int_0^L \left(q_1(x) S_1^2(t,x) + q_2(x) S_2^2(t,x) \right) dx, \qquad (5.31)$$

where $q_1 \in C^1([0,L]; (0,+\infty))$ and $q_2 \in C^1([0,L]; (0,+\infty))$ have to be determined. The time derivative of $\mathbf{V}$ along the trajectories of (5.28), (5.30) is

$$\frac{d\mathbf{V}}{dt}(\mathbf{S}, \mathbf{U}) = \int_0^L \left(2q_1 S_1 \partial_t S_1 + 2q_2 S_2 \partial_t S_2\right) dx$$

$$= -\int_0^L \left(2q_1 S_1(\lambda_1 \partial_x S_1 + aS_2) + 2q_2 S_2(-\lambda_2 \partial_x S_2 + bS_1)\right) dx \qquad (5.32)$$

$$= -B - \int_0^L I dx,$$

with

$$B \triangleq \lambda_1(L)q_1(L)S_1^2(t, L) - \lambda_2(L)q_2(L)u_2^2 - \lambda_1(0)q_1(0)u_1^2 + \lambda_2(0)q_2(0)S_2^2(t, 0),$$
$$(5.33)$$

$$I \triangleq (-(\lambda_1 q_1)_x)S_1^2 + 2(q_2 b + q_1 a)S_1 S_2 + ((\lambda_2 q_2)_x)S_2^2. \qquad (5.34)$$

We introduce the following definition.

Definition 5.6. A function $\mathbf{V}(\mathbf{S})$ with given q_1 and q_2 is a basic control Lyapunov function for the control system (5.28), (5.30) if and only if

$$\forall \mathbf{S} \in H^1(0, L)^2, \ \exists \mathbf{U} \in \mathbb{R}^2 \text{ such that } \frac{d\mathbf{V}}{dt}(\mathbf{S}, \mathbf{U}) \leqslant 0. \qquad (5.35)$$

It is a *strict* basic control Lyapunov function if and only if

$$\forall \mathbf{S} \in H^1(0, L)^2 \setminus \{(0, 0)^\mathsf{T}\}, \ \exists \mathbf{U} \in \mathbb{R}^2 \text{ such that } \frac{d\mathbf{V}}{dt}(\mathbf{S}, \mathbf{U}) < 0. \qquad (5.36)$$

We then have the following theorem.

Theorem 5.7. There exists a basic quadratic strict control Lyapunov function for the control system (5.28), (5.30) if and only if the maximal solution η of the Cauchy problem

$$\eta' = \left| \frac{a}{\lambda_1} + \frac{b}{\lambda_2}\eta^2 \right|, \qquad \eta(0) = 0. \qquad (5.37)$$

is defined on $[0, L]$. $\square$

The proof of this theorem will be based on the following preliminary proposition.

Proposition 5.8. Let $L > 0$, let $\alpha \in C^0([0, L])$ and $\beta \in C^0([0, L])$. If there exist $f \in C^1([0, L])$ and $g \in C^1([0, L])$ such that

$$f > 0 \text{ in } [0, L], \qquad (5.38)$$

$$g > 0 \text{ in } [0, L], \qquad (5.39)$$

$$f' \leqslant 0 \text{ in } [0, L], \tag{5.40}$$

$$g' \geqslant 0 \text{ in } [0, L], \tag{5.41}$$

$$-f'g' \geqslant (\alpha f + \beta g)^2 \text{ in } [0, L], \tag{5.42}$$

then the maximal solution η of the Cauchy problem

$$\eta' = \left| \alpha + \beta \eta^2 \right|, \ \eta(0) = 0, \tag{5.43}$$

is defined on $[0, L]$. Conversely, if the maximal solution of the Cauchy problem (5.43) is defined on $[0, L]$, there exist $f \in C^1([0, L])$ and $g \in C^1([0, L])$ such that (5.38) and (5.39) hold while (5.40), (5.41), and (5.42) are *strict* inequalities.

□

The proof of this proposition can be found in Bastin and Coron (2011). Let us remark that the function

$$(x, s) \in [0, L] \times \mathbb{R} \mapsto \left| \alpha(x) + \beta(x)s^2 \right| \in \mathbb{R}$$

is continuous in $[0, L] \times \mathbb{R}$ and locally Lipschitz with respect to s. Hence the Cauchy problem (5.43) has a unique maximal solution.

Proof of Theorem 5.7. a) *"Only if" condition.* A necessary condition for $\mathbf{V(S)}$ to be a (strict) control Lyapunov function is that I is a strictly positive quadratic form with respect to (S_1, S_2) for almost every x in $[0, L]$, i.e.

$$-(\lambda_1 q_1)_x \geqslant 0 \text{ in } [0, L], \tag{5.44}$$

$$(\lambda_2 q_2)_x \geqslant 0 \text{ in } [0, L], \tag{5.45}$$

$$-(\lambda_1 q_1)_x (\lambda_2 q_2)_x \geqslant (q_1 a + q_2 b)^2 \text{ in } [0, L]. \tag{5.46}$$

We define the functions $f \in C^1([0, L])$ and $g \in C^1([0, L])$ such that

$$f(x) \triangleq \lambda_1(x) q_1(x), \ \forall x \in [0, L], \tag{5.47}$$

$$g(x) \triangleq \lambda_2(x) q_2(x), \ \forall x \in [0, L]. \tag{5.48}$$

The quadratic form $\mathbf{V(S)}$ is coercive with respect to $(S_1, S_2)^\mathsf{T} \in L^2(0, L)^2$, i.e.

$$\exists \sigma > 0 \text{ such that } \mathbf{V(S)} \geqslant \sigma \int_0^L (S_1^2 + S_2^2) dx,$$

if and only if (5.38) and (5.39) hold. Note that (5.44) is equivalent to (5.40) and that (5.45) is equivalent to (5.41) for almost every x in $[0, L]$. Property (5.46) is equivalent to (5.42) for every x in $[0, L]$ with α and β defined by

$$\alpha(x) \triangleq \frac{a(x)}{\lambda_1(x)}, \ \beta(x) \triangleq \frac{b(x)}{\lambda_2(x)}, \ \forall x \in [0, L]. \tag{5.49}$$

Following Proposition 5.8, we consider the maximal solution η of the Cauchy problem

$$\eta' = \left| \frac{a}{\lambda_1} + \frac{b}{\lambda_2} \eta^2 \right|, \ \eta(0) = 0.$$

It follows from Proposition 5.8 that a necessary condition for the existence of a control Lyapunov function $\mathbf{V(S)}$ of the form (5.31) is that η is defined on $[0, L]$.

b) *"If" condition.* Let us assume that η is defined on $[0, L]$. Then there is a strict control Lyapunov function $V(y)$ of the form (5.31). Indeed, by Proposition 5.8, there exist $q_1 \in C^1([0, L]; (0, +\infty))$ and $q_2 \in C^1([0, L]; (0, +\infty))$ such that (5.44), (5.45), and (5.46) are *strict* inequalities in $[0, L]$. Let us define the following decentralized feedback control laws

$$u_1(t) \triangleq k_1 S_2(t, 0), \ u_2(t) \triangleq k_2 S_1(t, L). \tag{5.50}$$

Then for any constant k_1 and k_2 selected such that

$$k_1^2 \leq \frac{\lambda_2(0)q_2(0)}{\lambda_1(0)q_1(0)}, \qquad k_2^2 \leq \frac{\lambda_1(L)q_1(L)}{\lambda_2(L)q_2(L)}, \tag{5.51}$$

we have

$$\frac{d\mathbf{V}}{dt} \leq -\theta(|S_1|_{L^2(0,L)}^2 + |S_1|_{L^2(0,L)}^2), \tag{5.52}$$

for some $\theta > 0$ independent of (S_1, S_2). This leads to exponential stability with a rate depending on θ and σ, themselves depending on q_1 and q_2.

$\square$

Remark 5.9. One could believe that more general stabilizability conditions could be obtained by considering a more general Lyapunov function (with an additional cross-term) of the form

$$\mathbf{V(S)} \triangleq \int_0^L \left(q_1(x)S_1^2 + q_2(x)S_2^2 + q_3(x)S_1 S_2 \right) dx. \tag{5.53}$$

In fact, this is not true because it can be shown that, for the control system (5.28)–(5.30), if (5.53) is a control Lyapunov function then $q_3(x)$ must be zero. The proof of this assertion can be found in Bastin and Coron (2011).

In the next section, Theorem 5.7 is illustrated with an application to the control of an open channel represented by linearized Saint-Venant equations.

5.3.1   Application to the Control of an Open Channel

We consider a pool of a prismatic *horizontal* open channel with a rectangular cross-section and a unit width. The dynamics of the system are described by the Saint-Venant equations

$$\partial_t H + \partial_x(HV) = 0,$$
$$\partial_t V + \partial_x\left(\frac{V^2}{2} + gH\right) + C\frac{V^2}{H} = 0, \tag{5.54}$$

with the state variables $H(t, x)$ = water depth and $V(t, x)$ = water velocity. C is a friction coefficient and g is the gravity acceleration.

As illustrated in Fig. 5.3, the channel is provided with hydraulic control devices (pumps, valves, mobile spillways, sluice gates, etc.) which are located at the two extremities and allow to assign the values of the flow-rate on both sides:

$$Q_0(t) = H(t, 0)V(t, 0), \qquad Q_L(t) = H(t, L)V(t, L). \tag{5.55}$$

The system (5.54), (5.55) is a control system with state $H(t, x)$, $V(t, x)$ and controls $Q_0(t)$ and $Q_L(t)$.

A steady state (or equilibrium profile), corresponding to the set-point Q^*, is a couple of time-invariant *nonuniform* (i.e., space-varying) state functions $H^*(x)$, $V^*(x)$ such that $H^*(x)V^*(x) = Q^*$ which satisfy the differential equations

$$V^*\partial_x H^* = -H^*\partial_x V^* = -\frac{CV^{*3}}{gH^* - V^{*2}}. \tag{5.56}$$

For any constant inputs $Q_0(t) = Q_L(t) \equiv Q^*$, the open-loop system has a continuum of non-isolated equilibria $H^*(x)$, $V^*(x)$ which are therefore not asymptotically stable. The objective is to design decentralized control laws, with $Q_0(t)$ function of $H(t, 0)$ and $Q_L(t)$ function of $H(t, L)$, in order to stabilize the system about a constant flow-rate set point Q^* and a constant level set-point $H^*(0)$.

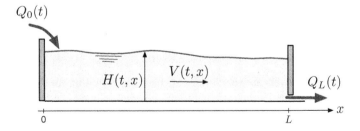

Fig. 5.3   Horizontal pool of an open channel controlled with boundary fluxes Q_0 and Q_L.

As we have seen in Section 1.4, using the definitions

$$R_1(t, x) = V(t, x) - V^*(x) + (H(t, x) - H^*(x))\sqrt{\frac{g}{H^*(x)}},$$

$$R_2(t, x) = V(t, x) - V^*(x) - (H(t, x) - H^*(x))\sqrt{\frac{g}{H^*(x)}},$$

(5.57)

the linearized Saint-Venant equations around the steady state, written in characteristic form, are:

$$\partial_t R_1 + \lambda_1(x)\partial_x R_1 + \gamma_1(x)R_1 + \delta_1(x)R_2 = 0,$$
$$\partial_t R_2 - \lambda_2(x)\partial_x R_2 + \gamma_2(x)R_1 + \delta_2(x)R_2 = 0,$$

with the characteristic velocities

$$\lambda_1(x) = V^*(x) + \sqrt{gH^*(x)}, \qquad -\lambda_2(x) = V^*(x) - \sqrt{gH^*(x)},$$

and the coefficients

$$\gamma_1(x) = \frac{CV^{*2}}{H^*}\left[-\frac{3}{4(\sqrt{gH^*} + V^*)} + \frac{1}{V^*} - \frac{1}{2\sqrt{gH^*}}\right],$$

$$\delta_1(x) = \frac{CV^{*2}}{H^*}\left[-\frac{1}{4(\sqrt{gH^*} + V^*)} + \frac{1}{V^*} + \frac{1}{2\sqrt{gH^*}}\right],$$

$$\gamma_2(x) = \frac{CV^{*2}}{H^*}\left[\frac{1}{4(\sqrt{gH^*} - V^*)} + \frac{1}{V^*} - \frac{1}{2\sqrt{gH^*}}\right],$$

$$\delta_2(x) = \frac{CV^{*2}}{H^*}\left[\frac{3}{4(\sqrt{gH^*} - V^*)} + \frac{1}{V^*} + \frac{1}{2\sqrt{gH^*}}\right].$$

The steady state flow is subcritical (or fluvial) if the following condition holds

$$gH^*(x) - V^{*2}(x) > 0 \quad \forall x \in [0, L].$$

(5.58)

Under this condition, the system is strictly hyperbolic with

$$-\lambda_2(x) < 0 < \lambda_1(x) \quad \forall x \in [0, L].$$

According to our analysis above, in order to check the condition for the existence of a basic quadratic control Lyapunov function, we need to solve the following third-order differential system on $[0, L]$:

$$\frac{dV^*}{dx} = \frac{C}{Q^*}\left(\frac{(V^*(x))^5}{gQ^* - (V^*(x))^3}\right) \qquad V^*(0) = \frac{Q^*}{H^*(0)},$$

$$\frac{d\psi}{dx} = \frac{\gamma_1(x)}{\lambda_1(x)} + \frac{\delta_2(x)}{\lambda_2(x)} \qquad \psi(0) = 0,$$

$$\frac{d\eta}{dx} = \frac{e^{\psi(x)}\delta_1(x)}{\lambda_1(x)} + \frac{\gamma_2(x)}{e^{\psi(x)}\lambda_2(x)}\eta^2(x) \qquad \eta(0) = 0.$$

The first equation computes the steady state profile $V^*(x)$. It is obtained from (5.56) and $Q^* = H^*(x)V^*(x)$. The solution ψ of the second equation is such that $\varphi = \exp(\psi)$ involved in the computation of $a(x)$ and $b(x)$ (see (5.28)). The third equation is the ODE (5.37) in the statement of Theorem 5.7.

As a matter of illustration, we compute the solution of this system with the following parameter values : $g = 9.81$ m/s², $C = 0.002$, $Q^* = 1$ m³/s, $H^*(0) = 2$ m. The function η exists over the interval $[0, L]$ with $L \simeq 18.795$ km which is the maximal length for which the flow remains subcritical, i.e. $V^*(x) < (gQ^*)^{1/3} \simeq 2.14$ m/sec. The functions $V^*(x)$ and $\eta(x)$ are shown in Fig. 5.4.

Let us now impose a boundary condition of the form

$$S_1(t,0) = k_1 S_2(t,0) \tag{5.59}$$

with

$$k_1^2 \leq \frac{\lambda_2(0)q_2(0)}{\lambda_1(0)q_1(0)}$$

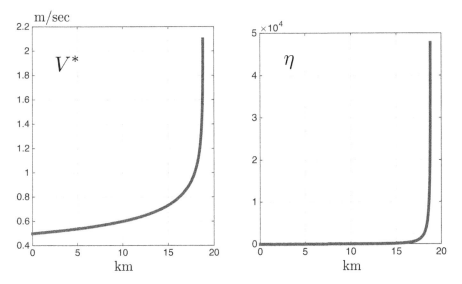

Fig. 5.4 Functions $V^*(x)$ and $\eta(x)$ computed for the example of an open channel.

to the system (5.28). Then, using the definition (5.27) of the S_i coordinates, the definition (5.57) of the R_i coordinates and the physical boundary condition (5.55), it is a matter of few calculations to get the physical stabilizing control law which implements the boundary condition (5.59)

$$Q_0(t) = \frac{H(t,0)}{H^*(0)} \left[Q^* - \frac{\varphi_1(0) + k_1\varphi_2(0)}{\varphi_1(0) - k_1\varphi_2(0)} \sqrt{gH^*(0)}(H(t,0) - H^*(0)) \right]$$

for the open channel represented by the Saint-Venant equations. We remark that this control law is a nonlinear feedback function of the water depth $H(t,0)$ although it is derived on the basis of a linearized model. Obviously, a similar derivation leads to a control law for $Q_L(t)$ at the other side of the channel.

In addition, it can also be emphasized that the implementation of the controls is particularly simple since only measurements of the levels $H(t,0)$ and $H(t,L)$ at the two boundaries are required. This means that the feedback implementation does not require neither level measurements inside the pool nor any velocity or flow rate measurements.

5.4 Boundary Control of Density-Flow Systems

In Section 2.2 we have analyzed the boundary control of linear density-flow systems described by conservation laws. In this section, we generalize the analysis to the case of linear density-flow systems described by balance laws with uniform steady states. We consider a system of two linear balance laws of the general form:

$$\partial_t H + \partial_x Q = 0,$$
$$\partial_t Q + \lambda_1\lambda_2\partial_x H + (\lambda_1 - \lambda_2)\partial_x Q - \alpha H + \beta Q = 0,$$
$$t \in [0, +\infty), x \in [0, L],$$
$$\tag{5.60}$$

where $\lambda_1 \in \mathbb{R}_+$, $\lambda_2 \in \mathbb{R}_+$, $\alpha \in \mathbb{R}$, $\beta \in \mathbb{R}$. The first equation can be interpreted as a mass conservation law with H the density and Q the flow density. The second equation can then be interpreted as a momentum balance law.

We are concerned with the solutions of the Cauchy problem for the system (5.60) under an initial condition:

$$H(0,x) = H_o(x), \qquad Q(0,x) = Q_o(x), \qquad x \in [0, L],$$

and two boundary conditions of the form:

$$Q(t,0) = Q_0(t), \qquad Q(t,L) = Q_L(t), \qquad t \in [0, +\infty). \tag{5.61}$$

Any pair of constant states H^*, Q^* such that $\alpha H^* = \beta Q^*$ is a potential steady state of the system. We assume that one of them has been selected as the desired steady state or *set point*.

The Riemann coordinates defined around the set point H^*, Q^* are:

$$R_1 = Q - Q^* + \lambda_2(H - H^*), \qquad R_2 = Q - Q^* - \lambda_1(H - H^*), \qquad (5.62)$$

with the inverse change of coordinates:

$$H = H^* + \frac{R_1 - R_2}{\lambda_1 + \lambda_2}, \qquad Q = Q^* + \frac{\lambda_1 R_1 + \lambda_2 R_2}{\lambda_1 + \lambda_2}. \qquad (5.63)$$

With these coordinates, the system (5.60) is written in characteristic form:

$$\partial_t R_1 + \lambda_1 \partial_x R_1 + \gamma R_1 + \delta R_2 = 0,$$
$$\qquad\qquad\qquad\qquad\qquad\qquad\qquad\qquad\qquad (5.64)$$
$$\partial_t R_2 - \lambda_2 \partial_x R_2 + \gamma R_1 + \delta R_2 = 0,$$

with

$$\gamma \triangleq \frac{\lambda_1 \beta - \alpha}{\lambda_1 + \lambda_2}, \qquad \delta \triangleq \frac{\lambda_2 \beta + \alpha}{\lambda_1 + \lambda_2}.$$

5.4.1   Transfer Functions

Using the Laplace transform, the transfer functions between the inputs Q_0, Q_L and the outputs $H(t, 0), H(t, L)$ are computed as follows (see, e.g., Litrico and Fromion (2009, Section 3.3)):

$$\begin{pmatrix} H(s, 0) \\ H(s, L) \end{pmatrix} = \begin{pmatrix} P_{00}(s) & P_{01}(s) \\ P_{10}(s) & P_{11}(s) \end{pmatrix} \begin{pmatrix} Q_0(s) \\ Q_L(s) \end{pmatrix},$$

with

$$P_{00}(s) = \frac{\sigma_2(s)e^{\sigma_1(s)L} - \sigma_1(s)e^{\sigma_2(s)L}}{s(e^{\sigma_2(s)L} - e^{\sigma_1(s)L})},$$

$$P_{01}(s) = \frac{\sigma_1(s) - \sigma_2(s)}{s(e^{\sigma_2(s)L} - e^{\sigma_1(s)L})},$$

$$P_{10}(s) = \frac{(\sigma_2(s) - \sigma_1(s))e^{(\sigma_1(s)+\sigma_2(s))L}}{s(e^{\sigma_2(s)L} - e^{\sigma_1(s)L})},$$

$$P_{11}(s) = \frac{\sigma_1(s)e^{\sigma_1(s)L} - \sigma_2(s)e^{\sigma_2(s)L}}{s(e^{\sigma_2(s)L} - e^{\sigma_1(s)L})},$$

$$\sigma_1(s) = \frac{(\lambda_1 - \lambda_2)s + \alpha - \sqrt{d(s)}}{2\lambda_1\lambda_2},$$

$$\sigma_2(s) = \frac{(\lambda_1 - \lambda_2)s + \alpha + \sqrt{d(s)}}{2\lambda_1\lambda_2},$$

$$d(s) = (\lambda_1 + \lambda_2)^2 s^2 + 2[(\lambda_1 - \lambda_2)\alpha + 2\lambda_1\lambda_2\beta]s + \alpha^2.$$

The poles of the system are the roots of the characteristic equation

$$s(e^{\sigma_2(s)L} - e^{\sigma_1(s)L}) = 0.$$

There is a pole at zero $p_0 = 0$ since the system is an integrator of the difference of flows $Q_0 - Q_L$. The other poles are given by (Litrico and Fromion 2009, Equ. 3.51):

$$p_{\pm k} = -(\lambda_1\gamma + \lambda_2\delta) \pm \frac{2\lambda_1\lambda_2}{\lambda_1 + \lambda_2} \sqrt{\frac{\gamma\delta}{\lambda_1\lambda_2} - \frac{k^2\pi^2}{L^2}}, \quad k \in \mathbb{Z} \setminus \{0\}.$$

In the special case where $\gamma = \delta = 0$ (i.e., a system of two linear conservation laws), the poles are given by

$$p_{\pm k} = -\pm \frac{2jk\pi}{L} \frac{\lambda_1\lambda_2}{\lambda_1 + \lambda_2}, \quad k \in \mathbb{Z}.$$

Thus they are all located on the imaginary axis (i.e., they have a zero real part) and we recover the property that the system is not asymptotically stable as in Section 2.2.

In the general case where $\gamma \neq 0$ and/or $\delta \neq 0$, let k_m be the greatest integer such that the radicand is positive. Then the poles obtained for $0 < k \leqslant k_m$ are real, and those obtained for $k > k_m$ are complex conjugate with a real part equal to

$$\Re(p_{\pm k}) = -(\lambda_1\gamma + \lambda_2\delta).$$

These poles can be stable or unstable depending of the respective signs and the relative values of γ and δ. In any case, the system is never asymptotically stable since there is always a pole at zero and the issue of boundary feedback stabilization is worth be addressed.

5.4.2 Boundary Feedback Stabilization with Two Local Controls

We consider the system (5.60) represented in Riemann coordinates by the model (5.64) with $\gamma \neq 0$ and/or $\delta \neq 0$ (the special case where $\gamma = \delta = 0$, i.e., a system of two linear conservation laws, has been treated extensively in Section 2.2). In order to define local boundary feedback controls, we use the property, introduced in Section 5.3, that the model (5.64) is equivalent to

$$\partial_t S_1 + \lambda_1 \partial_x S_1 + \delta e^{cx} S_2 = 0,$$

$$\partial_t S_2 - \lambda_2 \partial_x S_2 + \gamma e^{-cx} S_1 = 0,$$

(5.65)

where

$$c_1 \triangleq \frac{\gamma}{\lambda_1}, \quad c_2 \triangleq \frac{\delta}{\lambda_2}, \quad c \triangleq c_1 + c_2,$$

and the coordinates S_1 and S_2 are defined as

$$S_1(t, x) = e^{c_1 x} R_1(t, x), \quad S_2(t, x) = e^{-c_2 x} R_2(t, x).$$

(5.66)

From Theorem 5.7, we know that there exists a basic quadratic strict control Lyapunov function for the system (5.65) under boundary control if and only if the Cauchy problem

$$\eta' = \left| a e^{cx} + b e^{-cx} \eta^2 \right|, \quad a \triangleq \frac{\delta}{\lambda_1}, \quad b \triangleq \frac{\gamma}{\lambda_2}, \quad \eta(0) = 0,$$

(5.67)

has a solution on $[0, L]$. Let us define

$$\theta \triangleq e^{-cx} \eta.$$

Then (5.67) becomes

$$\theta' + c\theta = \left| a + b\theta^2 \right|, \quad \theta(0) = 0.$$

(5.68)

The Cauchy problem (5.68) has a solution for all $L \geq 0$ if and only if $c\theta = \left| a + b\theta^2 \right|$ for some $\theta \geq 0$, that is, if and only if $c \geq 0$ and therefore if and only if

$$\frac{\gamma}{\lambda_1} + \frac{\delta}{\lambda_2} \geq 0.$$

(5.69)

Let us assume that (5.69) holds. Then, using the analysis of Section 5.3, we know that, for any L, there is a quadratic Lyapunov function

$$\mathbf{V} = \int_0^L \left(q_1(x)S_1^2(t,x) + q_2(x)S_2^2(t,x)\right)dx,$$

such that the stability of the system (5.65) is obtained under condition (5.69) with boundary conditions of the form

$$S_1(t,0) = k_1 S_2(t,0), \quad k_1^2 \leqslant \frac{\lambda_2\, q_2(0)}{\lambda_1\, q_1(0)},$$

$$S_2(t,L) = k_2 S_1(t,L), \quad k_2^2 \leqslant \frac{\lambda_1\, q_1(L)}{\lambda_2\, q_2(L)}. \tag{5.70}$$

Using the definitions (5.62) and (5.66) of the coordinates R_i and S_i, we get the two local boundary feedback control laws that can be used to stabilize the density-flow system (5.60) at the set point H^*, Q^*:

$$Q_0(t) \triangleq Q^* + k_0(H^* - H(t,0)), \qquad k_0 \triangleq \frac{\lambda_1 k_1 + \lambda_2}{1 - k_1},$$

$$Q_L(t) \triangleq Q^* - k_L(H^* - H(t,L)), \qquad k_L \triangleq \frac{\lambda_2 k_2 e^{cL} + \lambda_1}{1 - k_2 e^{cL}}. \tag{5.71}$$

It follows from our analysis above that, under conditions (5.69), the steady state (H^*, Q^*) of the closed loop system (5.60), (5.61), (5.71) is exponentially stable if the control tuning parameters k_0 and k_L are chosen such that inequalities (5.70) hold.

5.4.3   Feedback-Feedforward Stabilization with a Single Control

Let us now consider the situation where $Q_0(t)$ is the only control input while the other input $Q_L(t)$ is a measurable disturbance. Then, from the previous section, a natural candidate control law is:

$$Q_0(t) \triangleq Q_L(t) + k_P(H^* - H(t,0)), \tag{5.72}$$

where k_P is a tuning parameter and H^* is the density set point. This control law involves a 'feedforward' term $Q_L(t)$ (which compensates for the measured disturbance) and a proportional feedback term $k_P(H^* - H(t,0))$ for the stabilization. Assuming a constant disturbance $Q_L(t) = Q^*$, if $k_P \neq 0$ the closed-loop system has a unique steady state (H^*, Q^*) and can be written in transformed Riemann coordinates (5.65) with boundary conditions

$$S_1(t, 0) = k_1 S_2(t, 0), \qquad k_1 = \frac{k_P - \lambda_2}{k_P + \lambda_1},$$

$$S_2(t, L) = -\frac{\lambda_1}{\lambda_2} e^{-cL} S_1(t, L).$$

Then, using again the analysis of Section 5.3, we know that the Cauchy problem (5.68) has a solution for all $L \geq 0$ if and only if inequality (5.69) holds. Using this inequality, it can be shown that the system (5.60) is exponentially stable with the control (5.72) for any L such that

$$\frac{\eta(L)}{e^{cL}} = \theta(L) < \frac{\lambda_2}{\lambda_1}, \tag{5.73}$$

provided the control tuning parameter k_P is selected such that

$$\left(\frac{k_P - \lambda_2}{k_P + \lambda_1}\right)^2 \leq \frac{\lambda_2 q_2(0)}{\lambda_1 q_1(0)}.$$

The value of $\theta(L)$ can be computed explicitly in the special case where γ and δ are both positive. In such case, we have $a > 0$, $b > 0$, $c > 0$ and (5.68) becomes

$$\theta' = a - c\theta + b\theta^2, \qquad \theta(0) = 0. \tag{5.74}$$

The discriminant of equation $b\theta^2 - c\theta + a = 0$ is

$$\Delta = (c_1 + c_2)^2 - 4c_1 c_2 = (c_1 - c_2)^2 \geq 0 \tag{5.75}$$

Therefore the two zeroes of the polynomial $b\theta^2 - c\theta + a$ are

$$\theta_1 \triangleq \frac{c_1}{b} = \frac{\lambda_2}{\lambda_1}, \qquad \theta_2 \triangleq \frac{c_2}{b} = \frac{\delta}{\gamma},$$

and (5.74) is equivalent to

$$\theta'\left(\frac{1}{\theta - \theta_2} - \frac{1}{\theta - \theta_1}\right) = (\theta_2 - \theta_1), \qquad \theta(0) = 0.$$

This gives the following function:

$$\theta(L) = \theta_1 \theta_2 \frac{e^{b\theta_2 L} - e^{b\theta_1 L}}{\theta_2 e^{b\theta_2 L} - \theta_1 e^{b\theta_1 L}}.$$

This function is monotonically increasing with

$$\lim_{L \to +\infty} \theta(L) = \min\left(\theta_1, \theta_2\right) = \min\left(\frac{\lambda_2}{\lambda_1}, \frac{\delta}{\gamma}\right).$$

Then, from (5.73) we see that the system (5.60) is stabilizable for any L.

5.4.4   Stabilization with Proportional-Integral Control

In the continuation of the previous paragraph, we now extend the stabilization analysis to the case where $Q_0(t)$ is provided with a Proportional-Integral (PI) controller and $Q_L(t)$ is a nonmeasurable disturbance (see Section 2.2 for further details and motivations on the use of PI control in this case).

The control law is defined as

$$Q_0(t) \triangleq Q_R + k_P(H^* - H(t, 0)) + k_I \int_0^t (H^* - H(\tau, 0)) d\tau, \tag{5.76}$$

where Q_R is a constant arbitrary reference value and H^* is the level set-point. For the closed-loop stability analysis there is no loss of generality in dealing with the special case $Q_L = Q_R = Q^* = \alpha H^*/\beta$. In the Riemann coordinates (5.62), the closed-loop system dynamics are given by equations (5.64) that we recall here:

$$\partial_t R_1 + \lambda_1 \partial_x R_1 + \gamma R_1 + \delta R_2 = 0,$$

$$\partial_t R_2 - \lambda_2 \partial_x R_2 + \gamma R_1 + \delta R_2 = 0. \tag{5.77}$$

Moreover, the boundary conditions $Q(t, 0) = Q_0(t)$ and $Q(t, L) = Q^*$ are written, using the transformation (5.63):

$$R_1(t, 0) = k_1 R_2(t, 0) + k_3 X(t), \quad k_1 \triangleq \frac{k_P - \lambda_2}{k_P + \lambda_1}, \quad k_3 \triangleq \frac{k_I}{k_P + \lambda_1},$$

$$R_2(t, L) = k_2 R_1(t, L), \quad k_2 \triangleq -\frac{\lambda_1}{\lambda_2}, \tag{5.78}$$

and

$$X(t) \triangleq \int_0^t \left(R_2(\tau, 0) - R_1(\tau, 0)\right) d\tau.$$

For the stability analysis, we use the quadratic Lyapunov function

$$V = \int_0^L \left(\frac{p_1}{\lambda_1} R_1^2 e^{-\mu x/\lambda_1} + \frac{p_2}{\lambda_2} R_2^2 e^{\mu x/\lambda_2} \right) dx + qX^2,$$

with positive constant coefficients p_1, p_2, q, and μ.

The time derivative of this function along the solutions of the system (5.77), (5.78) is

$$\frac{dV}{dt} = W_1 + W_2,$$

where

$$W_1 \triangleq - \int_0^L \begin{pmatrix} R_1 & R_2 \end{pmatrix} \Omega_1(\mu) \begin{pmatrix} R_1 \\ R_2 \end{pmatrix} dx,$$

$$\Omega_1(\mu) \triangleq \begin{pmatrix} \dfrac{p_1(\mu + 2\gamma)}{\lambda_1} e^{-\mu x/\lambda_1} & \dfrac{p_1\delta}{\lambda_1} e^{-\mu x/\lambda_1} + \dfrac{p_2\gamma}{\lambda_2} e^{\mu x/\lambda_2} \\ \dfrac{p_1\delta}{\lambda_1} e^{-\mu x/\lambda_1} + \dfrac{p_2\gamma}{\lambda_2} e^{\mu x/\lambda_2} & \dfrac{p_2(\mu + 2\delta)}{\lambda_2} e^{\mu x/\lambda_2} \end{pmatrix},$$

and

$$W_2 \triangleq - \begin{pmatrix} X(t) & R_2(t,0) & R_1(t,L) \end{pmatrix} \Omega_2(\mu) \begin{pmatrix} X(t) \\ R_2(t,0) \\ R_1(t,L) \end{pmatrix},$$

$$\Omega_2(\mu) \triangleq \begin{pmatrix} 2qk_3 - p_1 k_3^2 & -p_1 k_1 k_3 - q(1 - k_1) & 0 \\ -p_1 k_1 k_3 - q(1 - k_1) & p_2 - p_1 k_1^2 & 0 \\ 0 & 0 & p_1 e^{-\mu L/\lambda_1} - p_2 k_2^2 e^{\mu L/\lambda_2} \end{pmatrix}.$$

We are looking for conditions under which the matrices Ω_1 and Ω_2 are positive definite for all $x \in [0, L]$. We have the following proposition.

Proposition 5.10. If

$$\gamma > 0, \quad \delta > 0 \quad \text{and} \quad \frac{\lambda_1}{\lambda_2} < \frac{\gamma}{\delta} < \frac{\lambda_2}{\lambda_1}, \tag{5.79}$$

and if the control tuning parameters k_P and k_I are selected such that

$$|k_1| = \left| \frac{k_P - \lambda_2}{k_P + \lambda_1} \right| < \sqrt{\frac{\delta\lambda_2}{\gamma\lambda_1}}, \quad \text{and} \quad k_3 = \frac{k_I}{k_P + \lambda_1} > 0, \tag{5.80}$$

then there exist positive constant coefficients p_1, p_2, q, and μ such that the matrices $\Omega_1(\mu)$ and $\Omega_2(\mu)$ are positive definite for all $x \in [0, L]$, and consequently, such that the PI control law (5.76) exponentially stabilizes the system (5.60).

Proof. The determinant of the matrix $\Omega_1(\mu)$ is given by

$$D(\mu) = \frac{p_1(\mu + 2\gamma)}{\lambda_1} e^{-\mu x/\lambda_1} \frac{p_2(\mu + 2\delta)}{\lambda_2} e^{\mu x/\lambda_2} - \left(\frac{p_1\delta}{\lambda_1} e^{-\mu x/\lambda_1} + \frac{p_2\gamma}{\lambda_2} e^{\mu x/\lambda_2} \right)^2,$$

$$= \frac{p_1 p_2(\mu^2 + 2\mu(\gamma + \delta))}{\lambda_1 \lambda_2} e^{-\mu x/\lambda_1} e^{\mu x/\lambda_2} - \left(\frac{p_1\delta}{\lambda_1} e^{-\mu x/\lambda_1} - \frac{p_2\gamma}{\lambda_2} e^{\mu x/\lambda_2} \right)^2.$$

Let the coefficients $p_1 > 0$ and $p_2 > 0$ be selected such that

$$\frac{p_1\delta}{\lambda_1} = \frac{p_2\gamma}{\lambda_2}. \tag{5.81}$$

Then, under condition (5.79), the trace of $\Omega_1(\mu)$ is positive and, under conditions (5.79) and (5.81), the determinant of $\Omega_1(\mu)$ is positive for $\mu > 0$ sufficiently small since

$$D(0) = 0 \quad \text{and} \quad D'(0) = \frac{2p_1 p_2(\gamma + \delta)}{\lambda_1 \lambda_2} > 0.$$

The matrix $\Omega_2(\mu)$ is positive definite if and only if the three following inequalities hold:

(i) $p_1 e^{-\mu L/\lambda_1} - p_2 k_2^2 e^{\mu L/\lambda_2} > 0$;
(ii) $2qk_3 - p_1 k_3^2 + p_2 - p_1 k_1^2 > 0$;
(iii) $(2qk_3 - p_1 k_3^2)(p_2 - p_1 k_1^2) - (p_1 k_1 k_3 + q(1 - k_1))^2 > 0$.

Using the definition of k_2 in (5.78) and the conditions (5.79) and (5.81), it is obvious that

$$k_2^2 = \frac{\lambda_1^2}{\lambda_2^2} < \frac{\gamma\lambda_1}{\delta\lambda_2} = \frac{p_1}{p_2},$$

and therefore that there exists $\mu > 0$ sufficiently small such that inequality (i) is satisfied.

Under conditions (5.80) and (5.81) we have

$$\frac{p_1}{p_2} = \frac{\gamma\lambda_1}{\delta\lambda_2} < \frac{1}{k_1^2}, \tag{5.82}$$

and therefore inequality **(ii)** holds if

$$0 < k_3 < \frac{2q}{p_1}. \tag{5.83}$$

The left-hand side of **(iii)** is a second order polynomial in k_3:

$$P(k_3) = -p_1 p_2 k_3^2 + 2q(p_2 - p_1 k_1)k_3 - q^2(1 - k_1)^2.$$

The discriminant of $P(k_3)$ is

$$\Delta \triangleq 4q^2(p_2 - p_1 k_1^2)(p_2 - p_1).$$

In view of (5.79) and (5.82), it is clear that $\Delta > 0$. Then, the polynomial $P(k_3)$ has two real roots given by

$$k_3^{\pm} = \frac{q}{p_1 p_2}\left[(p_2 - p_1 k_1) \pm \sqrt{\Delta}\right].$$

Under conditions (5.79) and (5.83), it can be checked that

$$k_3^+ > 0 \quad \text{and} \quad k_3^- < \frac{2q}{p_1}.$$

It follows that, for any $k_3 > 0$, $q > 0$ can be selected such that $k_3 \in (0, 2q/p_1) \cap (k_3^-, k_3^+)$ and therefore that conditions **(ii)** and **(iii)** are satisfied. $\qquad\square$

5.5 Proportional-Integral Control in Navigable Rivers

In Section 1.15.2.2 (p. 48), we have seen that a navigable river is a string of pools separated by hydraulic control gates as illustrated in Fig. 1.20. An example of the control of a real life navigable river will be presented in detail in Chapter 8. In this section, for simplicity, we consider the special case of a channel having n rectangular pools with the same length L and the same width W. Using Saint-Venant equations (1.72), the dynamics of the system are described by the following set of balance laws:

$$\partial_t \begin{pmatrix} H_i \\ V_i \end{pmatrix} + \partial_x \begin{pmatrix} H_i V_i \\ \frac{1}{2}V_i^2 + gH_i \end{pmatrix} + \begin{pmatrix} 0 \\ g[CV_i^2 H_i^{-1} - S_i] \end{pmatrix} = \mathbf{0}, \; i = 1\ldots, n,$$

and the following set of boundary conditions:

$$H_i(t, L)V_i(t, L) = H_{i+1}(t, 0)V_{i+1}(t, 0), \quad i = 1, \ldots, n - 1, \tag{5.84a}$$

$$H_i(t, L)V_i(t, L) = \left(k_G\sqrt{2g}\right)\sqrt{[H_i(t, L) - U_i(t)]^3}, \qquad i = 1, \ldots, n, \qquad (5.84\text{b})$$

$$Q_0(t) = WH_1(t, 0)V_1(t, 0), \qquad (5.84\text{c})$$

where H_i and V_i denote the water level and the water velocity in the ith pool, U_i is the position of the ith gate which is used as control action, S_i is the constant slope of the ith pool, k_G and C are constant shape and friction coefficients respectively, $Q_0(t)$ is the inflow rate considered as an external disturbance.

For a constant inflow rate $Q_0(t) = Q^*$, a steady state of the system is a constant state H_i^*, V_i^* $(i = 1, \ldots, n)$ which satisfies the relations

$$Q^* = WH_i^* V_i^*, \qquad S_i H_i^* = C(V_i^*)^2.$$

From Section 1.4, we know that the Riemann coordinates may be defined as

$$R_i = (V_i - V_i^*) + (H_i - H_i^*)\sqrt{\frac{g}{H_i^*}},$$
$$\qquad\qquad\qquad\qquad i = 1, \ldots, n, \qquad (5.85)$$
$$R_{n+i} = (V_i - V_i^*) - (H_i - H_i^*)\sqrt{\frac{g}{H_i^*}},$$

and that the Saint-Venant equations, linearized about the steady state, are written as follows in these Riemann coordinates:

$$\partial_t R_i + \lambda_i \partial_x R_i + \gamma_i R_i + \delta_i R_{n+i} = 0,$$
$$\qquad\qquad\qquad\qquad i = 1, \ldots, n, \qquad (5.86)$$
$$\partial_t R_{n+i} - \lambda_{n+i} \partial_x R_{n+i} + \gamma_i R_i + \delta_i R_{n+i} = 0,$$

with the characteristic velocities

$$\lambda_i = V_i^* + \sqrt{gH_i^*}, \qquad -\lambda_{n+i} = V_i^* - \sqrt{gH_i^*}, \qquad i = 1, \ldots, n$$

and the parameters

$$\gamma_i = gS_i\left(\frac{1}{V_i^*} - \frac{1}{2\sqrt{gH_i^*}}\right), \qquad \delta_i = gS_i\left(\frac{1}{V_i^*} + \frac{1}{2\sqrt{gH_i^*}}\right), \qquad i = 1, \ldots, n,$$

such that $0 < \lambda_{n+i} < \lambda_i$ and $0 < \gamma_i < \delta_i$.

We assume that the control gates are provided with Proportional-Integral controllers that aim at regulating the water levels $H_i(t, L)$ at the steady state set-point values H_i^*. The control laws are as follows:

$$U_i(t) = U_R - k_{Pi}(H_i(t, L) - H_i^*) - k_{Ii}\int_0^t (H_i(\tau, L) - H_i^*)d\tau.$$

Then the linearization of the boundary conditions (5.84) gives the following relations in Riemann coordinates:

$$R_1(t,0) = -\frac{\lambda_{n+1}}{\lambda_1} R_{n+1}(t,0), \tag{5.87a}$$

$$R_{i+1}(t,0) = \frac{(\lambda_i + \lambda_{n+i})(\lambda_i + k_i\lambda_{n+i})}{\lambda_{i+1}(\lambda_{i+1} + \lambda_{n+i+1})} R_i(t,L) - \frac{\lambda_{n+i+1}}{\lambda_{i+1}} R_{n+i+1}(t,0)$$

$$+ \frac{\lambda_{n+i}(\lambda_i + \lambda_{n+i})k_{n+i}}{\lambda_{i+1}(\lambda_i + \lambda_{n+i})} X_i(t), \quad i = 1, \ldots, n-1, \tag{5.87b}$$

$$R_{n+i}(t,L) = k_i R_i(t,L) + k_{n+i} X_i(t), \quad i = 1, \ldots, n, \tag{5.87c}$$

with

$$X_i(t) \triangleq \int_0^t (R_i(\tau,L) - R_{n+i}(\tau,L)) \, d\tau, \quad i = 1, \ldots, n,$$

and

$$k_i \triangleq \frac{k_{Pi} - \lambda_i}{k_{Pi} + \lambda_{n+i}}, \qquad k_{n+i} \triangleq \frac{k_{Ii}}{k_{Pi} + \lambda_{n+i}}.$$

5.5.1 Dissipative Boundary Condition

Theorem 5.11. If the control tuning parameters k_{Pi} and k_{Ii} are selected such that the dissipativity conditions

$$|k_i| < \sqrt{\frac{\gamma_i}{\delta_i} \frac{\lambda_i}{\lambda_{n+i}}}, \qquad k_{n+i} > 0$$

are satisfied, then the solutions of the system (5.86), (5.87) exponentially converge to zero for the L^2-norm.

Proof. See Appendix E. ☐

5.5.2 Control Error Propagation

In the decentralized control structure of a navigable river, the role of the PI controller in a given pool is to regulate the water level at its set-point while rejecting load disturbances. As we shall see in this paragraph, the decentralized structure where the PI controllers are tuned separately may lead to problems of global performance because of the spatial propagation of water level control errors and the possible amplification of the control actions in the downstream direction. To analyze that phenomenon, we consider a string of n pools with PI controllers. From Section 1.4, we know that the Saint-Venant equations, linearized about the steady state, have the structure of linear density-flow systems

$$\partial_t h_i + \partial_x q_i = 0,$$
$$\qquad\qquad\qquad\qquad\qquad\qquad\qquad\qquad\qquad i = 1, \ldots, n,$$
$$\partial_t q_i + \lambda_i \lambda_{n+i} \partial_x h_i + (\lambda_i - \lambda_{n+i}) \partial_x q_i - \alpha_i h_i + \beta_i q_i = 0,$$

where $h_i \triangleq H_i - H_i^*$, $q_i \triangleq H_i V_i - H_i^* V_i^*$, $\alpha_i \triangleq CV_i^*/H_i^{*2}$, $\beta_i \triangleq 2CV_i^*/H_i^*$.

From Section 5.4, we deduce that the system dynamics are represented as follows in the frequency domain:

$$h_i(s) = -P_i(s)q_i(s) + P_{n+i}(s)q_{i-1}(s), \qquad i = 1, \ldots, n, \qquad (5.88)$$

with the transfer functions

$$P_i(s) \triangleq \frac{\sigma_{n+i}(s)e^{\sigma_{n+i}(s)L} - \sigma_i(s)e^{\sigma_i(s)L}}{s(e^{\sigma_{n+i}(s)L} - e^{\sigma_i(s)L})},$$

$$P_{n+i}(s) \triangleq \frac{(\sigma_{n+i}(s) - \sigma_i(s))e^{(\sigma_i(s)+\sigma_{n+i}(s))L}}{s(e^{\sigma_{n+i}(s)L} - e^{\sigma_i(s)L})},$$

$$\sigma_i(s) \triangleq \frac{(\lambda_i - \lambda_{n+i})s + \alpha_i - \sqrt{d_i(s)}}{2\lambda_i \lambda_{n+i}},$$

$$\sigma_{n+i}(s) \triangleq \frac{(\lambda_i - \lambda_{n+i})s + \alpha_i + \sqrt{d_i(s)}}{2\lambda_i \lambda_{n+i}},$$

$$d_i(s) \triangleq (\lambda_i + \lambda_{n+i})^2 s^2 + 2[(\lambda_i - \lambda_{n+i})\alpha_i + 2\lambda_i \lambda_{n+i}\beta_i]s + \alpha_i^2.$$

Moreover, the PI controls are defined as

$$q_i(s) = C_i(s)h_i(s) \quad \text{with} \quad C_i(s) \triangleq k_{Pi} + \frac{k_{Ii}}{s}, \qquad (5.89)$$

where k_{Pi} and k_{Ii} are the proportional and integral gains respectively. Then, from (5.88) and (5.89), we see that the control errors propagate according to the transfer function

$$T_i(s) \triangleq \frac{h_i(s)}{h_{i-1}(s)} = \frac{P_{n+i}(s)C_{i-1}(s)}{1 + P_i(s)C_i(s)}. \tag{5.90}$$

In particular, it can be verified that the static gain of this transfer function is

$$T_i(0) = \frac{k_{I(i-1)}}{k_{Ii}}.$$

Hence we see that $T_i(0) = 1$ if $k_{I(i-1)} = k_{Ii}$ (i.e., if the two integral gains are identically tuned), meaning that the level control error due to load disturbances will be propagated without attenuation (nor amplification). An error propagation with a significant attenuation (i.e., $T_i(0) \ll 1$) requires that the two controllers be differently tuned with $k_{I(i-1)} \ll k_{Ii}$. But, obviously this is necessarily detrimental for the performance quality in terms of set-point tracking in each pool.

5.6 Limit of Stabilizability

In this section we shall show that, for systems of balance laws, there is an intrinsic limit of stabilizability under local boundary control. We consider the simplest possible case of a control system with a single boundary control. The system is written in Riemann coordinates:

$$\partial_t S_1 + \partial_x S_1 + c S_2 = 0,$$

$$\partial_t S_2 - \partial_x S_2 + c S_1 = 0, \qquad\qquad t \in [0, +\infty), \, x \in [0, L], \tag{5.91}$$

$$S_1(t, 0) = u(t), \qquad S_2(t, L) = S_1(t, L).$$

This is a control system where the state is $(S_1, S_2)^\mathsf{T} \in L^2(0, L)^2$ and the control is $u \in \mathbb{R}$. This system is controllable (see Russell (1978), and also Li (2010) for the nonlinear case). In the special case where $c = 0$, the system (5.91) is a control system of conservation laws and we know from Chapter 2 that it is stabilizable with a boundary feedback local control of the form $u(t) = kS_2(t, 0)$ regardless of the value of L.

In contrast, it is shown in (Bastin and Coron 2011, Section 5) that for $c > 0$ and for L large enough the control system (5.91) has no control Lyapunov of the form

$$V \triangleq \int_0^L q_1(x)(S_1(x))^2 + q_2(x)(S_2(x))^2 + q_3(x)S_1(x)S_2(x)dx.$$

In the next proposition, we show that this limitation is intrinsic and that, in fact, the system (5.91) with $c > 0$ can absolutely not be stabilized by means of linear boundary feedback laws of the form $u(t) = kS_2(t, 0)$ if L is too large.

Proposition 5.12. If

$$L \geq \frac{\pi}{c} > 0 \tag{5.92}$$

there is no $k \in \mathbb{R}$ such that the equilibrium $(0,0)^{\mathsf{T}} \in L^2(0,L)^2$ is exponentially stable for the closed loop system

$$\partial_t S_1 + \partial_x S_1 + cS_2 = 0,$$

$$\partial_t S_2 - \partial_x S_2 + cS_1 = 0, \qquad\qquad t \in [0, +\infty), \ x \in [0, L], \tag{5.93}$$

$$S_1(t, 0) = kS_2(t, 0), \qquad S_2(t, L) = S_1(t, L).$$

Proof. Let us look at the real part of the eigenvalues of the generator associated with (5.93) (as in finite dimension). Let $\sigma \in \mathbb{C}$. We look for a solution $(S_1, S_2)^{\mathsf{T}}$ of (5.93) of the form

$$S_1(t, x) = e^{\sigma t} f(x), \ S_2(t, x) = e^{\sigma t} g(x), \ t \in [0, +\infty), \ x \in [0, L].$$

Such a $(S_1, S_2)^{\mathsf{T}}$ is a solution of (5.93) if and only if

$$\sigma f + f_x + cg = 0, \tag{5.94}$$

$$\sigma g - g_x + cf = 0, \tag{5.95}$$

$$f(L) = g(L), f(0) = kg(0). \tag{5.96}$$

From (5.94), we have

$$g = -\frac{1}{c}(\sigma f + f_x). \tag{5.97}$$

From (5.95) and (5.97), we have

$$f_{xx} + (c^2 - \sigma^2)f = 0. \tag{5.98}$$

From now on we assume that

$$\sigma^2 \neq c^2. \tag{5.99}$$

Let $\xi \in \mathbb{C}$ be such that

$$\xi^2 = \sigma^2 - c^2. \tag{5.100}$$

From (5.98), we get the existence of $A \in \mathbb{C}$ and $B \in \mathbb{C}$ such that

$$f = Ae^{\xi x} + Be^{-\xi x}. \tag{5.101}$$

Then (5.97) gives

$$g = -\frac{1}{c}\left(A(\sigma + \xi)e^{\xi x} + B(\sigma - \xi)e^{-\xi x}\right). \tag{5.102}$$

From now on we assume that (5.101) and (5.102) hold. It is easily checked that (5.94) and (5.95) hold. Equation (5.96) is equivalent to

$$A\left(c + k\left(\sigma + \xi\right)\right) + B\left(c + k\left(\sigma - \xi\right)\right) = 0, \tag{5.103}$$

$$A\left(c + \sigma + \xi\right)e^{\xi L} + B\left(c + \sigma - \xi\right)e^{-\xi L} = 0. \tag{5.104}$$

There exists $(A, B) \in \mathbb{C} \times \mathbb{C} \setminus \{(0,0)\}$ such that (5.103) and (5.104) hold if and only if

$$\left(c + k\left(\sigma + \xi\right)\right)\left(c + \sigma - \xi\right) = \left(c + k\left(\sigma - \xi\right)\right)\left(c + \sigma + \xi\right)e^{2\xi L}. \tag{5.105}$$

It follows from Lichtner (2008), Neves et al. (1986), and Renardy (1993) that if the equilibrium $(0,0)^{\mathsf{T}} \in L^2(0,L)^2$ is exponentially stable for the closed loop system (5.93), then

$$|k| < 1. \tag{5.106}$$

From (5.105), we have

$$k = \frac{\xi\left(e^{2L\xi} + 1\right) + (\sigma + c)\left(e^{2L\xi} - 1\right)}{\xi\left(e^{2L\xi} + 1\right) - (\sigma + c)\left(e^{2L\xi} - 1\right)},$$

and therefore

$$k - 1 = \frac{2\left(\sigma + c\right)\left(e^{2L\xi} - 1\right)}{\xi\left(e^{2L\xi} + 1\right) - (\sigma + c)\left(e^{2L\xi} - 1\right)},$$

$$k + 1 = \frac{2\xi\left(e^{2L\xi} + 1\right)}{\xi\left(e^{2L\xi} + 1\right) - (\sigma + c)\left(e^{2L\xi} - 1\right)},$$

which imply

$$\frac{k-1}{k+1} = \frac{2\left(\sigma + c\right)\left(e^{2L\xi} - 1\right)}{2\xi\left(e^{2L\xi} + 1\right)} = (\sigma + c)\frac{\operatorname{sh}(L\xi)}{\xi}\frac{1}{\operatorname{ch}(L\xi)}. \tag{5.107}$$

Conversely, if (5.107) holds, then σ is an eigenvalue.

We remark that the quantities

$$\frac{\text{sh}\,(L\xi)}{\xi}, \ \text{ch}\,(L\xi)$$

are not changed if ξ is substituted by $-\xi$. The functions

$$\sigma \in \mathbb{C} \mapsto \begin{cases} \dfrac{\text{sh}\,(L\xi)}{\xi} & \text{if } \sigma \neq c, \\ L & \text{if } \sigma = c, \end{cases}$$

$$\sigma \in \mathbb{C} \mapsto \text{ch}\,(L\xi),$$

are holomorphic functions of σ.

We now take $\sigma \in (0, c)$. Then (5.107) is equivalent to

$$F(\sigma) = \frac{k-1}{k+1}, \tag{5.108}$$

where $F : [0, c) \to \mathbb{R}$ is defined by

$$F(\sigma) := \sqrt{\frac{\sigma + c}{c - \sigma}}\,\text{tg}\left(L\sqrt{c^2 - \sigma^2}\right). \tag{5.109}$$

We now assume that (5.92) holds. Let $\sigma_1 \in [0, c)$ and $\sigma_2 \in (\sigma_1, c)$ be defined by

$$\sigma_1 := \sqrt{c^2 - \frac{\pi^2}{L^2}}, \ \sigma_2 := \sqrt{c^2 - \frac{\pi^2}{4L^2}}. \tag{5.110}$$

Then F is continuous on $[\sigma_1, \sigma_2)$ and

$$F(\sigma_1) = 0, \tag{5.111}$$

$$\lim_{\sigma \to \sigma_2^-} F(\sigma) = -\infty. \tag{5.112}$$

From (5.106), we get that

$$\frac{k-1}{k+1} \in (-\infty, 0). \tag{5.113}$$

From (5.111), (5.112), and (5.113), there exists $\sigma \in (\sigma_1, \sigma_2)$ such that (5.108) holds. This concludes the proof of Proposition 5.12. $\qquad\qquad\square$

Remark 5.13. Our proof of Proposition 5.12 shows that, if the inequality (5.92) is strict, then for every $k \in \mathbb{R}$, the equilibrium $(0, 0)^\top \in L^2(0, L)^2$ is exponentially unstable for the closed loop system (5.93).

5.7 References and Further Reading

Various interesting generalizations of the results presented in this chapter can be found in the following references.

- Tchousso et al. (2009) show how the Lyapunov approach can be extended to linear systems of balance laws of higher spatial dimension.
- Prieur and Mazenc (2012) show how time varying strict Lyapunov functions can be defined to get input-to-state stability for time varying linear systems of balance laws.
- For a linearized Saint-Venant-Exner system, Diagne and Sène (2013) use the Faedo-Galerkin method to prove the exponential convergence of the solutions with a Lyapunov function quadratic in the physical coordinates. Furthermore this reference presents simulations of the control system carried out with a finite volume method based on a Roe's scheme.
- The stability of switching systems of linear balance laws is addressed by Amin et al. (2012) with the method of characteristics for the L^∞-norm, and by Lamare et al. (2013) and Prieur et al. (2014) with the Lyapunov method for the L^2-norm.
- The issue of singular perturbations in linear systems of balance laws is addressed in Tang et al. (2015a).
- The propagation of control errors in strings of density-flow systems under nonlocal boundary control is discussed by Cantoni et al. (2007) and Li and De Schutter (2010).

Chapter 6
Quasi-Linear Hyperbolic Systems

I N THIS CHAPTER, we continue to explore the use of Lyapunov functions for the
stability analysis of quasi-linear hyperbolic systems under dissipative boundary
conditions. We address the most general case of systems that cannot be transformed
into Riemann coordinates.

A balance law can be viewed as a conservation law which is perturbed by a so-
called *source term*. In the first section, we shall see that, if the perturbation is not
too big, the hyperbolic systems of balance laws with uniform steady states inherit
of the stability properties of the corresponding hyperbolic systems of conservation
laws.

On the other hand, we remember that, in Chapter 5, *linear* hyperbolic systems are exponentially stable for the L^2-norm under the matrix inequalities of
Proposition 5.1. In the second section, we shall see that, for *quasi-linear* systems,
exactly the same matrix inequalities are sufficient to have the exponential stability
of the steady state for the H^2-norm, in a way which is reminiscent to nonlinear
conservation laws.

6.1 Stability of Systems with Uniform Steady States

Let us first consider the special case of a quasi-linear hyperbolic system with a
uniform steady state (see page 4). The system is written:

$$\mathbf{Y}_t + F(\mathbf{Y})\mathbf{Y}_x + G(\mathbf{Y}) = \mathbf{0}, \quad t \in [0, +\infty), \quad x \in [0, L], \tag{6.1}$$

$$\begin{pmatrix} \mathbf{Y}^+(t,0) \\ \mathbf{Y}^-(t,L) \end{pmatrix} = \mathcal{H} \begin{pmatrix} \mathbf{Y}^+(t,L) \\ \mathbf{Y}^-(t,0) \end{pmatrix}, \quad t \in [0, +\infty), \tag{6.2}$$

© Springer International Publishing Switzerland 2016
G. Bastin, J.-M. Coron, *Stability and Boundary Stabilization of 1-D Hyperbolic
Systems*, Progress in Nonlinear Differential Equations and Their Applications 88,
DOI 10.1007/978-3-319-32062-5_6

with $\mathbf{Y} : [0, +\infty) \times [0, L] \to \mathbb{R}^n$ and where $F : \mathcal{Y} \to \mathcal{M}_{n,n}(\mathbb{R})$, $G : \mathcal{Y} \to \mathbb{R}^n$ and $\mathcal{H} : \mathcal{Y} \to \mathbb{R}^n$ are of class C^1. We assume that the system is *strictly* hyperbolic which means that for each $\mathbf{Y} \in \mathcal{Y}$, the matrix $F(\mathbf{Y})$ has distinct real eigenvalues. Let us recall also that we assume that those eigenvalues do not vanish in $\mathcal{Y}$.

In this section, we assume furthermore that the system (6.1), (6.2) has a *uniform* steady state $\mathbf{Y}^*$ which is constant with respect to t and x and can therefore be assumed to be $\mathbf{0}$ without loss of generality. In that case, we necessarily have $G(\mathbf{0}) = \mathbf{0}$ and $\mathcal{H}(\mathbf{0}) = \mathbf{0}$.

As before, the matrix $\mathbf{K}$ is defined as the linearization of the map $\mathcal{H}$ at the steady state:

$$\mathbf{K} \triangleq \mathcal{H}'(\mathbf{0}) \in \mathcal{M}_{n,n}(\mathbb{R}).$$

The system (6.1), (6.2) is considered under an initial condition

$$\mathbf{Y}(0, x) = \mathbf{Y}_0(x), \quad x \in [0, L], \tag{6.3}$$

which satisfies the following compatibility conditions of order 1 and are extensions of conditions (4.7), (4.8):

$$\begin{pmatrix} \mathbf{Y}_0^+(0) \\ \mathbf{Y}_0^-(L) \end{pmatrix} = \mathcal{H} \begin{pmatrix} \mathbf{Y}_0^+(L) \\ \mathbf{Y}_0^-(0) \end{pmatrix}, \tag{6.4}$$

$$F^+(\mathbf{Y}_0(0))\partial_x\mathbf{Y}_0(0) + G^+(\mathbf{Y}_0(0)) =$$

$$\left[\frac{\partial \mathcal{H}^+}{\partial \mathbf{Y}^+}\begin{pmatrix} \mathbf{Y}_0^+(L) \\ \mathbf{Y}_0^-(0) \end{pmatrix}\right]\left(F^+(\mathbf{Y}_0(L))\partial_x\mathbf{Y}_0(L) + G^+(\mathbf{Y}_0(L))\right)$$

$$+ \left[\frac{\partial \mathcal{H}^+}{\partial \mathbf{Y}^-}\begin{pmatrix} \mathbf{Y}_0^+(L) \\ \mathbf{Y}_0^-(0) \end{pmatrix}\right]\left(F^-(\mathbf{Y}_0(0))\partial_x\mathbf{Y}_0(0) + G^-(\mathbf{Y}_0(0))\right), \tag{6.5}$$

$$F^-(\mathbf{Y}_0(L))\partial_x\mathbf{Y}_0(L) + G^-(\mathbf{Y}_0(L)) =$$

$$\left[\frac{\partial \mathcal{H}^-}{\partial \mathbf{Y}^+}\begin{pmatrix} \mathbf{Y}_0^+(L) \\ \mathbf{Y}_0^-(0) \end{pmatrix}\right]\left(F^+(\mathbf{Y}_0(L))\partial_x\mathbf{Y}_0(L) + G^+(\mathbf{Y}_0(L))\right)$$

$$+ \left[\frac{\partial \mathcal{H}^-}{\partial \mathbf{Y}^-}\begin{pmatrix} \mathbf{Y}_0^+(L) \\ \mathbf{Y}_0^-(0) \end{pmatrix}\right]\left(F^-(\mathbf{Y}_0(0))\partial_x\mathbf{Y}_0(0) + G^-(\mathbf{Y}_0(0))\right), \tag{6.6}$$

where $F^+ \in \mathcal{M}_{m,n}(\mathbb{R})$, $F^- \in \mathcal{M}_{n-m,n}(\mathbb{R})$, $G^+ \in \mathbb{R}^m$, $G^- \in \mathbb{R}^{n-m}$, $\mathcal{H}^+ \in \mathbb{R}^m$, $\mathcal{H}^- \in \mathbb{R}^{n-m}$ are defined such that

$$F \triangleq \begin{pmatrix} F^+ \\ F^- \end{pmatrix}, \quad G \triangleq \begin{pmatrix} G^+ \\ G^- \end{pmatrix}, \quad \mathcal{H} \triangleq \begin{pmatrix} \mathcal{H}^+ \\ \mathcal{H}^- \end{pmatrix}.$$

Our purpose in this section is to show that the dissipativity conditions $\rho_2(\mathbf{K}) < 1$ and $\rho_\infty(\mathbf{K}) < 1$ which are sufficient for the stability of systems of conservation laws (see Chapter 4, Theorems 4.3, 4.11, 4.18, 4.22, 4.24), remain valid for the exponential stability of the steady state of general quasi-linear systems (6.1), (6.2) provided $\|G'(\mathbf{0})\|$ is sufficiently small. We have the following stability theorem.

Theorem 6.1. For given functions F and $\mathcal{H}$, if $\rho_\infty(\mathbf{K}) < 1$, there exists $\varepsilon_0 > 0$, $\varepsilon_1 > 0$, $C_1 > 0$, and $\nu > 0$, such that, for every function G of class C^1 such that $G(\mathbf{0}) = \mathbf{0}$ and $\|G'(\mathbf{0})\| < \varepsilon_0$, for every $\mathbf{Y}_\circ \in C^1([0, L]; \mathbb{R}^n)$ such that $|\mathbf{Y}_\circ|_1 < \varepsilon_1$ and satisfying the compatibility conditions (6.4), (6.5),(6.6), the C^1-solution of the Cauchy problem (6.1), (6.2), (6.3) satisfies

$$|\mathbf{Y}(t, .)|_1 \le C_1 e^{-\nu t} |\mathbf{Y}_\circ|_1, \quad \forall t \in [0, +\infty). \qquad \Box$$

Proof. The well-posedness of the Cauchy problem results from a straightforward extension of Theorem 4.1. The proof of the exponential stability is a direct extension of the proofs of Theorems 4.3 and 4.18. It relies on the well-known robustness of the Lyapunov analysis with respect to small perturbations of the system dynamics. An alternative proof using the method of characteristics can be found in Prieur et al. (2008) (see also Dos Santos Martins and Prieur (2008)). $\qquad \Box$

Remark 6.2. It is worth noting that the proof of this theorem gives an explicit estimate of ε_0 which depends on $F'(\mathbf{0})$ and $\mathbf{K} = \mathcal{H}'(\mathbf{0})$. In the special case where G is quadratic with respect to $\mathbf{Y}$ (i.e., $G'(\mathbf{0}) = \mathbf{0}$), the stability becomes independent of F and $\mathbf{K}$ as stated in the next corollary. This property was repeatedly used in the stability proofs of Chapter 4 for systems of nonlinear conservation laws.

Corollary 6.3. If $\rho_\infty(\mathbf{K}) < 1$, for every function G of class C^1 such that $G(\mathbf{0}) = \mathbf{0}$ and $G'(\mathbf{0}) = \mathbf{0}$, the steady state $\mathbf{Y}(t, x) \equiv \mathbf{0}$ of the system (6.1), (6.2) is exponentially stable for the C^1-norm. $\qquad \Box$

Using the same approach as in Section 4.5 of Chapter 4, stability theorems similar to Theorem 6.1 can be stated for any C^p-norm, $p \in \mathbb{N} \smallsetminus \{0\}$, under the dissipativity condition $\rho_\infty(\mathbf{K}) < 1$ and for any H^p-norm, $p \in \mathbb{N} \smallsetminus \{0, 1\}$, under the dissipativity condition $\rho_2(\mathbf{K}) < 1$, provided the initial condition satisfies compatibility conditions of order p and $p - 1$ respectively.

6.2 Stability of General Quasi-Linear Hyperbolic Systems

We now consider the case of a general quasi-linear hyperbolic system

$$\mathbf{Y}_t + F(\mathbf{Y})\mathbf{Y}_x + G(\mathbf{Y}) = \mathbf{0}, \quad t \in [0, +\infty), \quad x \in [0, L], \tag{6.7}$$

$$\mathcal{B}\big(\mathbf{Y}(t, 0), \mathbf{Y}(t, L)\big) = \mathbf{0}, \quad t \in [0, +\infty), \tag{6.8}$$

with $\mathbf{Y} : [0, +\infty) \times [0, L] \to \mathbb{R}^n$ and where $F : \mathcal{Y} \to \mathcal{M}_{n,n}(\mathbb{R})$, $G : \mathcal{Y} \to \mathbb{R}^n$ and $\mathcal{B} : \mathcal{Y} \times \mathcal{Y} \to \mathbb{R}^n$ are sufficiently smooth functions (see property (6.16)). We assume that the system is *strictly* hyperbolic which means that for each $\mathbf{Y} \in \mathcal{Y}$, the matrix $F(\mathbf{Y})$ has distinct real eigenvalues. Furthermore, those eigenvalues are supposed to not vanish in $\mathcal{Y}$.

6.2.1 Stability Condition for the H^2-Norm for Systems with Positive Characteristic Velocities

In this subsection, for simplicity, we treat only the case where all eigenvalues of $F(\mathbf{Y})$ are strictly positive for all $\mathbf{Y} \in \mathcal{Y}$. Therefore, for all $x \in [0, L]$, with $\mathbf{Y}^*(x)$ the steady state such that $F(\mathbf{Y}^*(x))\mathbf{Y}_x^*(x) + G(\mathbf{Y}^*(x)) = \mathbf{0}$, the matrix $F(\mathbf{Y}^*(x))$ can be diagonalized:

$$\exists N(x) \in \mathcal{M}_{n,n}(\mathbb{R}) \quad \text{such that} \quad N(x)F(\mathbf{Y}^*(x)) = \Lambda(x)N(x),$$

$$\text{with} \quad \Lambda(x) \triangleq \text{diag}\left\{\lambda_1\left(F(\mathbf{Y}^*(x))\right), \dots, \lambda_n\left(F(\mathbf{Y}^*(x))\right)\right\} \in \mathcal{D}_n^+,$$

where $\lambda_i(F)$ is the i-th eigenvalue of F. We define the following change of coordinates:

$$\mathbf{Z}(t, x) \triangleq N(x)\left(\mathbf{Y}(t, x) - \mathbf{Y}^*(x)\right), \quad \mathbf{Z} = (Z_1, \dots, Z_n)^\mathsf{T}.$$

In the $\mathbf{Z}$ coordinates, the system (6.7), (6.8) is rewritten

$$\mathbf{Z}_t + A(\mathbf{Z}, x)\mathbf{Z}_x + B(\mathbf{Z}, x) = \mathbf{0}, \tag{6.9}$$

$$\mathcal{B}\left(N(0)^{-1}\mathbf{Z}(t, 0) + \mathbf{Y}^*(0), N(L)^{-1}\mathbf{Z}(t, L) + \mathbf{Y}^*(L)\right) = \mathbf{0}, \tag{6.10}$$

where

$$A(\mathbf{Z}, x) \triangleq N(x)F(N(x)^{-1}\mathbf{Z} + \mathbf{Y}^*(x))N(x)^{-1} \text{ with } A(\mathbf{0}, x) = \Lambda(x),$$

$$B(\mathbf{Z}, x) \triangleq N(x)\Big[F(N^{-1}(x)\mathbf{Z} + \mathbf{Y}^*(x))(\mathbf{Y}_x^*(x) - N(x)^{-1}N'(x)N(x)^{-1}\mathbf{Z})$$

$$+ G(N(x)^{-1}\mathbf{Z} + \mathbf{Y}^*(x))\Big].$$

Since, by definition of the steady state,

$$B(\mathbf{0}, x) = N(x)\Big[F(\mathbf{Y}^*(x))\mathbf{Y}_x^*(x) + G(\mathbf{Y}^*(x))\Big] = \mathbf{0},$$

it follows that there exists a matrix $M(\mathbf{Z}, x) \in \mathcal{M}_{n \times n}(\mathbb{R})$ such that (6.9) may be rewritten as

$$\mathbf{Z}_t + A(\mathbf{Z}, x)\mathbf{Z}_x + M(\mathbf{Z}, x)\mathbf{Z} = \mathbf{0}, \tag{6.11}$$

with

$$M(\mathbf{0}, x) \triangleq \frac{\partial B}{\partial \mathbf{Z}}(\mathbf{0}, x).$$

As motivated in Chapter 1, it is assumed that the boundary condition (6.10) can be solved for $\mathbf{Z}(t, 0)$ and therefore written into the form

$$\mathbf{Z}(t, 0) = \mathcal{H}\big(\mathbf{Z}(t, L)\big). \tag{6.12}$$

Our concern is to analyze the exponential stability of the steady state $\mathbf{Z}(t, x) \equiv \mathbf{0}$ of the system (6.11) under the boundary condition (6.12) and under an initial condition

$$\mathbf{Z}(0, x) = \mathbf{Z}_\mathrm{o}(x), \quad x \in [0, L]. \tag{6.13}$$

which satisfies the compatibility conditions

$$\mathbf{Z}_\mathrm{o}(0) = \mathcal{H}\big(\mathbf{Z}_\mathrm{o}(L)\big), \tag{6.14}$$

$$A(\mathbf{Z}_\mathrm{o}(0), 0)\partial_x \mathbf{Z}_\mathrm{o}(0) + M(\mathbf{Z}_\mathrm{o}(0), 0)\mathbf{Z}_\mathrm{o}(0) =$$
$$\mathcal{H}'\big(\mathbf{Z}_\mathrm{o}(L)\big)\big(A(\mathbf{Z}_\mathrm{o}(L), L)\partial_x \mathbf{Z}_\mathrm{o}(L) + M(\mathbf{Z}_\mathrm{o}(L), L)\mathbf{Z}_\mathrm{o}(L)\big), \tag{6.15}$$

where $\mathcal{H}'$ denotes the Jacobian matrix of the map $\mathcal{H}$. The analysis is carried on under the property

$$A, M, \mathcal{H} \text{ are of class } C^2. \tag{6.16}$$

The matrix $\mathbf{K}$ is defined as the linearization of the map $\mathcal{H}$ at the steady state:

$$\mathbf{K} \triangleq \mathcal{H}'(\mathbf{0}).$$

The well-posedness of the Cauchy problem and the existence of a unique classical solution results from the following theorem which is a direct extension of Theorem 4.9 (see Appendix B for details).

Theorem 6.4. There exists $\delta_0 > 0$ such that, for every $\mathbf{Z}_\mathrm{o} \in H^2((0, L); \mathbb{R}^n)$ satisfying

$$\|\mathbf{Z}_\mathrm{o}\|_{H^2((0,L);\mathbb{R}^n)} \leqslant \delta_0$$

and the compatibility conditions (6.14) and (6.15), the Cauchy problem (6.9), (6.12), (6.13) has a unique maximal classical solution

$$\mathbf{Z} \in C^0([0, T), H^2((0, L); \mathbb{R}^n))$$

with $T \in (0, +\infty]$. Moreover, if

$$\|\mathbf{Z}(t, \cdot)\|_{H^2((0,L);\mathbb{R}^n)} \leqslant \delta_0, \ \forall t \in [0, T),$$

then $T = +\infty$. $\square$

The definition of the exponential stability is as follows.

Definition 6.5. The steady state $\mathbf{Z}(t, x) \equiv 0$ of the system (6.11), (6.12) is exponentially stable (for the H^2-norm) if there exist $\delta > 0$, $\nu > 0$, and $C > 0$ such that, for every $\mathbf{Z}_o \in H^2((0, L); \mathbb{R}^n)$ satisfying $\|\mathbf{Z}_o\|_{H^2((0,L);\mathbb{R}^n)} \leqslant \delta$ and the compatibility conditions (6.14), (6.15), the (unique) solution $\mathbf{Z}$ of the Cauchy problem (6.11), (6.12), (6.13) is defined on $[0, +\infty) \times [0, L]$ and satisfies

$$\|\mathbf{Z}(t, .)\|_{H^2((0,L);\mathbb{R}^n)} \leqslant Ce^{-\nu t}\|\mathbf{Z}_o\|_{H^2((0,L);\mathbb{R}^n)}, \ \ \forall t \in [0, +\infty).$$

$\square$

We then have the following stability theorem.

Theorem 6.6. The steady state $\mathbf{Z}(t, x) \equiv 0$ of the system (6.11), (6.12) is exponentially stable for the H^2-norm if there exists a map $Q \in C^1([0, L]; \mathcal{D}_n^+)$ such that the following *Matrix Inequalities* hold:

(i) the matrix

$$Q(L)\Lambda(L) - \mathbf{K}^T Q(0)\Lambda(0)\mathbf{K}$$

is positive semi-definite;

(ii) the matrix

$$-\big(Q(x)\Lambda(x)\big)_x + Q(x)M(\mathbf{0}, x) + M^T(\mathbf{0}, x)Q(x)$$

is positive definite $\forall x \in [0, L]$.

$\square$

In order to define an appropriate Lyapunov function, we need the following lemma which is a straightforward generalization of Assumption 4.14.

Lemma 6.7. Let $D(\mathbf{Z}, x)$ be the diagonal matrix whose diagonal entries are the eigenvalues $\lambda_i(\mathbf{Z}, x)$, $i = 1, \ldots, n$, of the matrix $A(\mathbf{Z}, x)$. There exist a positive real number η and a map $E : \mathcal{B}_\eta \times [0, L] \to \mathcal{M}_{n,n}(\mathbb{R})$ of class C^2 such that

$$E(\mathbf{Z}, x)A(\mathbf{Z}, x) = D(\mathbf{Z}, x)E(\mathbf{Z}, x), \ \forall \mathbf{Z} \in \mathcal{B}_\eta, \ \ \forall x \in [0, L], \tag{6.17}$$

$$E(\mathbf{0}, x) = \mathbf{I}_n, \ \ \forall x \in [0, L], \tag{6.18}$$

where $\mathbf{I}_n$ is the identity matrix of $\mathcal{M}_{n,n}(\mathbb{R})$. $\square$

In order to prove Theorem 6.6, we define the following candidate Lyapunov function:

$$\mathbf{V} \triangleq \mathbf{V}_1 + \mathbf{V}_2 + \mathbf{V}_3, \tag{6.19}$$

with

$$\mathbf{V}_1 \triangleq \int_0^L \mathbf{Z}^\top E^\top(\mathbf{Z}, x) Q(x) E(\mathbf{Z}, x) \mathbf{Z} dx, \tag{6.20}$$

$$\mathbf{V}_2 \triangleq \int_0^L \mathbf{Z}_t^\top E^\top(\mathbf{Z}, x) Q(x) E(\mathbf{Z}, x) \mathbf{Z}_t dx, \tag{6.21}$$

$$\mathbf{V}_3 \triangleq \int_0^L \mathbf{Z}_{tt}^\top E^\top(\mathbf{Z}, x) Q(x) E(\mathbf{Z}, x) \mathbf{Z}_{tt} dx. \tag{6.22}$$

The proof of Theorem 6.6 will then be based on estimates of the time derivatives $(d\mathbf{V}_i/dt)$ $(i = 1,2,3)$ along the system solutions. As usual (see, e.g., Comment 4.6), we assume that the solutions $\mathbf{Z}$ are of class C^3. Indeed, using a density argument similar to Comment 4.6, the estimates of $d\mathbf{V}_i/dt$ given below remain valid, in the distribution sense with $\mathbf{Z} \in C^0([0, T], H^2((0, L); \mathbb{R}^n))$ (see the statement of Theorem 6.4).

Estimate of $d\mathbf{V}_1/dt$
The time derivative of $\mathbf{V}_1$ along the solutions of (6.11), (6.12) is

$$\frac{d\mathbf{V}_1}{dt} = \int_0^L 2\mathbf{Z}^\top E^\top(\mathbf{Z}, x) Q(x) \big[E(\mathbf{Z}, x)\mathbf{Z} \big]_t dx$$

$$= \int_0^L 2\mathbf{Z}^\top E^\top(\mathbf{Z}, x) Q(x) \Big(\big[E(\mathbf{Z}, x) \big]_t \mathbf{Z} + E(\mathbf{Z}, x)\mathbf{Z}_t \Big) dx$$

$$= \int_0^L 2\mathbf{Z}^\top E^\top(\mathbf{Z}, x) Q(x) \Big(\big[E(\mathbf{Z}, x) \big]_t \mathbf{Z}$$
$$- E(\mathbf{Z}, x) A(\mathbf{Z}, x)\mathbf{Z}_x - E(\mathbf{Z}, x) B(\mathbf{Z}, x) \Big) dx.$$

Using equality (6.17), we have

$$\frac{d\mathbf{V}_1}{dt} = \int_0^L 2\mathbf{Z}^\top E^\top(\mathbf{Z}, x) Q(x) \Big(\big[E(\mathbf{Z}, x) \big]_t \mathbf{Z}$$
$$- D(\mathbf{Z}, x) E(\mathbf{Z}, x)\mathbf{Z}_x - E(\mathbf{Z}, x) B(\mathbf{Z}, x) \Big) dx.$$

Then, using integrations by parts, we get

$$\frac{d\mathbf{V}_1}{dt} = \mathcal{T}_{11} + \mathcal{T}_{12},$$

with

$$T_{11} \triangleq \left[- \mathbf{Z}^\mathsf{T} E^\mathsf{T}(\mathbf{Z}, x) Q(x) D(\mathbf{Z}, x) E(\mathbf{Z}, x) \mathbf{Z} \right]_0^L, \tag{6.23}$$

$$T_{12} \triangleq \int_0^L \mathbf{Z}^\mathsf{T} E^\mathsf{T}(\mathbf{Z}, x) Q'(x) D(\mathbf{Z}, x) E(\mathbf{Z}, x) \mathbf{Z}$$
$$- \mathbf{Z}^\mathsf{T} [E^\mathsf{T}(\mathbf{Z}, x) Q(x) D(\mathbf{Z}, x) E(\mathbf{Z}, x)]_x \mathbf{Z}$$
$$+ 2 \mathbf{Z}^\mathsf{T} E^\mathsf{T}(\mathbf{Z}, x) Q(x) \Big([E(\mathbf{Z}, x)]_t \mathbf{Z} - E(\mathbf{Z}, x) B(\mathbf{Z}, x) \Big) dx, \tag{6.24}$$

From (6.23), we have

$$T_{11} = -\mathbf{Z}^\mathsf{T}(t, L) E^\mathsf{T}(\mathbf{Z}(t, L), L) Q(L) D(\mathbf{Z}(t, L), L) E(\mathbf{Z}(t, L), L) \mathbf{Z}(t, L)$$
$$+ \mathbf{Z}^\mathsf{T}(t, 0) E^\mathsf{T}(\mathbf{Z}(t, 0), 0) Q(0) D(\mathbf{Z}(t, 0), 0) E(\mathbf{Z}(t, 0), 0) \mathbf{Z}(t, 0). \tag{6.25}$$

Let us introduce a notation in order to deal with estimates on "higher order terms." We denote by $\mathcal{O}(X; Y)$, with $X \geq 0$ and $Y \geq 0$, quantities for which there exist $C > 0$ and $\varepsilon > 0$, independent of $\mathbf{Z}$, $\mathbf{Z}_t$ and $\mathbf{Z}_{tt}$, such that

$$(Y \leq \varepsilon) \Rightarrow (|\mathcal{O}(X; Y)| \leq C X).$$

Then from (6.18) and (6.25), using the boundary condition (6.12), we have

$$T_{11} = -\mathbf{Z}^\mathsf{T}(t, L) \Big[Q(L) \Lambda(L) - \mathbf{K}^\mathsf{T} Q(0) \Lambda(0) \mathbf{K} \Big] \mathbf{Z}(t, L) + \mathcal{O}(|\mathbf{Z}(t, L)|^3; |\mathbf{Z}(t, L)|), \tag{6.26}$$

and from (6.18), (6.24) we have

$$T_{12} = -\int_0^L \mathbf{Z}^\mathsf{T} \Big[-\big(Q(x)\Lambda(x)\big)_x + M^\mathsf{T}(\mathbf{0}, x) Q(x) + Q(x) M(\mathbf{0}, x) \Big] \mathbf{Z} \, dx$$
$$+ \mathcal{O}\Big(\int_0^L (|\mathbf{Z}|^3 + |\mathbf{Z}||\mathbf{Z}_t|^2) dx; |\mathbf{Z}(t, .)|_0 \Big). \tag{6.27}$$

(Recall that for $f \in C^0([0, L]; \mathbb{R}^n)$, we denote $|f|_0 = \max\{|f(x)|; x \in [0, L]\}$, see Section 4.1.)

Estimate of $d\mathbf{V}_2/dt$

By time differentiation of the system equations (6.11), (6.12), $\mathbf{Z}_t$ can be shown to satisfy the following hyperbolic dynamics:

$$\mathbf{Z}_{tt} + A(\mathbf{Z}, x)\mathbf{Z}_{tx} + \text{diag}\left[\frac{\partial A}{\partial \mathbf{Z}}(\mathbf{Z}, x)\mathbf{Z}_t\right]\mathbf{Z}_x + \frac{\partial B}{\partial \mathbf{Z}}(\mathbf{Z}, x)\mathbf{Z}_t = \mathbf{0}, \qquad (6.28)$$

$$\mathbf{Z}_t(t, 0) = \mathcal{H}'(\mathbf{Z}(t, L))\mathbf{Z}_t(t, L). \qquad (6.29)$$

In (6.28), the matrix $\partial A/\partial \mathbf{Z}$ is defined as the matrix where the i, j entry is $\partial A_{ij}/\partial Z_j$, while the matrix $\text{diag}[(\partial A/\partial \mathbf{Z})\mathbf{Z}_t]$ stands for the diagonal matrix whose diagonal entries are the components of the vector $(\partial A/\partial \mathbf{Z})\mathbf{Z}_t$.

The time derivative of $\mathbf{V}_2$ along the solutions of (6.11), (6.12), (6.28), (6.29) is

$$
\begin{aligned}
\frac{d\mathbf{V}_2}{dt} &= \int_0^L 2\mathbf{Z}_t^\mathsf{T} E^\mathsf{T}(\mathbf{Z}, x)Q(x)\left[E(\mathbf{Z}, x)\mathbf{Z}_t\right]_t dx \\
&= \int_0^L 2\mathbf{Z}_t^\mathsf{T} E^\mathsf{T}(\mathbf{Z}, x)Q(x)\left(\left[E(\mathbf{Z}, x)\right]_t \mathbf{Z}_t + E(\mathbf{Z}, x)\mathbf{Z}_{tt}\right)dx \\
&= \int_0^L 2\mathbf{Z}_t^\mathsf{T} E^\mathsf{T}(\mathbf{Z}, x)Q(x)\left(\left[E(\mathbf{Z}, x)\right]_t \mathbf{Z}_t - E(\mathbf{Z}, x)A(\mathbf{Z}, x)\mathbf{Z}_{tx}\right. \\
&\quad \left. - E(\mathbf{Z}, x)\left(\text{diag}\left[\frac{\partial A}{\partial \mathbf{Z}}(\mathbf{Z}, x)\mathbf{Z}_t\right]\mathbf{Z}_x + \frac{\partial B}{\partial \mathbf{Z}}(\mathbf{Z}, x)\mathbf{Z}_t\right)\right)dx.
\end{aligned}
$$

Using equality (6.17), we have

$$
\begin{aligned}
\frac{d\mathbf{V}_2}{dt} &= \int_0^L 2\mathbf{Z}_t^\mathsf{T} E^\mathsf{T}(\mathbf{Z}, x)Q(x)\left(\left[E(\mathbf{Z}, x)\right]_t \mathbf{Z}_t - D(\mathbf{Z}, x)E(\mathbf{Z}, x)\mathbf{Z}_{tx}\right. \\
&\quad \left. - E(\mathbf{Z}, x)\left(\text{diag}\left[\frac{\partial A}{\partial \mathbf{Z}}(\mathbf{Z}, x)\mathbf{Z}_t\right]\mathbf{Z}_x + \frac{\partial B}{\partial \mathbf{Z}}(\mathbf{Z}, x)\mathbf{Z}_t\right)\right)dx.
\end{aligned}
$$

Then, using integrations by parts, we get

$$\frac{d\mathbf{V}_2}{dt} = \mathcal{T}_{21} + \mathcal{T}_{22},$$

with

$$\mathcal{T}_{21} \triangleq \left[-\mathbf{Z}_t^\mathsf{T} E^\mathsf{T}(\mathbf{Z}, x)Q(x)D(\mathbf{Z}, x)E(\mathbf{Z}, x)\mathbf{Z}_t\right]_0^L, \qquad (6.30)$$

$$T_{22} \triangleq \int_0^L \mathbf{Z}_t^\mathsf{T} E^\mathsf{T}(\mathbf{Z}, x) Q'(x) D(\mathbf{Z}, x) E(\mathbf{Z}, x) \mathbf{Z}_t$$

$$- \mathbf{Z}_t^\mathsf{T} [E^\mathsf{T}(\mathbf{Z}, x) Q(x) D(\mathbf{Z}, x) E(\mathbf{Z}, x)]_x \mathbf{Z}_t e^{-\mu x}$$

$$+ 2\mathbf{Z}_t^\mathsf{T} E^\mathsf{T}(\mathbf{Z}, x) Q(x) \Big(\big[E(\mathbf{Z}, x) \big]_t \mathbf{Z}_t - E(\mathbf{Z}, x) \big(\mathrm{diag} \Big[\frac{\partial A}{\partial \mathbf{Z}} (\mathbf{Z}, x) \mathbf{Z}_t \Big] \mathbf{Z}_x$$

$$+ \frac{\partial B}{\partial \mathbf{Z}} (\mathbf{Z}, x) \mathbf{Z}_t \big) \Big) dx. \qquad (6.31)$$

From (6.30), we have

$$T_{21} = -\mathbf{Z}_t^\mathsf{T}(t, L) E^\mathsf{T}(\mathbf{Z}(t, L), L) Q(L) D(\mathbf{Z}(t, L), L) E(\mathbf{Z}(t, L), L) \mathbf{Z}_t(t, L)$$

$$+ \mathbf{Z}_t^\mathsf{T}(t, 0) E^\mathsf{T}(\mathbf{Z}(t, 0), 0) Q(0) D(\mathbf{Z}(t, 0), 0) E(\mathbf{Z}(t, 0), 0) \mathbf{Z}_t(t, 0). \qquad (6.32)$$

Then, using the boundary condition (6.29), we get

$$T_{21} = -\mathbf{Z}_t^\mathsf{T}(t, L) \Big[Q(L) \Lambda(L) - \mathbf{K}^\mathsf{T} Q(0) \Lambda(0) \mathbf{K} \Big] \mathbf{Z}_t(t, L)$$

$$+ \mathcal{O}(|\mathbf{Z}_t(t, L)|^2 |\mathbf{Z}(t, L)|; |\mathbf{Z}(t, L)|). \qquad (6.33)$$

Moreover T_{22} is written

$$T_{22} = -\int_0^L \mathbf{Z}_t^\mathsf{T} \Big[-\big(Q(x) \Lambda(x) \big)_x + M^\mathsf{T}(\mathbf{0}, x) Q(x) + Q(x) M(\mathbf{0}, x) \Big] \mathbf{Z}_t \, dx$$

$$+ \mathcal{O}\Big(\int_0^L |\mathbf{Z}_t|^2 (|\mathbf{Z}_t| + |\mathbf{Z}|) dx; |\mathbf{Z}(t, .)|_0 \Big). \qquad (6.34)$$

Estimate of $d\mathbf{V}_3/dt$

By time differentiation of the system equations (6.28), (6.29), $\mathbf{Z}_{tt}$ can be shown to satisfy the following hyperbolic dynamics:

$$\mathbf{Z}_{ttt} + A(\mathbf{Z}, x) \mathbf{Z}_{ttx} + 2\mathrm{diag} \Big[\frac{\partial A}{\partial \mathbf{Z}} (\mathbf{Z}, x) \mathbf{Z}_t \Big] \mathbf{Z}_{tx} + \mathrm{diag} \Big[\frac{\partial A}{\partial \mathbf{Z}} (\mathbf{Z}, x) \mathbf{Z}_t \Big]_t \mathbf{Z}_x$$

$$+ \frac{\partial B}{\partial \mathbf{Z}} (\mathbf{Z}, x) \mathbf{Z}_{tt} + \Big[\frac{\partial B}{\partial \mathbf{Z}} (\mathbf{Z}, x) \Big]_t \mathbf{Z}_t = \mathbf{0}, \qquad (6.35)$$

$$\mathbf{Z}_{tt}(t, 0) = \mathcal{H}'(\mathbf{Z}(t, L)) \mathbf{Z}_{tt}(t, L) + \mathcal{H}''(\mathbf{Z}(t, L)) (\mathbf{Z}_t(t, L), \mathbf{Z}_t(t, L)). \qquad (6.36)$$

The time derivative of $\mathbf{V}_3$ along the C^3-solutions of (6.11), (6.12), (6.28), (6.29), (6.35), (6.36) is

$$\frac{d\mathbf{V}_3}{dt} = \int_0^L 2\mathbf{Z}_{tt}^\top E^\top(\mathbf{Z}, x)Q(x)\big[E(\mathbf{Z}, x)\mathbf{Z}_{tt}\big]_t dx$$

$$= \int_0^L 2\mathbf{Z}_{tt}^\top E^\top(\mathbf{Z}, x)Q(x)\Big(\big[E(\mathbf{Z}, x)\big]_t\mathbf{Z}_{tt} + E(\mathbf{Z}, x)\mathbf{Z}_{ttt}\Big) dx$$

$$= \int_0^L 2\mathbf{Z}_{tt}^\top E^\top(\mathbf{Z}, x)Q(x)\Big(\big[E(\mathbf{Z}, x)\big]_t\mathbf{Z}_{tt} - E(\mathbf{Z}, x)A(\mathbf{Z}, x)\mathbf{Z}_{ttx}$$

$$- E(\mathbf{Z}, x)\Big(2\mathrm{diag}\Big[\frac{\partial A}{\partial \mathbf{Z}}(\mathbf{Z}, x)\mathbf{Z}_t\Big]\mathbf{Z}_{tx} + \mathrm{diag}\Big[\frac{\partial A}{\partial \mathbf{Z}}(\mathbf{Z}, x)\mathbf{Z}_t\Big]_t\mathbf{Z}_x$$

$$+ \frac{\partial B}{\partial \mathbf{Z}}(\mathbf{Z}, x)\mathbf{Z}_{tt} + \Big[\frac{\partial B}{\partial \mathbf{Z}}(\mathbf{Z}, x)\Big]_t\mathbf{Z}_t\Big)\Big) dx.$$

Using equality (6.17), we have

$$\frac{d\mathbf{V}_3}{dt} = \int_0^L 2\mathbf{Z}_{tt}^\top E^\top(\mathbf{Z}, x)Q(x)\Big(\big[E(\mathbf{Z}, x)\big]_t\mathbf{Z}_{tt} - D(\mathbf{Z}, x)E(\mathbf{Z}, x)\mathbf{Z}_{ttx}$$

$$- E(\mathbf{Z}, x)\Big(2\mathrm{diag}\Big[\frac{\partial A}{\partial \mathbf{Z}}(\mathbf{Z}, x)\mathbf{Z}_t\Big]\mathbf{Z}_{tx} + \mathrm{diag}\Big[\frac{\partial A}{\partial \mathbf{Z}}(\mathbf{Z}, x)\mathbf{Z}_t\Big]_t\mathbf{Z}_x$$

$$+ \frac{\partial B}{\partial \mathbf{Z}}(\mathbf{Z}, x)\mathbf{Z}_{tt} + \Big[\frac{\partial B}{\partial \mathbf{Z}}(\mathbf{Z}, x)\Big]_t\mathbf{Z}_t\Big)\Big) dx.$$

Then, using integration by parts, we get

$$\frac{d\mathbf{V}_3}{dt} = \mathcal{T}_{31} + \mathcal{T}_{32}, \tag{6.37}$$

with

$$\mathcal{T}_{31} \triangleq \Big[-\mathbf{Z}_{tt}^\top E^\top(\mathbf{Z}, x)Q(x)E(\mathbf{Z}, x)A(\mathbf{Z}, x)\mathbf{Z}_{tt}\Big]_0^L, \tag{6.38}$$

$$\mathcal{T}_{32} \triangleq \int_0^L \mathbf{Z}_{tt}^\mathsf{T} E^\mathsf{T}(\mathbf{Z}, x) Q'(x) E(\mathbf{Z}, x) A(\mathbf{Z}, x) \mathbf{Z}_{tt}$$

$$- \mathbf{Z}_{tt}^\mathsf{T} [E^\mathsf{T}(\mathbf{Z}, x) Q(x) E(\mathbf{Z}, x) A(\mathbf{Z}, x)]_x \mathbf{Z}_{tt} e^{-\mu x}$$

$$+ 2\mathbf{Z}_{tt}^\mathsf{T} E^\mathsf{T}(\mathbf{Z}, x) Q(x) \left([E(\mathbf{Z}, x)]_t \mathbf{Z}_{tt} - E(\mathbf{Z}, x) \left(2\mathrm{diag} \left[\frac{\partial A}{\partial \mathbf{Z}}(\mathbf{Z}, x) \mathbf{Z}_t \right] \mathbf{Z}_{tx} \right. \right.$$

$$\left. \left. + \mathrm{diag} \left[\frac{\partial A}{\partial \mathbf{Z}}(\mathbf{Z}, x) \mathbf{Z}_t \right]_t \mathbf{Z}_x + \frac{\partial B}{\partial \mathbf{Z}}(\mathbf{Z}, x) \mathbf{Z}_{tt} + \left[\frac{\partial B}{\partial \mathbf{Z}}(\mathbf{Z}, x) \right]_t \mathbf{Z}_t \right) \right) dx. \qquad (6.39)$$

From (6.38), we have

$$\mathcal{T}_{31} = -\mathbf{Z}_{tt}^\mathsf{T}(t, L) E^\mathsf{T}(\mathbf{Z}(t, L), L) Q(L) E(\mathbf{Z}(t, L), L) A(\mathbf{Z}(t, L), L) \mathbf{Z}_{tt}(t, L)$$

$$+ \mathbf{Z}_{tt}^\mathsf{T}(t, 0) E^\mathsf{T}(\mathbf{Z}(t, 0), 0) Q(0) E(\mathbf{Z}(t, 0), 0) A(\mathbf{Z}(t, 0), 0) \mathbf{Z}_{tt}(t, 0).$$

Then, using the boundary condition (6.36), $\mathcal{T}_{31}$ is written

$$\mathcal{T}_{31} = -\mathbf{Z}_{tt}^\mathsf{T}(t, L) \left[Q(L) \Lambda(L) - \mathbf{K}^\mathsf{T} Q(0) \Lambda(0) \mathbf{K} \right] \mathbf{Z}_{tt}(t, L)$$

$$+ \mathcal{O}(|\mathbf{Z}_{tt}(t, L)|^2 |\mathbf{Z}(t, L)| + |\mathbf{Z}_{tt}(t, L)| |\mathbf{Z}_t(t, L)|^2 + |\mathbf{Z}_t(t, L)|^4; |\mathbf{Z}(t, L)|). \qquad (6.40)$$

Moreover $\mathcal{T}_{32}$ is written

$$\mathcal{T}_{32} = -\int_0^L \mathbf{Z}_{tt}^\mathsf{T} \left[-(Q(x) \Lambda(x))_x + M^\mathsf{T}(0, x) Q(x) + Q(x) M(0, x) \right] \mathbf{Z}_{tt} \, dx$$

$$+ \mathcal{O}\left(\int_0^L \left(|\mathbf{Z}_{tt}|^2 (|\mathbf{Z}_t| + |\mathbf{Z}|) + |\mathbf{Z}_{tt}| |\mathbf{Z}_t|^2 \right) dx; |\mathbf{Z}(t, .)|_0 + |\mathbf{Z}_t(t, .)|_0 \right). \qquad (6.41)$$

In the next lemma, we shall now use these estimates to show that the Lyapunov function exponentially decreases along the system trajectories.

Lemma 6.8. There exist positive real constants α, β, and δ such that, for every $\mathbf{Z}$ such that $|\mathbf{Z}|_0 + |\mathbf{Z}_t|_0 \leq \delta$, we have

$$\frac{1}{\beta} \int_0^L (|\mathbf{Z}|^2 + |\mathbf{Z}_t|^2 + |\mathbf{Z}_{tt}|^2) dx \leq \mathbf{V} \leq \beta \int_0^L (|\mathbf{Z}|^2 + |\mathbf{Z}_t|^2 + |\mathbf{Z}_{tt}|^2) dx, \qquad (6.42)$$

$$\frac{d\mathbf{V}}{dt} \leq -\alpha \mathbf{V}. \qquad (6.43)$$

Proof. Inequalities (6.42) follow directly from the definition of $\mathbf{V}$ and straightforward estimations.

Let us introduce the following compact matrix notations:

$$\mathcal{K} \triangleq Q(L)\Lambda(L) - \mathbf{K}^{\mathsf{T}}Q(0)\Lambda(0)\mathbf{K}, \tag{6.44}$$

$$\mathcal{L}(x) \triangleq \left[-\left(Q(x)\Lambda(x)\right)_x + M^{\mathsf{T}}(\mathbf{0}, x)Q(x) + Q(x)M(\mathbf{0}, x) \right]. \tag{6.45}$$

Then, it follows from (6.26), (6.27), (6.33), (6.34), (6.40), (6.41) that

$$
\begin{aligned}
\frac{d\mathbf{V}}{dt} = &-\mathbf{Z}^{\mathsf{T}}(t, L)\mathcal{K}\,\mathbf{Z}(t, L) - \mathbf{Z}_t^{\mathsf{T}}(t, L)\mathcal{K}\,\mathbf{Z}_t(t, L) - \mathbf{Z}_{tt}^{\mathsf{T}}(t, L)\mathcal{K}\,\mathbf{Z}_{tt}(t, L) \\
&+ \mathcal{O}(|\mathbf{Z}(t, L)|(|\mathbf{Z}(t, L)|^2 + |\mathbf{Z}_t(t, L)|^2 + |\mathbf{Z}_{tt}(t, L)|^2) \\
&+ |\mathbf{Z}_{tt}(t, L)||\mathbf{Z}_t(t, L)|^2 + |\mathbf{Z}_t(t, L)|^4; |\mathbf{Z}(t, L)|) \\
&- \int_0^L \left(\mathbf{Z}^{\mathsf{T}}\mathcal{L}(x)\,\mathbf{Z} + \mathbf{Z}_t^{\mathsf{T}}\mathcal{L}(x)\,\mathbf{Z}_t + \mathbf{Z}_{tt}^{\mathsf{T}}\mathcal{L}(x)\,\mathbf{Z}_{tt} \right)dx \\
&+ \mathcal{O}\Big(\int_0^L \Big(|\mathbf{Z}|^2|\mathbf{Z}| + |\mathbf{Z}|^2|\mathbf{Z}_t| + |\mathbf{Z}_t|^2|\mathbf{Z}| \\
&\qquad\qquad + |\mathbf{Z}_t|^2|\mathbf{Z}_t| + |\mathbf{Z}_{tt}|^2|\mathbf{Z}_t| \\
&\qquad\qquad + |\mathbf{Z}_t|^2|\mathbf{Z}_{tt}| + |\mathbf{Z}_{tt}|^2|\mathbf{Z}_t| \Big)dx; |\mathbf{Z}(t, .)|_0 + |\mathbf{Z}_t(t, .)|_0\Big). \tag{6.46}
\end{aligned}
$$

Then by assumption (i) of Theorem 6.6 and from (6.44), there exists $\delta_1 > 0$ such that if $|\mathbf{Z}(t, L)| + |\mathbf{Z}_t(t, L)| < \delta_1$ then

$$
\begin{aligned}
&-\mathbf{Z}^{\mathsf{T}}(t, L)\mathcal{K}\,\mathbf{Z}(t, L) - \mathbf{Z}_t^{\mathsf{T}}(t, L)\mathcal{K}\,\mathbf{Z}_t(t, L) - \mathbf{Z}_{tt}^{\mathsf{T}}(t, L)\mathcal{K}\,\mathbf{Z}_{tt}(t, L) \\
&+ \mathcal{O}(|\mathbf{Z}(t, L)|(|\mathbf{Z}(t, L)|^2 + |\mathbf{Z}_t(t, L)|^2 + |\mathbf{Z}_{tt}(t, L)|^2) \\
&+ |\mathbf{Z}_{tt}(t, L)||\mathbf{Z}_t(t, L)|^2 + |\mathbf{Z}_t(t, L)|^4; |\mathbf{Z}(t, L)|) \leqslant 0. \tag{6.47}
\end{aligned}
$$

Let us recall the following Sobolev inequality (see, e.g., Brezis (1983)): for a function $\varphi \in C^1([0, L]; \mathbb{R}^n)$, there exits $C_1 > 0$ such that

$$|\varphi|_0 \leqslant C_1 \int_0^L (|\varphi(x)|^2 + |\varphi'(x)|^2)dx. \tag{6.48}$$

Moreover, from (6.9) and (6.28), we know also that there exist $\delta_2 > 0$ and $C_2 > 0$ such that, if $|\mathbf{Z}(t, x)| + |\mathbf{Z}_t(t, x)| < \delta_2$, then

$$|\mathbf{Z}_t(t,x)| \leqslant C_2\big(|\mathbf{Z}(t,x)| + |\mathbf{Z}_x(t,x)|\big), \tag{6.49}$$

$$|\mathbf{Z}_{tt}(t,x)| \leqslant C_2\big(|\mathbf{Z}(t,x)| + |\mathbf{Z}_x(t,x)| + |\mathbf{Z}_{xx}(t,x)|\big), \tag{6.50}$$

$$|\mathbf{Z}_x(t,x)| \leqslant C_2\big(|\mathbf{Z}(t,x)| + |\mathbf{Z}_t(t,x)|\big), \tag{6.51}$$

$$|\mathbf{Z}_{xx}(t,x)| \leqslant C_2\big(|\mathbf{Z}(t,x)| + |\mathbf{Z}_t(t,x)| + |\mathbf{Z}_{tt}(t,x)|\big). \tag{6.52}$$

By using repeatedly inequalities (6.48) to (6.52), it follows that there exists $\delta_3 > 0$ and $C_3 > 0$ such that, if $|\mathbf{Z}(t,.)|_0 + |\mathbf{Z}_t(t,.)|_0 < \delta_3$, then

$$\mathcal{O}\big(\int_0^L \big(|\mathbf{Z}|^2|\mathbf{Z}| + |\mathbf{Z}|^2|\mathbf{Z}_t| + |\mathbf{Z}_t|^2|\mathbf{Z}| + |\mathbf{Z}_t|^2|\mathbf{Z}_t|$$

$$+ |\mathbf{Z}_{tt}|^2|\mathbf{Z}_t| + |\mathbf{Z}_t|^2|\mathbf{Z}_{tt}| + |\mathbf{Z}_{tt}|^2|\mathbf{Z}_t|\big)dx; |\mathbf{Z}(t,.)|_0 + |\mathbf{Z}_t(t,.)|_0\big)$$

$$\leqslant C_3(|\mathbf{Z}(t,.)|_0 + |\mathbf{Z}_t(t,.)|_0)\mathbf{V}. \tag{6.53}$$

Using assumption (ii) of Theorem 6.6, there exists $\gamma > 0$ such that

$$-\int_0^L \Big(\mathbf{Z}^\mathsf{T}\mathcal{L}(x)\,\mathbf{Z} + \mathbf{Z}_t^\mathsf{T}\mathcal{L}(x)\,\mathbf{Z}_t + \mathbf{Z}_{tt}^\mathsf{T}\mathcal{L}(x)\,\mathbf{Z}_{tt}\Big)dx$$

$$\leqslant -2\gamma\mathbf{V}(\mathbf{Z}(t,.), \mathbf{Z}_t(t,.), \mathbf{Z}_{tt}(t,.)).$$

It follows from (6.46) that, if $\delta < \min(\delta_1, \delta_2, \delta_3)$ is taken sufficiently small, then $\alpha > 0$ can be selected such that

$$\frac{d\mathbf{V}}{dt} = (-2\gamma + C_3(|\mathbf{Z}(t,.)|_0 + |\mathbf{Z}_t(t,.)|_0)\mathbf{V} \leqslant -\alpha\mathbf{V},$$

for every $\mathbf{Z}$ such that $|\mathbf{Z}|_0 + |\mathbf{Z}_t|_0 \leq \delta$. This concludes the proof of Lemma 6.8.   □

Proof of Theorem 6.6. The proof follows from Lemma 6.8 exactly as the proof of Theorem 4.11 follows from Lemma 4.13.   □

Remark 6.9 (Semi-linear Systems). Here above, we have analyzed the exponential stability of general quasi-linear hyperbolic systems of the form

$$\mathbf{Z}_t + A(\mathbf{Z},x)\mathbf{Z}_x + B(\mathbf{Z},x) = \mathbf{0}.$$

In this equation, A depends on $\mathbf{Z}$ and it is necessary to address the exponential stability for the H^2-norm. It is however interesting to point out that there are many examples where the system is *semi-linear*, i.e., A is constant or depends only on x

but not on $\mathbf{Z}$, as for instance in the models of Raman amplifier, plug flow reactors or chemotaxis presented in Chapter 1. In that special case of a semi-linear system of the form

$$\mathbf{Z}_t + \Lambda(x)\mathbf{Z}_x + B(\mathbf{Z}, x) = 0,$$

it is possible to establish the exponential stability in H^1-norm under assumptions (i) and (ii) of Theorem 6.6. The details of the analysis when Λ is constant can be found in Bastin and Coron (2016).

6.2.2 Stability Condition for the H^p-Norm for Any $p \in \mathbb{N} \smallsetminus \{0, 1\}$

We now consider the most general class of quasi-linear hyperbolic systems with m positive and $n - m$ negative characteristic velocities represented by the equations

$$\mathbf{Z}_t + A(\mathbf{Z}, x)\mathbf{Z}_x + B(\mathbf{Z}, x) = 0, \quad t \in [0, +\infty), \quad x \in [0, L], \tag{6.54}$$

$$\begin{pmatrix} \mathbf{Z}^+(t, 0) \\ \mathbf{Z}^-(t, L) \end{pmatrix} = \mathcal{H} \begin{pmatrix} \mathbf{Z}^+(t, L) \\ \mathbf{Z}^-(t, 0) \end{pmatrix}, \quad t \in [0, +\infty), \tag{6.55}$$

$$\mathbf{Z}(0, x) = \mathbf{Z}_o(x), \quad x \in [0, L], \tag{6.56}$$

with

$$A(\mathbf{0}, x) \triangleq \begin{pmatrix} \Lambda^+(x) & 0 \\ 0 & -\Lambda^-(x) \end{pmatrix}, \quad B(\mathbf{0}, x) = \mathbf{0}, \quad x \in [0, L].$$

Assumption 6.16 is now replaced by

$$A, M, \mathcal{H} \text{ are of class } C^p. \tag{6.57}$$

Using the same approach as in Section 4.5 of Chapter 4, the conditions for the H^2-norm given in Theorem 6.6 of the previous section can be generalized to the stability for any H^p-norm if the definition of exponential stability involves an appropriate extension of the compatibility conditions of order $p - 1$ (see page 153).

With the usual notation $\mathbf{K} \triangleq \mathcal{H}'(\mathbf{0})$, the stability theorem may then be stated as follows.

Theorem 6.10. The steady state $\mathbf{Z}(t, x) \equiv \mathbf{0}$ of the system (6.54), (6.55) is exponentially stable for the H^p-norm if there exists a map $Q \triangleq \text{diag} \{Q^+, Q^-\}$ with $Q^+ \in C^1([0, L]; \mathcal{D}_m^+)$ and $Q^- \in C^1([0, L]; \mathcal{D}_{n-m}^+)$ such that the following *Matrix Inequalities* hold:

(i) the matrix

$$\begin{pmatrix} Q^+(L)\Lambda^+(L) & 0 \\ 0 & Q^-(0)\Lambda^-(0) \end{pmatrix} - \mathbf{K}^\top \begin{pmatrix} Q^+(0)\Lambda^+(0) & 0 \\ 0 & Q^-(L)\Lambda^-(L) \end{pmatrix} \mathbf{K} \tag{6.58}$$

is positive semi-definite;

(ii) the matrix

$$-Q'(x)\Lambda + Q(x)M(\mathbf{0}, x) + M^\top(\mathbf{0}, x)Q(x)$$

is positive definite $\forall x \in [0, L]$.

$\square$

The proof of this theorem is omitted.

6.3   References and Further Reading

The generalization of Theorem 6.10 to the case of a nonautonomous coefficient matrix $A(\mathbf{Z}, x, t)$ explicitly depending on both x and t has been carried out by Diagne and Drici (2012). Moreover the issue of the *input-to-state* stability is addressed by Prieur and Mazenc (2012).

An interesting application of boundary control of the Raman amplifier system (1.21) is presented by Pavel and Chang (2012). In this paper, the authors use an entropy Lyapunov function in order to prove the *global* existence of a C^0-solution and its exponential convergence to the desired steady state. Extremum seeking control of cascaded Raman amplifiers is also addressed by Dower et al. (2008).

This chapter has dealt with the stability of the smooth C^p or H^p-solutions of systems of nonlinear balance laws. Some results for *entropic* solutions have also been recently obtained. The boundary feedback stabilization of entropic solutions of scalar nonlinear balance laws is addressed by Perollaz (2013). Moreover, in Coron et al. (2015), the condition $\rho_1(\mathbf{K}) < 1$ is shown to be sufficient for the exponential stability of the steady state of systems of two conservation laws in the framework of BV entropic solutions when the two characteristic velocities are positive.

Chapter 7
Backstepping Control

IN THIS CHAPTER, we address the problem of boundary stabilization of hyperbolic systems of balance laws by *full state feedback* and by dynamic output feedback in *observer-controller form*. We consider only the case of systems of two balance laws as in Section 5.3. The control design problem is solved by using a 'backstepping' method where the gains of the feedback laws are solutions of an associated system of linear hyperbolic PDEs. The backstepping method for hyperbolic PDEs was initially introduced by Krstic and Smyshlyaev (2008a), Krstic and Smyshlyaev (2008b), and Smyshlyaev et al. (2010). This chapter is essentially based on Vazquez et al. (2011) and Coron et al. (2013).

7.1 Motivation and Problem Statement

We have seen in Section 5.3 that there is always a proper coordinate transformation such that any system of two linear balance laws can be written in the general form

$$\partial_t S_1 + \lambda_1(x)\partial_x S_1 + a(x)S_2 = 0,$$
$$\partial_t S_2 - \lambda_2(x)\partial_x S_2 + b(x)S_1 = 0, \qquad t \in [0, +\infty), \quad x \in [0, L], \qquad (7.1)$$

where λ_1, λ_2 are in $C^1([0, L]; \mathbb{R}_+)$ and a, b are in $C^1([0, L]; \mathbb{R})$. This system is here considered under the boundary conditions

$$S_1(t, 0) = u(t), \qquad S_2(t, L) = \gamma S_1(t, L), \qquad t \in [0, +\infty), \qquad (7.2)$$

where γ is a real constant. This is a control system where $u(t) \in \mathbb{R}$ is the command signal. From Section 5.6, we know that, if this system is open-loop unstable, there is a limitation to the stabilization with a *static boundary output feedback* (i.e., a

© Springer International Publishing Switzerland 2016
G. Bastin, J.-M. Coron, *Stability and Boundary Stabilization of 1-D Hyperbolic Systems*, Progress in Nonlinear Differential Equations and Their Applications 88, DOI 10.1007/978-3-319-32062-5_7

feedback of the state values at the boundaries only): there is a maximal length L above which the stabilization by a static boundary output feedback is impossible.

In this chapter, we show how this limitation can be bypassed by using a dynamic boundary output feedback in the so-called *observer-controller form*. In a first step, we design a full-state feedback control law. Then we design a boundary feedback state observer. Finally the stabilizing output feedback controller is built by combining both designs.

We define a linear *reference model* of the following form:

$$\partial_t S_1^\star + \lambda_1(x)\partial_x S_1^\star = 0,$$
$$\qquad\qquad\qquad\qquad\qquad\qquad t \in [0, +\infty), \quad x \in [0, L], \qquad (7.3)$$
$$\partial_t S_2^\star - \lambda_2(x)\partial_x S_2^\star = 0,$$

with boundary conditions

$$S_1^\star(t, 0) = k S_2^\star(t, 0), \qquad S_2^\star(t, L) = \gamma S_1^\star(t, L), \qquad t \in [0, +\infty), \qquad (7.4)$$

where k is a tuning parameter selected such that $|k\gamma| < 1$. According to Theorem 2.4, this reference model is exponentially stable for the L^2-norm. The design method is then to seek a feedback control law which transforms the closed loop system into this reference model (which, for this reason, is also called *target system* in the literature).

7.2 Full-State Feedback

We introduce the vector and matrix notations

$$\mathbf{S} \triangleq \begin{pmatrix} S_1 \\ S_2 \end{pmatrix}, \qquad \mathbf{S}^\star \triangleq \begin{pmatrix} S_1^\star \\ S_2^\star \end{pmatrix},$$

$$\Lambda(x) \triangleq \begin{pmatrix} \lambda_1(x) & 0 \\ 0 & -\lambda_2(x) \end{pmatrix}, \qquad M(x) \triangleq \begin{pmatrix} 0 & a(x) \\ b(x) & 0 \end{pmatrix}.$$

With these notations, the system (7.1) is written

$$\mathbf{S}_t + \Lambda(x)\mathbf{S}_x + M(x)\mathbf{S} = 0. \qquad (7.5)$$

The state transformation (called *backstepping transformation*) is then defined as follows:

$$\mathbf{S}^\star(t, x) \triangleq \mathbf{S}(t, x) - \int_x^L P(x, \xi)\mathbf{S}(t, \xi)d\xi, \qquad (7.6)$$

where the map $P : \mathcal{X} \to \mathcal{M}_{2,2}$, with $\mathcal{X} \triangleq \{(x, \xi); 0 \leqslant x \leqslant \xi \leqslant L\}$. The map $\mathbf{S}(t, .) \mapsto \mathbf{S}^{\star}(t, .)$ is a Volterra transformation of the second kind. The transformation is linear, continuous, and bijective from $L^2((0, L); \mathbb{R}^2)$ into $L^2((0, L); \mathbb{R}^2)$, see Volterra (1896) and Evans (1910).

Using (7.5) and (7.6), we have

$$\mathbf{S}_t(t, x) = \mathbf{S}_t^{\star}(t, x) + \int_x^L P(x, \xi) \mathbf{S}_t(t, \xi) d\xi$$

$$= \mathbf{S}_t^{\star}(t, x) - \int_x^L P(x, \xi) \Big(\Lambda(\xi) \mathbf{S}_x(t, \xi) + M(\xi) \mathbf{S}(t, \xi) \Big) d\xi$$

$$= \mathbf{S}_t^{\star}(t, x) + \int_x^L \Big(P_{\xi}(x, \xi) \Lambda(\xi) + P(x, \xi) \Lambda_x(\xi) - P(x, \xi) M(\xi) \Big) \mathbf{S}(t, \xi) d\xi$$

$$- P(x, L) \Lambda(L) \mathbf{S}(t, L) + P(x, x) \Lambda(x) \mathbf{S}(t, x).$$
$$(7.7)$$

Moreover, using again (7.6), we have

$$\mathbf{S}_x(t, x) = \mathbf{S}_x^{\star}(t, x) + \int_x^L P_x(x, \xi) \mathbf{S}(t, \xi) d\xi - P(x, x) \mathbf{S}(t, x). \qquad (7.8)$$

Hence, from (7.5) and (7.7), (7.8), we get

$$\mathbf{S}_t^{\star}(t, x) + \Lambda(x) \mathbf{S}_x^{\star}(t, x) = -E_1(x) \mathbf{S}(t, x) + E_2(t, x) - \int_x^L E_3(x, \xi) \mathbf{S}(t, \xi) d\xi,$$

with

$$E_1(x) \triangleq M(x) + P(x, x) \Lambda(x) - \Lambda(x) P(x, x),$$

$$E_2(t, x) \triangleq P(x, L) \Lambda(L) \mathbf{S}(t, L),$$

$$E_3(x, \xi) \triangleq P_{\xi}(x, \xi) \Lambda(\xi) + \Lambda(x) P_x(x, \xi) + P(x, \xi) \Lambda_x(\xi) - P(x, \xi) M(\xi).$$

We denote

$$P \triangleq \begin{pmatrix} p_{00} & p_{01} \\ p_{10} & p_{11} \end{pmatrix}.$$

Let us now show how the map P can be selected such that $E_1 = 0$, $E_2 = 0$, and $E_3 = 0$. We see that $E_1 = 0$ is equivalent to

$$p_{01}(x, x) = \frac{a(x)}{\lambda_1(x) + \lambda_2(x)}, \quad p_{10}(x, x) = \frac{-b(x)}{\lambda_1(x) + \lambda_2(x)}. \qquad (7.9)$$

Using the boundary condition $S_2(t, L) = \gamma S_1(t, L)$, we have $E_2(t, x) = 0$ if

$$p_{00}(x, L)\lambda_1(L) - \gamma p_{01}(x, L)\lambda_2(L) = 0,$$
$$p_{10}(x, L)\lambda_1(L) - \gamma p_{11}(x, L)\lambda_2(L) = 0. \tag{7.10}$$

Then, in order to have $E_3(t, x) = 0$, the matrix function $P(x, \xi)$ is defined as the solution, in the domain $\mathcal{X} \triangleq \{(x, \xi); 0 \leq x \leq \xi \leq L\}$, of the matrix hyperbolic partial differential equation

$$P_\xi(x, \xi)\Lambda(\xi) + \Lambda(x)P_x(x, \xi) + P(x, \xi)\Lambda_x(\xi) - P(x, \xi)M(\xi) = 0 \tag{7.11}$$

under the boundary conditions (7.9) and (7.10). The well-posedness of the system (7.9), (7.10), and (7.11) is established in Vazquez et al. (2011) for $\gamma \neq 0$. In the special case where $\gamma = 0$, the reference model is modified as follows:

$$\partial_t S_1^\star + \lambda_1(x)\partial_x S_1^\star - g(x)S_2^\star(L, t) = 0,$$
$$\partial_t S_2^\star - \lambda_2(x)\partial_x S_2^\star = 0,$$

where the function $g(x)$ has to be selected adequately, see Vazquez et al. (2011). From now on, we assume that $\gamma \neq 0$.

Let us now look at the boundary conditions for $\mathbf{S}^\star$. From the definition (7.6) of the state transformation and using the boundary condition (7.2), we have

$$S_2^\star(t, L) = \gamma S_1^\star(t, L),$$

$$S_1^\star(t, 0) = u(t) - \int_0^L \Big(p_{00}(0, \xi)S_1(t, \xi) + p_{01}(0, \xi)S_2(t, \xi)\Big)d\xi,$$

$$S_2^\star(t, 0) = S_2(t, 0) - \int_0^L \Big(p_{10}(0, \xi)S_1(t, \xi) + p_{11}(0, \xi)S_2(t, \xi)\Big)d\xi.$$

In order to realize the dissipative boundary condition (7.4), the feedback control law is defined as

$$u(t) = kS_2(t, 0) + \int_0^L \Big(\big(p_{00}(0, \xi) - kp_{10}(0, \xi)\big)S_1(t, \xi)$$
$$+ \big(p_{01}(0, \xi) - kp_{11}(0, \xi)\big)S_2(t, \xi)\Big)d\xi, \tag{7.12}$$

where k is a tuning parameter selected such that $|k\gamma| < 1$. Hence this control law exponentially stabilizes the control system (7.1), (7.2). In the special case of a deadbeat control (i.e., $k = 0$), the steady state is reached in finite time

$$t_F = \int_0^L \left(\frac{1}{\lambda_1(\xi)} + \frac{1}{\lambda_2(\xi)}\right)d\xi.$$

Obviously, from (7.12), the practical implementation of this feedback control law needs the on-line knowledge of the full state $S(t, x)$ on $[0, L]$. In the next section, we shall see how this knowledge can be provided by a state-observer that uses a boundary on-line measurement of $S_2(t, 0)$ only.

7.3   Observer Design and Output Feedback

The objective is now to design an observer for the on-line estimation of $S(t, x)$. Assuming that the output $S_2(t, 0)$ is measured on-line, the observer is a copy of the system (7.1) with additional so-called *output injection* terms:

$$\widehat{S}_t(t, x) + \Lambda(x)\widehat{S}_x(t, x) + M(x)\widehat{S} + \begin{pmatrix} \upsilon_1(x) \\ \upsilon_2(x) \end{pmatrix} \left(S_2(t, 0) - \widehat{S}_2(t, 0)\right) = 0 \qquad (7.13)$$

with the boundary conditions

$$\widehat{S}_1(t, 0) = u(t), \qquad \widehat{S}_2(t, L) = \gamma\widehat{S}_1(t, L). \qquad (7.14)$$

In these equations, the estimates are denoted by a hat accent while $\upsilon_1(x)$ and $\upsilon_2(x)$ are the output injection gains. We define the estimation errors

$$\widetilde{S}_1 \triangleq S_1 - \widehat{S}_1, \qquad \widetilde{S}_2 \triangleq S_2 - \widehat{S}_2.$$

Then the so-called error system is obtained by subtraction of the observer equations (7.13) from the system equations (7.5):

$$\widetilde{S}_t(t, x) + \Lambda(x)\widetilde{S}_x(t, x) + M(x)\widetilde{S} - N(x)\widetilde{S}(t, 0) = 0, \qquad (7.15)$$

with boundary conditions

$$\widetilde{S}_1(t, 0) = 0, \qquad \widetilde{S}_2(t, L) = \gamma\widetilde{S}_1(t, L). \qquad (7.16)$$

The matrix $N(x)$ is defined as

$$N(x) \triangleq \begin{pmatrix} 0 & \upsilon_1(x) \\ 0 & \upsilon_2(x) \end{pmatrix}. \qquad (7.17)$$

In order to find the output injection gains $\upsilon_i(x)$, we use a backstepping transformation of the following form:

$$\widetilde{S}(t, x) \triangleq S^\sharp(t, x) + \int_0^x \widetilde{P}(x, \xi)S^\sharp(t, \xi)d\xi. \qquad (7.18)$$

where $\mathbf{S}^\sharp(t,x)$ is the state of the following reference model:

$$\partial_t S_1^\sharp + \lambda_1(x)\partial_x S_1^\sharp = 0,$$

$$\partial_t S_2^\sharp - \lambda_2(x)\partial_x S_2^\sharp = 0,$$

$$S_1^\sharp(t,0) = 0, \qquad S_2^\sharp(t,L) = \gamma S_1(t,L).$$

Using (7.15) and (7.18), we have

$$\widetilde{\mathbf{S}}_t(t,x) = \mathbf{S}_t^\sharp(t,x) + \int_0^x \widetilde{P}(x,\xi)\mathbf{S}_t^\sharp(t,\xi)d\xi$$

$$= \mathbf{S}_t^\sharp(t,x) - \int_0^x \widetilde{P}(x,\xi)\Lambda(\xi)\mathbf{S}_x^\sharp(t,\xi)d\xi. \tag{7.19}$$

Moreover, using again (7.18), we have

$$\widetilde{\mathbf{S}}_x(t,x) = \mathbf{S}_x^\sharp(t,x) + \int_0^x \widetilde{P}_x(x,\xi)\mathbf{S}^\sharp(t,\xi)d\xi + \widetilde{P}(x,x)\mathbf{S}^\sharp(t,x). \tag{7.20}$$

Hence, from (7.15), (7.19), and (7.20), we get

$$A(t,x) + \int_0^x \widetilde{P}(x,\xi)A(t,\xi)d\xi = E_1^\sharp(x)\mathbf{S}^\sharp(t,x) + E_2^\sharp(t,x) - \int_0^x E_3^\sharp(x,\xi)\mathbf{S}^\sharp(t,\xi)d\xi,$$

with

$$A(t,x) \triangleq \mathbf{S}_t^\sharp(t,x) + \Lambda(x)\mathbf{S}_x^\sharp(t,x)$$

$$E_1^\sharp(x) \triangleq \widetilde{P}(x,x)\Lambda(x) - \Lambda(x)\widetilde{P}(x,x) - M(x),$$

$$E_2^\sharp(t,x) \triangleq (N(x) - \widetilde{P}(x,0)\Lambda(0))\mathbf{S}^\sharp(t,0),$$

$$E_3^\sharp(x,\xi) \triangleq P_\xi(x,\xi)\Lambda(\xi) + \widetilde{P}(x,\xi)\Lambda_x(\xi) + M(x)\widetilde{P}(x,\xi) + \Lambda(x)\widetilde{P}_x(x,\xi).$$

Denoting

$$\widetilde{P} \triangleq \begin{pmatrix} \tilde{p}_{00} & \tilde{p}_{01} \\ \tilde{p}_{10} & \tilde{p}_{11} \end{pmatrix},$$

let us now show how the map $\widetilde{P}$ and the output gains $\upsilon_i(x)$ can be selected such that $E_1^\sharp = 0$, $E_2^\sharp = 0$, and $E_3^\sharp = 0$. We see that $E_1^\sharp = 0$ is equivalent to

$$p_{01}(x,x) = \frac{-a(x)}{\lambda_1(x) + \lambda_2(x)}, \qquad p_{10}(x,x) = \frac{b(x)}{\lambda_1(x) + \lambda_2(x)}. \tag{7.21}$$

Using the boundary condition $S_1^{\sharp}(t,0) = 0$, we see that $E_2^{\sharp}(t,x) = 0$ if

$$(N(x) - \widetilde{P}(x,0)\Lambda(0)) \begin{pmatrix} 0 \\ S_2(t,0) \end{pmatrix} = 0 \quad \text{for any } S_2(t,0),$$

and therefore if the output injection gains are chosen such that

$$\upsilon_1(x) = -\lambda_2(0)\tilde{p}_{01}(x,0), \qquad \upsilon_2(x) = -\lambda_2(0)\tilde{p}_{11}(x,0). \tag{7.22}$$

Moreover, in order to satisfy the boundary condition $S_2^{\sharp}(t,L) = \gamma S_1^{\sharp}(t,L)$, we impose

$$\tilde{p}_{10}(L,\xi) = \gamma\tilde{p}_{00}(L,\xi), \qquad \tilde{p}_{11}(L,\xi) = \gamma\tilde{p}_{01}(L,\xi). \tag{7.23}$$

Then, in order to have $E_3^{\sharp}(t,x) = 0$, the matrix function $\widetilde{P}(x,\xi)$ is defined as the solution, in the domain $\mathcal{X} \triangleq \{(x,\xi); 0 \leqslant x \leqslant \xi \leqslant L\}$, of the matrix hyperbolic partial differential equation

$$\widetilde{P}_\xi(x,\xi)\Lambda(\xi) + \widetilde{P}(x,\xi)\Lambda_x(\xi) - M(x)\widetilde{P}(x,\xi) + \Lambda(x)\widetilde{P}_x(x,\xi) = 0 \tag{7.24}$$

under the boundary conditions (7.21) and (7.23). The well-posedness of the system (7.21), (7.23), (7.24) is established in Vazquez et al. (2011).

The exponential stability of the error system (7.15) follows. This implies that the state estimate $\widehat{S}(t,x)$ exponentially converges to the real state $S(t,x)$, and even in finite time in the special case where $\gamma = 0$.

A stabilizing output feedback controller is then obtained by combining the full state feedback controller (7.12) and the observer (7.13) as

$$u(t) = kS_2(t,0) + \int_0^L \Big(\big(p_{00}(0,\xi) - kp_{10}(0,\xi)\big)\widehat{S}_1(t,\xi)$$

$$+ \big(p_{01}(0,\xi) - kp_{11}(0,\xi)\big)\widehat{S}_2(t,\xi) \Big)d\xi, \tag{7.25}$$

and we have the following stability theorem.

Theorem 7.1. Consider the system (7.1) with boundary condition (7.2), control law (7.25), and initial condition $S_o \in L^2((0,L);\mathbb{R}^2)$. Then, for any k such that $|k\gamma| < 1$, there exist $v > 0$ and $C_0 > 0$ such that

$$\|S(t,.)\|_{L^2((0,L);\mathbb{R}^2)} \leqslant C_0 e^{-vt}\|S_o\|_{L^2((0,L);\mathbb{R}^2)}, \quad \forall t \in [0,+\infty).$$

Furthermore, the equilibrium $S = 0$ is reached in finite time when $\gamma = 0$.   □

7.4 Backstepping Control of Systems of Two Balance Laws

In this section, we briefly examine how the backstepping control approach can be extended to the nonlinear case, with a local stability property. We consider a general system of two balance laws in quasilinear form:

$$\mathbf{Z}_t + A(\mathbf{Z}, x)\mathbf{Z}_x + B(\mathbf{Z}, x) = \mathbf{0}, \tag{7.26}$$

where $\mathbf{Z} : [0, L] \times [0, +\infty) \to \mathbb{R}^2$ and

$$A : \mathbb{R}^2 \times [0, L] \to \mathcal{M}_{2,2}(\mathbb{R}), \qquad B : \mathbb{R}^2 \times [0, L] \to \mathbb{R}^2,$$

$A(\mathbf{Z}, x)$ and $B(\mathbf{Z}, x)$ are twice continuously differentiable w.r.t. $\mathbf{Z}$ and x,

$A(\mathbf{0}, x) = \text{diag}\{\lambda_1(x), -\lambda_2(x)\}$ with $\lambda_i(x) > 0$ and $B(\mathbf{0}, x) = \mathbf{0} \ \forall x \in [0, L]$.

Remark that these assumptions imply that $\mathbf{Z} \equiv \mathbf{0}$ may be a steady state of the system. With the notation $\mathbf{Z} \triangleq (Z_1, Z_2)^\mathsf{T}$, the system (7.26) is considered under boundary conditions of the following form:

$$Z_1(t, 0) = u(t), \qquad Z_2(t, L) = G(Z_1(t, L)), \tag{7.27}$$

where the map G is assumed to be twice differentiable with $G(0) = 0$. The system (7.26), (7.27) is an open-loop control system when $u(t) \in \mathbb{R}$ is an exogenous command signal.

In order to design a backstepping observer-controller, we first rewrite the quasilinear system in a form which is, up to the nonlinear terms, identical to (7.1), (7.2). For that, we introduce the notation

$$\frac{\partial B}{\partial \mathbf{Z}}(\mathbf{0}, x) \triangleq \begin{bmatrix} \gamma_1(x) & \delta_1(x) \\ \gamma_2(x) & \delta_2(x) \end{bmatrix} = \Gamma(x)$$

and we use the transformation (5.27):

$$\mathbf{S}(t, x) = \Phi(x)\mathbf{Z}(t, x) \quad \text{with} \quad \Phi(x) \triangleq \begin{pmatrix} \varphi_1(x) & 0 \\ 0 & \varphi_2(x) \end{pmatrix} \tag{7.28}$$

where the functions φ_1 and φ_2 are given by (5.26). Then, it can be shown (for details see Coron et al. (2013)) that the quasilinear system (7.26) is written as follows in the $\mathbf{S}$ coordinates:

$$\mathbf{S}_t + \Lambda(x)\mathbf{S}_x + M(x)\mathbf{S} + f(\mathbf{S}, x) = 0 \tag{7.29}$$

where

$$\Lambda(x) \triangleq \begin{pmatrix} \lambda_1(x) & 0 \\ 0 & -\lambda_2(x) \end{pmatrix}, \quad M(x) \triangleq \begin{pmatrix} 0 & a(x) \\ b(x) & 0 \end{pmatrix},$$

$$a(x) \triangleq \frac{\varphi_1(x)}{\varphi_2(x)} \delta_1(x), \quad b(x) \triangleq \frac{\varphi_2(x)}{\varphi_1(x)} \gamma_2(x),$$

and

$$f(\mathbf{S}, x) \triangleq \Phi(x) \left[A(\Phi^{-1}(x)\mathbf{S}, x) - \Lambda(x) \right] \partial_x \left(\Phi^{-1}(x)\mathbf{S} \right)$$
$$+ B(\Phi^{-1}(x)\mathbf{S}, x) - \Gamma(x)\Phi^{-1}(x)\mathbf{S}.$$

In the $\mathbf{S}$ coordinates, the boundary conditions (7.27) are then written

$$S_1(t, 0) = u(t), \quad S_2(t, L) = \gamma S_1(t, L) + g(S_1(t, L)), \tag{7.30}$$

with

$$\gamma \triangleq \frac{\varphi_2(L)}{\varphi_1(L)} G'(0) \quad \text{and} \quad g(S_1) \triangleq \varphi_2(L) G\left(\frac{S_1}{\varphi_1(L)}\right) - \gamma S_1.$$

It is clear that, in these equations, the nonlinear terms satisfy $f(\mathbf{0}, x) = \mathbf{0}$, $g(0) = 0$, and $g'(0) = 0$. Then we see that the linear parts of (7.29) and (7.30) are indeed identical to the linear system equations (7.1) and (7.2) in Section 7.1.

Therefore, it is quite natural to apply the linear observer-controller designed in the previous sections to the quasi-linear system case and to get a local stability property in H^2-norm in the same vein as what has been done in Section 6.2. The details of the analysis can be found in Coron et al. (2013).

7.5 References and Further Reading

Backstepping is a technique originally developed from 1990 for designing stabilizing controls for nonlinear dynamical systems, see, e.g., Coron (2007, Section 12.5) and the tutorial textbook by Krstic et al. (1995).

The first extensions to PDEs were published by Coron and d'Andréa-Novel (1998) for the beam equation and by Liu and Krstic (2000) for discretized PDEs. Later on, Krstic and his collaborators introduced a modification of the method by means of an integral Volterra transformation of the second kind. This invertible transformation maps the original PDE control system into a reference system which is exponentially stable (and the exponential decay can be made arbitrarily fast by choosing suitably the reference system). In this framework, the first backstepping designs were proposed for the heat equation in Liu (2003) and Smyshlyaev and Krstic (2004). The applications to wave equations appeared later in Krstic et al. (2008), Smyshlyaev and Krstic (2009), Smyshlyaev et al. (2010). An excellent introduction to the backstepping method for PDEs is the book "Boundary control of PDEs: A course on backstepping designs" by Krstic and Smyshlyaev (2008b).

In this chapter we have only considered systems of two hyperbolic balance laws. More general systems have also been considered in the literature: linear systems with only one negative characteristic velocity in Di Meglio et al. (2013), systems of three linear balance laws in Hu and Di Meglio (2015), systems of n linear balance laws in Hu et al. (2016), and systems of n nonlinear balance laws in Hu et al. (2015b). Note that in all these papers the reference models (target systems) need to be more complicated than in this chapter.

Let us finally mention the following interesting contributions on the backstepping approach for hyperbolic systems.

- An extension of backstepping to Navier-Stokes two-dimensional time-varying problems can be found in Vazquez et al. (2008)
- In some cases it is useful to use more general integral transformations than integral Volterra transformations of the second kind: see Smyshlyaev et al. (2009) for an Euler-Bernoulli beam equation and Coron and Lü (2014) for a Korteweg-de Vries equation.
- In the special case of systems of two linear balance laws, it is possible to find the exact analytical solution to a Goursat PDE system governing the kernels of a backstepping-based boundary control law that stabilizes the system, see Vazquez and Krstic (2013).
- The backstepping method can also be extended to design *adaptive* output feedback controllers for hyperbolic systems with the goal of disturbance rejection, see Aamo (2013), or to account for unknown parameters, see Bernard and Krstic (2014).
- In Lamare et al. (2015a), the backstepping method is used for trajectory generation and the design of PI controllers for 2×2 linear hyperbolic systems with nonuniform coefficients.

Chapter 8
Case Study: Control of Navigable Rivers

T HE OBJECTIVE of this chapter is to emphasize the main technological features that may occur in real live applications of boundary feedback control of hyperbolic systems of balance laws. The issue is presented through the specific case study of the control of navigable rivers with a particular focus on the Meuse river in Wallonia (south of Belgium).

8.1 Geographic and Technical Data

According to Wikipedia "the Meuse (see Fig. 8.1) is a major European river, rising in France and flowing through Belgium and the Netherlands before draining into the North Sea. It has a total length of 925 km." In this chapter we shall mainly report on the control implementation in a stretch of the Meuse river which is called 'Haute Meuse' and is located between the city of Givet (at the Belgian-French border) and the city of Namur (in Belgium). A map of the basin of the Meuse river is shown in Fig. 8.2. As in most natural navigable rivers, the riverbed has been transformed such that the water flows through successive pools separated by weirs. The locations and the names of the Haute Meuse weirs can also be seen in Fig. 8.2 while the profile of the river is shown in Fig. 8.3.

Each weir is provided with three or four parallel automated gates that are used to regulate the water level in the pools. This is illustrated in Fig. 8.4 with the last weir of the Haute Meuse (La Plante). The gates can be operated in both overflow and underflow modes as sketched in Fig. 8.5 and illustrated in Fig. 8.6. The overflow mode is used in case of low flow rates (less than 300 m^3/sec) while the underflow mode is used in the range 300–800 m^3/sec.

© Springer International Publishing Switzerland 2016
G. Bastin, J.-M. Coron, *Stability and Boundary Stabilization of 1-D Hyperbolic Systems*, Progress in Nonlinear Differential Equations and Their Applications 88, DOI 10.1007/978-3-319-32062-5_8

Fig. 8.1 The Meuse river in the city of Liege (Belgium). From http://www.studentsoftheworld.info

The river bathymetry has been recorded by using a swath sonar system (see Fig. 8.7 and Dal Cin et al. (2005) for more details). A typical example of a cross-section of the riverbed obtained with swath bathymetry is shown in Fig. 8.8.

8.2 Modeling and Simulation

The dynamics of each pool are represented by the following general Saint-Venant equations (see Equations (1.29) in Section 1.4):

$$\partial_t A + \partial_x Q = 0,$$

$$\partial_t Q + \partial_x \frac{Q^2}{A} + gA\left[(\frac{\partial H}{\partial A}\partial_x A - S_b) + S_f\right] = 0,$$

where $A(t,x)$ is the cross-sectional area of the water in the channel, $Q(t,x)$ is the flow rate (or discharge), $H(A)$ is the water depth, S_f is the friction term, $S_b(x)$ is the bottom slope, and g is the constant gravity acceleration. The friction term S_f is represented by the Manning (1891) formula

$$S_f \triangleq \left(\frac{Q(P(A))^{2/3}}{\nu}A^{5/3}\right)^2$$

where ν is the so-called Strickler constant coefficient and $P(A)$ is the perimeter of the cross-sectional area.

For simulation purpose, the Saint-Venant equations are integrated numerically using a standard Preissman scheme (see, e.g., Litrico and Fromion (2009, Section 2.2.2)) with a spatial step size $\Delta x = 1\,\mathrm{m}$, a time step $\Delta t = 1\,\mathrm{s}$ and a Strickler coefficient $\nu = 33\,\mathrm{m}^{1/3}/\mathrm{s}$.

For the simulations, the boundary conditions are given by the hydraulic model of the gates (see Equations (1.25) and (1.26) in Section 1.4) written here in a form that covers both underflow and overflow modes in a single expression:

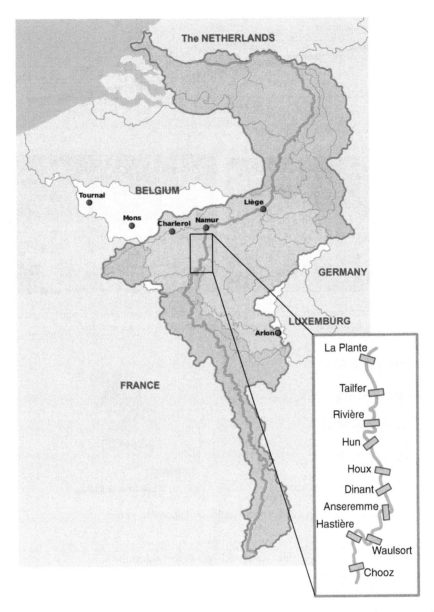

Fig. 8.2 Meuse river basin and zoom on the 'Haute-Meuse' stretch with the weir names and locations. (The figure is adapted from en.wikipedia.org/wiki/Meuse)

$$Q_g(t,L) = w\sqrt{2g}\left[\underbrace{k_{G1}\sqrt{\left(\max\{0, H(t,L) - U_{og}\}\right)^3}}_{\text{overflow}} + \underbrace{k_{G2}U_{ug}\sqrt{\Delta H}}_{\text{underflow}}\right]$$

where (see Fig. 8.9)

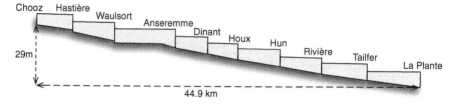

Fig. 8.3 Profile of the Haute Meuse stretch

Fig. 8.4 La Plante weir with four parallel gates. Top view on the left (from Google maps) and panoramic view from the Namur fortress on the right (http://www.meusenamuroise.be)

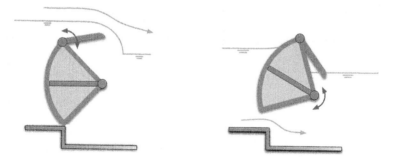

Fig. 8.5 Each gate can be operated in both overflow and underflow mode

- $H(t, L)$ is the regulated water level and ΔH is the waterfall,
- w is the width of the gate,
- U_{og} is the elevation of the top the gate and U_{ug} is the depth of the underflow aperture of the gate,
- $k_{G1} = 0.39$ and $k_{G2} = 0.7$ are the adimensional discharge coefficients,
- $Q_g(t, L)$ is the flow rate of one gate which has to be multiplied by the number of gates to get the total discharge at the boundary (i.e., at the corresponding weir).

In the next sections, we shall see that the simulation model is a fundamental tool for the practical set-up of the control system. In particular, it will be used to rationalize the selection of the set-points, of the control tuning parameters, and of the time step for the digital implementation of the control law.

Overflow mode

Underflow mode

Fig. 8.6 Automated control gates in the Meuse river (Belgium)

8.3 Control Implementation

In this section, we address some practical issues that have to be considered for the control implementation.

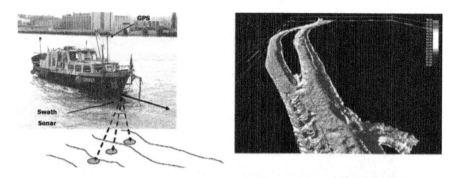

Fig. 8.7 Bathymetric recording with the swath sonar system (from Dal Cin et al. (2005))

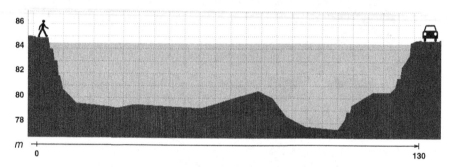

Fig. 8.8 Example of a Meuse cross-section obtained with swath bathymetry

Fig. 8.9 The hydraulic gates determine the boundary conditions

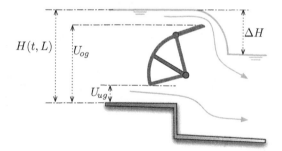

8.3.1 Local or Nonlocal Control?

Two basic configurations to achieve set-point level regulation with PI boundary controllers are shown in Fig. 8.10.

Local control, which is the usual structure in navigable rivers, means that the flow over a gate is used to regulate the water level immediately upstream the same gate. In contrast, nonlocal control means that the flow over the upstream gate of a pool is used to regulate the water level at the downstream gate of the pool. With nonlocal

Fig. 8.10 PI control
configurations for a pool of
an open channel

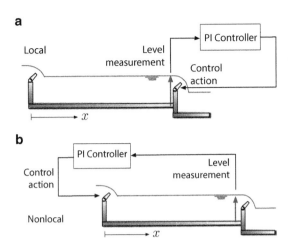

control, the effect of the control action is delayed which limits the achievable
performance. However, in arid regions, this structure of control is generally used
in irrigation networks to enhance water savings.

8.3.2 Steady State and Set-Point Selection

For a constant flow rate Q^*, the profile of the steady state cross-section $A^*(x)$ of the
water flow is the solution of the differential equation

$$Q^{*2}\partial_x A^* - A^{*3}\left[g(\frac{\partial H}{\partial A^*}\partial_x A^* - S_b) + S_f\right] = 0.$$

In order to set up the control system, the first task is to use the simulation model
to compute the steady state profile in all pools of the river. This is illustrated in
Fig. 8.11 for the pool between Houx and Hun weirs. It can be seen that for low
flow rates (less than 200 m³/sec) the steady state profile is almost horizontal, while
for high flow rates (above 800 m³/sec) it is quasi-parallel to the bottom of the river
(albeit with a slight curvature).

At each weir, the level set point must therefore be carefully selected in order to
meet three conflicting objectives

 (i) to guarantee a sufficient draft for the boats, that is a sufficient depth of water
      for a safe navigation (see Fig. 8.12);
 (ii) to guarantee a sufficient air draft allowing the boats to pass safely under the
      bridges and other obstacles such as power lines;
(iii) to avoid water overflow on the banks of the river in the upper parts of the pools
      in case of high flow rates.

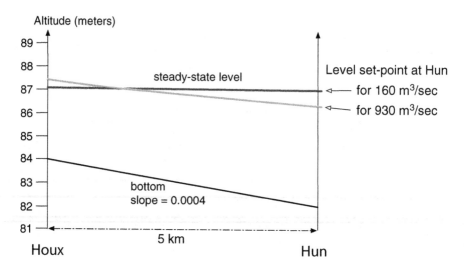

Fig. 8.11 Steady state levels for low and high flow rates in the pool between Houx and Hun weirs

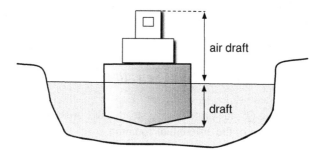

Fig. 8.12 The level set point is selected to guarantee a safe navigation

For these reasons, as shown in Fig. 8.11, the level set point is generally lower when the flow rate is higher. Therefore, a schedule for the level set point is provided at each weir as a piecewise constant decreasing function of the flow rate. This also explains the counterintuitive observation that, in a regulated navigable river, the water level is often lower when the flow is faster!

8.3.3   Choice of the Time Step for Digital Control

As already said above, in the Meuse river, the level set-point regulation is achieved with PI controllers. In practice, each weir is provided with a local *Programmable Logic Controller* (PLC) wherein a numerical incremental PI control algorithm is implemented. The starting point is the following continuous time expression for PI control (see Equation (2.25) in Section 2.2):

$$U(t) \triangleq U_R + k_P e(t) + k_I \int_0^t e(\tau)d\tau.$$

In this equation, $U(t)$ denotes the actual gate position and U_R is a constant reference value. Moreover, $e(t) \triangleq H^* - H(t, L)$ is the regulation error with H^* the level set point and $H(t, L)$ the on-line measurement of the actual water level immediately upstream of the gate (see Fig. 8.10(a)). k_P and k_I are the tuning parameters. For the computer implementation, the continuous variables are approximated by their sampled counterparts, see, e.g., Åström and Murray (2009, Section 10.5):

$$U(t_k) \triangleq U_R + k_P e(t_k) + k_I I(t_k),$$

where t_k denotes the sampling instant. The sampling period is denoted $\Delta t \triangleq t_k - t_{k-1}$. In this expression, the integral term $I(t_k)$ is given by approximating the integral with a sum:

$$I(t_k) = I(t_{k-1}) + \Delta t\, k_I e(t_k).$$

The incremental form is obtained by considering the time difference of the controller output as follows:

$$U(t_k) = U(t_k - \Delta t) + k_P\big(e(t_k) - e(t_k - \Delta t)\big) + k_I e(t_k).$$

Regarding the choice of the time step value, it is well known that, for the efficiency and the robustness of the control, it is important to use the largest time step which is compatible with the system dynamics. A standard rule of thumb is to select Δt between one-tenth and one-fifth of the step response time of the system. Experimental step responses recorded at the weir of Dinant are shown in Fig. 8.13 where it can be seen that the response time is around 200 minutes.

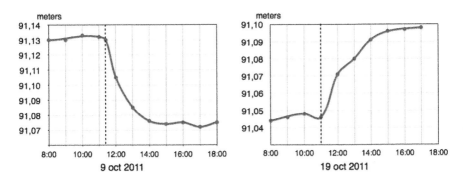

Fig. 8.13 Water level step responses recorded at Dinant on 9 and 19 October 2011 for a flow rate of about $20\, \mathrm{m}^3/\mathrm{sec}$

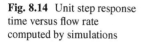

Fig. 8.14 Unit step response
time versus flow rate
computed by simulations

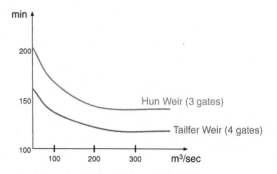

However, the response time may be different from one place to another and, in addition, it is highly dependent on the flow rate value and on the number of gates that are in operation in a weir. This issue was investigated by intensive simulations for the various pools. For instance, the dependence of the step response time on the flow rate is shown in Fig. 8.14 for the weir of Tailfer in the case where four gates are used, and also for the case where only three gates are in operation at the weir of Hun. It can be seen that, depending on the number of gates and the flow rate, the step response time ranges from 120 to 200 minutes.

Finally, it was decided to use a time step $\Delta t = 20$ minutes. This means that the gates move, during a short time, once every 20 minutes at most. This value has also the advantage not to exceed the limits of the patience of the human operators who are supervising the weir operation.

8.4 Control Tuning and Performance

The role of PI control is to achieve set-point regulation, off-set errors cancellation, and attenuation of load disturbances in a robust way (see, e.g., Åström and Murray (2009, Chapter 10) for a detailed motivation of PI controllers). In the Meuse river, the load disturbances mainly come from the river tributaries, the operation of electrical power plants along the river, or, sometimes, from inadvertent actions of the human operators at the weirs.

For each pool, a procedure in three steps is followed for the tuning of the control parameters k_P and k_I:

1. A simple first order model is identified from open loop unit step simulations. The PI controller is tuned to cancel the pole of this first order model and have a closed loop response time identical to the open loop. This gives a first guess for the values of the parameters k_P and k_I.
2. The controller is implemented on the numerical simulator. The parameters are then optimized by trial and error in order to achieve a good performance in simulation.

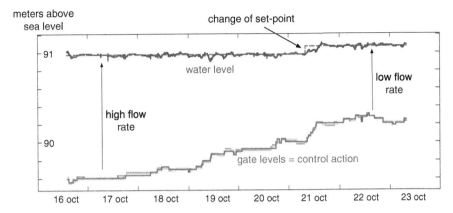

Fig. 8.15 Water level regulation at Dinant from 16 to 23 October 2012 with gates in overflow modes

3. The controller is set up in the PLC and the real life closed loop system is started. If necessary, a final adjustment of the parameters is done after a few weeks of system monitoring.

A typical experimental illustration of the performance of the water level regulation is shown in Fig. 8.15.

As it is well documented by Cantoni et al. (2007), the achievable performance is limited by two fundamental constraints:

1. First it must be ensured that the high frequency wave dynamics of the pool are not excited by a too tight tuning of the regulator. In equivalent terms, this means that the closed-loop bandwidth must be smaller than the dominant wave frequency by one decade at least. In practice, in our case, this is achieved by having a dominant time constant for the closed loop system of the same order of magnitude than for the open loop.

2. More important is the error-propagation phenomenon in strings of open channels under local control. The control actions to regulate the water level in a pool automatically induce flow disturbances at the input of the next pool. This in turn will result in water level errors and control actions in the next pool which will continue to propagate along the string of pools. As we have shown in Section 5.5 (p. 195) this phenomenon is unavoidable with local PI controllers and can lead to excessive errors if the regulators are not properly tuned. Hence there is a tuning trade-off between the performance of the local water level regulation, on one hand, and the performance of the water level error propagation on the other hand. For a big disturbance that occurred on October 15, 2014, Fig. 8.16 illustrates how the water level error propagation is controlled with the current tuning of the PI regulators.

Fig. 8.16 Propagation of
level control errors observed
on 15 October 2014 at the
four successive weirs of
Houx, Riviere, Tailfer, and La
Plante, for a flow rate of
about $100\,\mathrm{m^3/sec}$

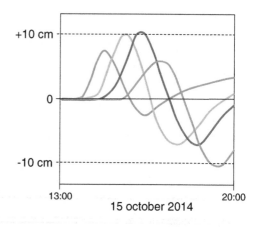

8.5   References and Further Reading

In this chapter, we have addressed the closed-loop control of open channels when
the control design is based on Saint-Venant equations models.

By testing the model against real experimental data, Ooi et al. (2005) show that
the Saint-Venant equations can adequately capture the dynamics of real channels
and constitute therefore a relevant tool for the control design (see also the interesting
identification study of van Overloop and Bombois (2012)).

In addition to closed-loop feedback control, it is evident that open-loop feedforward control is another issue of great interest for rivers and irrigation canals. Bastin
et al. (2005) describe a method to determine the gate motion at the downstream of
a pool in order to exactly cancel the influence of the variations of the flow of the
upstream discharge. The relevance of the control is illustrated with a simulation
experiment regarding the water level control in an open channel disturbed by a
time varying inflow rate. The computation of the feedforward control law is based
on linearized Saint-Venant equations for a horizontal pool with negligible friction
effects, but the simulation results show that it can be robustly applied to a channel
described by nonlinear Saint-Venant equations with slope and friction.

Similarly, Di Meglio et al. (2008) and Rabbani et al. (2010) present a feedforward
control of the flow of the upstream discharge of a pool that exactly produces a
desired downstream discharge. Using differential flatness (e.g., Fliess et al. (1995)),
the control derivation is based on the Hayami model, a parabolic partial differential
equation resulting from a simplification of the Saint-Venant equations. Here also,
numerical simulations show the effectiveness of the approach when the open-loop
controller is applied to irrigation canals modeled by the full Saint-Venant equations.

Litrico and Fromion (2006) investigate the control of oscillating modes occurring
in open channels, due to the reflection of propagating waves on the boundaries.
These modes are well captured by linearized Saint-Venant equations. They use a
distributed transfer function approach to compute a dynamic boundary controller
that cancels the oscillating modes over all the canal pools.

Various methods for the detection and the localization of leaks in open channels are presented and evaluated in Bedjaoui et al. (2009) and Bedjaoui and Weyer (2011). Some of the proposed methods make use of Saint Venant equation models with the leak modeled as a lateral outflow. The size of the leak is estimated with the aid of Luenberger type observers based on the Saint-Venant equations.

Fovet et al. (2010) study the control of algae detachment in regulated canal networks. An original strategy to manage these algae developments consists in flushing the fixed algae. The design objective consists in maximizing the algae detachment using as little water as possible, and without overcoming a maximal turbidity level. In this work, Saint Venant equations are complemented with an advection-dispersion equation based on mass conservation and Fick's law.

Let us finally mention that Van Pham et al. (2010) study the application of infinite-dimensional receding horizon optimal control techniques for an open channel.

Appendix A
Well-Posedness of the Cauchy Problem for Linear Hyperbolic Systems

In this appendix we study a Cauchy problem associated with the linear hyperbolic partial differential equation

$$\mathbf{R}_t + \Lambda(x)\mathbf{R}_x + M(x)\mathbf{R} = 0, \quad t \in [0, +\infty), \quad x \in [0, L], \tag{A.1}$$

with $\mathbf{R} : [0, +\infty) \times [0, L] \to \mathbb{R}^n$, $M : [0, L] \to \mathcal{M}_{n,n}(\mathbb{R})$ and $\Lambda(x) \triangleq \text{diag}\{\Lambda^+(x), \Lambda^-(x)\}$ such that

$$\Lambda^+(x) \triangleq \text{diag}\{\lambda_1(x), \dots, \lambda_m(x)\},$$
$$\Lambda^-(x) \triangleq -\text{diag}\{\lambda_{m+1}(x), \dots, \lambda_n(x)\}, \qquad \text{with } \lambda_i(x) > 0, \ \forall x \in [0, L]. \tag{A.2}$$

Concerning the regularity of Λ and M, we assume that

$$\lambda_i \text{ is Lipschitzian in } [0, L], \tag{A.3}$$

$$M \in L^\infty((0, L); \mathcal{M}_{n,n}(\mathbb{R})). \tag{A.4}$$

The boundary conditions are defined as

$$\begin{pmatrix} \mathbf{R}^+(t, 0) \\ \mathbf{R}^-(t, L) \end{pmatrix} = \mathbf{K} \begin{pmatrix} \mathbf{R}^+(t, L) \\ \mathbf{R}^-(t, 0) \end{pmatrix} \tag{A.5}$$

with

$$\mathbf{K} \triangleq \begin{pmatrix} K_{00} & K_{01} \\ K_{10} & K_{11}, \end{pmatrix} \tag{A.6}$$

© Springer International Publishing Switzerland 2016
G. Bastin, J.-M. Coron, *Stability and Boundary Stabilization of 1-D Hyperbolic Systems*, Progress in Nonlinear Differential Equations and Their Applications 88, DOI 10.1007/978-3-319-32062-5

such that

$$K_{00} \in \mathcal{M}_{m,m}(\mathbb{R}), \ K_{01} \in \mathcal{M}_{m,n-m}(\mathbb{R}), \ K_{10} \in \mathcal{M}_{n-m,m}(\mathbb{R}), \ K_{11} \in \mathcal{M}_{n-m,n-m}(\mathbb{R}).$$

We consider the following Cauchy problem

$$\mathbf{R}_t + \Lambda(x)\mathbf{R}_x + M(x)\mathbf{R} = 0, \quad t \in [0, +\infty), \quad x \in [0, L], \tag{A.7}$$

$$\begin{pmatrix} \mathbf{R}^+(t,0) \\ \mathbf{R}^-(t,L) \end{pmatrix} = \mathbf{K} \begin{pmatrix} \mathbf{R}^+(t,L) \\ \mathbf{R}^-(t,0) \end{pmatrix}, \quad t \in [0, +\infty), \tag{A.8}$$

$$\mathbf{R}(0,x) = \mathbf{R}_o(x), \quad x \in (0, L), \tag{A.9}$$

where $\mathbf{R}_o \in L^2((0,L); \mathbb{R}^n)$ is given.

We start this appendix by proving the following theorem on the well-posedness of the Cauchy problem (A.7), (A.8), (A.9) for the special case where the initial data $\mathbf{R}_o \in H^1((0,L); \mathbb{R}^n)$ and satisfy the compatibility condition

$$\begin{pmatrix} \mathbf{R}_o^+(0) \\ \mathbf{R}_o^-(L) \end{pmatrix} = \mathbf{K} \begin{pmatrix} \mathbf{R}_o^+(L) \\ \mathbf{R}_o^-(0) \end{pmatrix}. \tag{A.10}$$

We have the following theorem.

Theorem A.1. For every $\mathbf{R}_o \in H^1((0,L); \mathbb{R}^n)$ satisfying the compatibility condition (A.10), there exists one and only one

$$\mathbf{R} \in C^1([0, +\infty); L^2((0,L); \mathbb{R}^n)) \cap C^0([0, +\infty); H^1((0,L); \mathbb{R}^n))$$

which is a solution to the Cauchy problem (A.7), (A.8), (A.9). Moreover, there exists $C_0 > 0$ such that, for every $\mathbf{R}_o \in H^1((0,L); \mathbb{R}^n)$ satisfying the compatibility condition (A.10), this unique solution $\mathbf{R}$ satisfies

$$\|\mathbf{R}(t,.)\|_{H^1((0,L); \mathbb{R}^n)} + \|\mathbf{R}_t(t,.)\|_{L^2((0,L); \mathbb{R}^n)} \leqslant C_0 e^{C_0 t} \|\mathbf{R}_o\|_{H^1((0,L); \mathbb{R}^n)}, \quad \forall t \in [0, +\infty),$$

$$\tag{A.11}$$

$$\|\mathbf{R}(t,.)\|_{L^2((0,L); \mathbb{R}^n)} \leqslant C_0 e^{C_0 t} \|\mathbf{R}_o\|_{L^2((0,L); \mathbb{R}^n)}, \quad \forall t \in [0, +\infty). \tag{A.12}$$

Proof. We are going to deduce Theorem A.1 from the Lumer-Philipps theorem (see, for example, (Pazy 1983, Corollary 4.4, Chapter 1, page 15)). We could alternatively use the Hille-Yosida theorem; see, for example, (Brezis 1983, Theorem VII.4, page 105) or (Pazy 1983, Theorem 3.1, Chapter 1, page 8 and Theorem 1.3 page 102). Let $\mathcal{J} \triangleq L^2((0,L); \mathbb{R}^n)$. Let $\mu > 0$. The vector space $\mathcal{J}$ is equipped with the scalar product

$$((\varphi_1, \ldots, \varphi_n)^\mathsf{T}, (\psi_1, \ldots, \psi_n)^\mathsf{T})_\mu \triangleq \sum_{i=1}^{m} \int_0^L \varphi_i \psi_i e^{\mu(x-L)} dx + \sum_{i=m+1}^{n} \int_0^L \varphi_i \psi_i e^{-\mu x} dx,$$

(A.13)

for every $(\varphi_1, \ldots, \varphi_n)^\mathsf{T} \in \mathcal{J}$ and for every $(\psi_1, \ldots, \psi_n)^\mathsf{T} \in \mathcal{J}$. With this scalar product, $\mathcal{J}$ is an Hilbert space. Let $A : D(A) \to \mathcal{I}$ be the linear operator defined by

$$D(A) \triangleq \left\{ \varphi \in H^1((0, L); \mathbb{R}^n); \begin{pmatrix} \varphi^+(0) \\ \varphi^-(L) \end{pmatrix} = \mathbf{K} \begin{pmatrix} \varphi^+(L) \\ \varphi^-(0) \end{pmatrix} \right\} \subset \mathcal{J}. \qquad (A.14)$$

$$(A\varphi)(x) \triangleq -\Lambda(x)\varphi_x(x) - M(x)\varphi(x). \qquad (A.15)$$

Since the set $\varphi \in C^1([0, L]; \mathbb{R}^n)$ such that $\varphi(0) = \varphi(L) = 0$ is dense in $L^2((0, L); \mathbb{R}^n)$ and contained in $D(A)$,

$$D(A) \text{ is dense in } \mathcal{J}. \qquad (A.16)$$

We say that the linear operator A is quasi-dissipative if there exists $C > 0$ such that

$$(A\varphi, \varphi)_\mu \leqslant C(\varphi, \varphi)_\mu, \ \forall \varphi \in D(A).$$

Let us check that A is quasi-dissipative. For every $\varphi = (\varphi_1, \ldots, \varphi_n)^\mathsf{T} \in D(A)$, we have, using integrations by parts

$$(A\varphi, \varphi)_\mu = -\sum_{i=1}^{m} \int_0^L \lambda_i \varphi_{ix} \varphi_i e^{\mu(x-L)} dx$$

$$+ \sum_{i=m+1}^{n} \int_0^L \lambda_i \varphi_{ix} \varphi_i e^{-\mu x} dx + (M\varphi, \varphi)_\mu \qquad (A.17)$$

$$= \tfrac{1}{2}(B + I),$$

with

$$B \triangleq -\sum_{i=1}^{m} \left(\lambda_i(L)\varphi_i^2(L) - \lambda_i(0)\varphi_i^2(0)e^{-\mu L} \right)$$

$$+ \sum_{i=m+1}^{n} \left(\lambda_i(L)\varphi_i^2(L)e^{-\mu L} - \lambda_i(0)\varphi_i^2(0) \right) \qquad (A.18)$$

$$I \triangleq \sum_{i=1}^{m} \int_{0}^{L} \varphi_i^2 \left(\lambda_i e^{\mu(x-L)} \right)_x dx - \sum_{i=m+1}^{m} \varphi_i^2 \left(\lambda_i e^{-\mu x} \right)_x dx + 2(M\varphi, \varphi)_\mu. \qquad \text{(A.19)}$$

By (A.2) and (A.3), there exists $\delta > 0$ such that

$$\lambda_i(L) \geqslant 2\delta, \, \forall i \in \{1, \dots, m\} \text{ and } \lambda_i(0) \geqslant 2\delta, \, \forall i \in \{m+1, \dots, n\}. \qquad \text{(A.20)}$$

Using (A.14), (A.18), and (A.20), we see that if $\mu > 0$ is large enough,

$$B \leqslant -\delta \left(\sum_{i=1}^{m} \varphi_i^2(L) + \sum_{i=m+1}^{n} \varphi_i^2(0) \right), \, \forall \varphi \in D(A). \qquad \text{(A.21)}$$

From now on, we assume that $\mu > 0$ is large enough so that (A.21) holds. Then, using (A.3), (A.4), and (A.19), there exists $C > 0$ (depending on μ) such that

$$I \leqslant C(\varphi, \varphi)_\mu, \, \forall \varphi \in \mathcal{J}. \qquad \text{(A.22)}$$

From (A.17), (A.21), and (A.22) we get that

$$(A\varphi, \varphi)_\mu \leqslant C(\varphi, \varphi)_\mu \, \forall \varphi \in D(A), \qquad \text{(A.23)}$$

i.e., that A is quasi-dissipative.

Let us now prove that

$$A \text{ is a closed operator.} \qquad \text{(A.24)}$$

Let us recall that (A.24) simply means that the graph of A is a closed subset of $\mathcal{J} \times \mathcal{J}$. Let $(\varphi^k)_{k \in \mathbb{N}}$ be a sequence of elements in $D(A)$, let $\varphi \in \mathcal{J}$ and $\psi \in \mathcal{J}$ be such that

$$\varphi^k \to \varphi \text{ in } \mathcal{J} \text{ as } k \to +\infty, \qquad \text{(A.25)}$$

$$A\varphi^k \to \psi \text{ in } \mathcal{J} \text{ as } k \to +\infty. \qquad \text{(A.26)}$$

From (A.15), (A.25), and (A.26), we get that

$$\Lambda \varphi_x^k \to -\psi - M\varphi \text{ as } k \to +\infty. \qquad \text{(A.27)}$$

From (A.2), (A.3), (A.4), (A.25), and (A.27), we get that

$$\varphi^k \to \varphi \text{ in } H^1((0, L); \mathbb{R}^n) \text{ as } k \to +\infty. \qquad \text{(A.28)}$$

From (A.27) and (A.28), we have

$$\Lambda \varphi_x = -\psi - M\varphi. \tag{A.29}$$

Since φ^k is in $D(A)$ for every $k \in \mathbb{N}$ we have

$$\begin{pmatrix} \varphi^{k+}(0) \\ \varphi^{k-}(L) \end{pmatrix} = \mathbf{K} \begin{pmatrix} \varphi^{k+}(L) \\ \varphi^{k-}(0) \end{pmatrix}. \tag{A.30}$$

Since $H^1((0, L); \mathbb{R}^n)$ is continuously embedded in $C^0([0, L]; \mathbb{R}^n)$, it follows from (A.28) and (A.30) that

$$\begin{pmatrix} \varphi^+(0) \\ \varphi^-(L) \end{pmatrix} = \mathbf{K} \begin{pmatrix} \varphi^+(L) \\ \varphi^-(0) \end{pmatrix}. \tag{A.31}$$

Hence $\varphi \in D(A)$. Then, using (A.29), we get that $A\varphi = \psi$, which concludes the proof of (A.24).

Let us now show that adjoint A^* of the operator A is quasi-dissipative. Let us recall that A^* is the linear operator defined by

$$D(A^*) \triangleq \{\psi \in \mathcal{J};\ \exists C > 0 \text{ such that } |(A\varphi, \psi)_\mu| \leqslant C|\varphi|_\mu|,\ \forall \varphi \in D(A)\}, \tag{A.32}$$

$$(A\varphi, \psi)_\mu = (\varphi, A^*\psi)_\mu,\ \forall \varphi \in D(A),\ \forall \psi \in D(A^*). \tag{A.33}$$

The description of A^* is given in the following lemma.

Lemma A.2. Let $D : [0, L] \to \mathcal{M}_{n,n}(\mathbb{R})$ be defined by

$$D(x) \triangleq \operatorname{diag}(D_1(x), \ldots, D_n(x)),\ \forall x \in [0, L], \tag{A.34}$$

with

$$D_i(x) \triangleq e^{\mu(x-L)},\ \forall i \in \{1, \ldots, m\},\ \forall x \in [0, L], \tag{A.35}$$

$$D_i(x) \triangleq e^{-\mu x},\ \forall i \in \{m + 1, \ldots, n\},\ \forall x \in [0, L]. \tag{A.36}$$

Let $D^+ : [0, L] \to \mathcal{M}_{m,m}(\mathbb{R})$ and $D^- : [0, L] \to \mathcal{M}_{n-m,n-m}(\mathbb{R})$ be defined by

$$D^+(x) \triangleq \operatorname{diag}(D_1(x), \ldots, D_m(x)),\ \forall x \in [0, L], \tag{A.37}$$

$$D^-(x) \triangleq \operatorname{diag}(D_{m+1}(x), \ldots, D_n(x)),\ \forall x \in [0, L]. \tag{A.38}$$

Then $D(A^*)$ is the set of $\psi \in H^1((0,L); \mathbb{R}^n)$ such that

$$\psi^+(L) = \Lambda^+(L)^{-1} K_{00}^\mathsf{T} \Lambda^+(0) D^+(0) \psi^+(0) - \Lambda^+(L)^{-1} K_{10}^\mathsf{T} \Lambda^-(L) D^-(L) \psi^-(L),$$
(A.39)

$$\psi^-(0) = -\Lambda^-(0)^{-1} K_{01}^\mathsf{T} \Lambda^+(0) D^+(0) \psi^+(0) + \Lambda^-(0)^{-1} K_{11}^\mathsf{T} \Lambda^-(L) D^-(L) \psi^-(L).$$
(A.40)

and, for every $\psi \in D(A^*)$,

$$(A^* \psi)(x) = \Lambda(x) \psi_x(x) + \Lambda_x(x) \psi_x(x) + \widetilde{M}(x) \psi(x), \quad x \in (0,L),$$
(A.41)

where $\widetilde{M}$ is defined by

$$\widetilde{M}(x) = -D(x)^{-1} M(x)^\mathsf{T} D(x) + \mu \begin{pmatrix} \Lambda^+(x) & 0 \\ 0 & -\Lambda^-(x) \end{pmatrix}, \quad x \in (0,L).$$
(A.42)

Proof of Lemma A.2. Let us first point out that, for every $\varphi \in C_0^\infty((0,L); \mathbb{R}^n) \subset D(A)$ and for every $\psi \in L^2((0,L); \mathbb{R}^n)$, we have

$$(A\varphi, \psi)_\mu = \int_0^L (-\varphi_x^\mathsf{T} + \varphi^\mathsf{T} M^\mathsf{T}) D\psi \, dx$$

$$= \langle \psi, D(-\varphi_x + M\varphi) \rangle_{\mathcal{D}', \mathcal{D}}$$

$$= \langle \psi, -(D\varphi)_x + D'\varphi + DM\varphi \rangle_{\mathcal{D}', \mathcal{D}}$$

$$= \langle D\psi_x + D'\psi + M^\mathsf{T} D\psi, \varphi \rangle_{\mathcal{D}', \mathcal{D}}.$$

Hence, if $\psi \in D(A^*)$ then

$$D\psi_x + D'\psi + M^\mathsf{T} D\psi \in L^2((0,L); \mathbb{R}^n)$$

and therefore $\psi \in H^1((0,L); \mathbb{R}^n)$. Hence

$$D(A^*) \subset H^1((0,L); \mathbb{R}^n).$$
(A.43)

Moreover, for every $\varphi \in D(A)$ and for every $\psi \in H^1((0,L); \mathbb{R}^n)$, a simple integration by parts shows that

$$(A\varphi, \psi)_\mu = \int_0^L (-\varphi_x^\mathsf{T} \Lambda + \varphi^\mathsf{T} M^\mathsf{T}) D\psi \, dx$$

$$= -[\varphi^\mathsf{T} \Lambda D\psi]_0^L + \int_0^L \varphi^\mathsf{T} D((\Lambda D^{-1} D_x + D^{-1} M^\mathsf{T} D)\psi + \Lambda \psi_x) \, dx.$$
(A.44)

Note that

$$\Lambda D^{-1} D_x = \mu \begin{pmatrix} \Lambda^+(x) & 0 \\ 0 & -\Lambda^-(x) \end{pmatrix}, \ \forall x \in [0, L].$$

Hence, taking once more $\varphi \in C^\infty((0, L); \mathbb{R}^n)$ in (A.44), we get (A.41) for every $\psi \in D(A^*)$. Let us now study the boundary term in the right-hand side of (A.44). For every $\varphi \in D(A)$ and for every $\psi \in H^1((0, L); \mathbb{R}^n)$, we have

$$\begin{aligned}
[\varphi^\mathsf{T} \Lambda D \psi]_0^L &= \varphi^+(L)^\mathsf{T} \Lambda^+(L) D^+(L) \psi^+(L) + \varphi^-(L)^\mathsf{T} \Lambda^-(L) D^-(L) \psi^-(L) \\
&\quad - \varphi^+(0)^\mathsf{T} \Lambda^+(0) D^+(0) \psi^+(0) - \varphi^-(0)^\mathsf{T} \Lambda^-(0) D^-(0) \psi^-(0) \\
&= \varphi^+(L)^\mathsf{T} X + \varphi^-(0)^\mathsf{T} \mathbf{R},
\end{aligned}$$

(A.45)

with

$$X \triangleq \Lambda^+(L) D^+(L) \psi^+(L) + K_{10}^\mathsf{T} \Lambda^-(L) D^-(L) \psi^-(L) - K_{00}^\mathsf{T} \Lambda^+(0) D^+(0) \psi^+(0),$$

(A.46)

$$\mathbf{R} \triangleq K_{11}^\mathsf{T} \Lambda^-(L) D^-(L) \psi^-(L) - K_{01}^\mathsf{T} \Lambda^+(0) D^+(0) - \Lambda^-(0) D^-(0) \psi^-(0).$$

(A.47)

Let us now assume that $\psi \in D(A^*)$. Then there exists $\theta > 0$ such that

$$|\varphi^+(L)^\mathsf{T} X + \varphi^-(0)^\mathsf{T} \mathbf{R}| \le \theta |\varphi|_\mu, \ \forall \varphi \in D(A),$$

which clearly implies that $X = 0$ and $\mathbf{R} = 0$, that is (A.2) and (A.40) hold. Conversely, it follows from (A.44), (A.45), (A.46), and (A.47) that, if $\psi \in H^1((0, L); \mathbb{R}^n)$ satisfies (A.2) and (A.40), then $\psi \in D(A^*)$. This concludes the proof of Lemma A.2. □

We go back to the proof of Theorem A.1. Note that A^* has a similar form than A. In particular, A^*, as A, is densely defined and closed (as a matter of fact, the adjoint of a densely defined operator is always closed). Let us now look at the quasi-dissipativity of A^*. For every $\psi \in D(A^*)$, a simple integration by parts gives

$$\begin{aligned}
(A^* \psi, \psi)_\mu &= \int_0^L (\psi_x^\mathsf{T} \Lambda + \psi^\mathsf{T} \widetilde{M}^\mathsf{T}) D \psi \, dx \\
&= \frac{1}{2} [\psi^\mathsf{T} \Lambda D \psi]_0^L + \int_0^L \psi^\mathsf{T} \left(\widetilde{M}^\mathsf{T} D - \frac{1}{2} D_x \right) \psi \, dx.
\end{aligned}$$

(A.48)

We have

$$\begin{aligned}
[\psi^\mathsf{T} \Lambda D \psi]_0^L &= \psi^-(L)^\mathsf{T} \Lambda^-(L) D^-(L) \psi^-(L) - \psi^+(0) \Lambda^+(0) D^+(0) \psi^+(0) \\
&\quad \psi^+(L)^\mathsf{T} \Lambda^+(L) \psi^+(L) - \psi^-(0) \Lambda^-(0) \psi^-(0).
\end{aligned}$$

(A.49)

Clearly, there exists $\nu > 0$ such that

$$\psi^-(L)^\mathsf{T} \Lambda^-(L) D^-(L) \psi^-(L) - \psi^+(0) \Lambda^+(0) D^+(0) \psi^+(0) \leqslant$$
$$- \nu e^{-\mu L} \left(|\psi^-(L)|^2 + |\psi^+(0)|^2 \right), \; \forall \mu \in (0 + \infty), \; \forall \psi^-(L) \in \mathbb{R}^{n-m}, \; \forall \psi^+(0) \in \mathbb{R}^m.$$
$$(A.50)$$

However, from (A.2) and (A.40), there exists $C > 0$ such that

$$|\psi^+(L)| + |\psi^-(0)| \leqslant C e^{-\mu L} \left(|\psi^+(0)| + |\psi^-(L)| \right), \; \forall \psi \in D(A^*). \qquad (A.51)$$

From (A.49), (A.50), and (A.51), one sees that, if $\mu > 0$ is large enough, then

$$[\psi^\mathsf{T} \Lambda D \psi]_0^L \leqslant -\frac{\nu}{2} e^{-\mu L} \left(|\psi^-(L)|^2 + |\psi^+(0)|^2 \right), \; \forall \mu \in (0 + \infty), \; \forall \psi \in D(A^*).$$
$$(A.52)$$

From (A.48) and (A.52), we get that A^* is quasi-dissipative. Hence A and A^* are densely defined closed and quasi-dissipative. With the Lumer-Philipps theorem, this concludes the proof of Theorem A.1. $\qquad \qquad \square$

We now deal with weak solutions to the Cauchy problem (A.7), (A.8), (A.9). In order to motivate the definition of a (weak) solution to the Cauchy problem (A.7), (A.8), (A.9) with $\mathbf{R}_o$ only in $L^2((0, L); \mathbb{R}^n)$, let us multiply (A.7) on the left by φ^T where $\varphi : [0, T] \times [0, L] \to \mathbb{R}^n$, where $T > 0$ is given, and integrate the obtained equation on $(0, T) \times (0, L)$. Assuming $\mathbf{R}$ and φ to be of class C^1 in $[0, T] \times [0, L]$, we have using integrations by parts and (A.9),

$$\begin{aligned}
0 &= \int_0^L \int_0^T \varphi^\mathsf{T} \mathbf{R}_t \, dt dx + \int_0^T \int_0^L \varphi^\mathsf{T} \Lambda \mathbf{R}_x \, dx dt + \int_0^T \int_0^L \varphi^\mathsf{T} M \mathbf{R} \, dx dt \\
&= \int_0^L \varphi(T, x)^\mathsf{T} \mathbf{R}(T, x) dx - \int_0^L \varphi(0, x)^\mathsf{T} \mathbf{R}_o(x)^\mathsf{T} dx \\
&\quad + \int_0^T \varphi^\mathsf{T}(t, L) \Lambda(t, L) \mathbf{R}(t, L) dt - \int_0^T \varphi^\mathsf{T}(t, 0) \Lambda(t, 0) \mathbf{R}(t, 0) dt \\
&\quad - \int_0^T \int_0^L (\varphi_t^\mathsf{T} + \varphi_x^\mathsf{T} \Lambda + \varphi^\mathsf{T}(\Lambda_x - M)) \mathbf{R} \, dx dt.
\end{aligned} \qquad (A.53)$$

From (A.6), (A.8), and (A.53), we get

$$\begin{aligned}
0 &= \int_0^L \varphi(T, x)^\mathsf{T} \mathbf{R}(T, x) dx - \int_0^L \varphi(0, x)^\mathsf{T} \mathbf{R}_o(x) dx \\
&\quad + \int_0^T (\varphi^+(t, L)^\mathsf{T} \Lambda^+(L) - \varphi^-(t, L)^\mathsf{T} \Lambda^-(L) K_{10} \\
&\quad - \varphi^+(t, 0)^\mathsf{T} \Lambda^+(0) K_{00}) \mathbf{R}^+(t, L) dt
\end{aligned}$$

$$-\int_0^T (\varphi^+(t,0)^\mathsf{T}\Lambda^+(0)K_{01} + \varphi^-(t,L)^\mathsf{T}\Lambda^-(L)K_{11}$$

$$-\varphi^-(t,0)^\mathsf{T}\Lambda^-(0))\mathbf{R}^-(t,0)dt$$

$$-\int_0^T \int_0^L (\varphi_t^\mathsf{T} + \varphi_x^\mathsf{T}\Lambda + \varphi^\mathsf{T}(\Lambda_x - M))\mathbf{R}dxdt. \tag{A.54}$$

In particular, if

$$\begin{pmatrix} \varphi^+(t,L) \\ \varphi^-(t,0) \end{pmatrix} = \begin{pmatrix} \Lambda^+(L)^{-1}K_{00}^\mathsf{T}\Lambda^+(0) & \Lambda^+(L)^{-1}K_{10}^\mathsf{T}\Lambda^-(L) \\ \Lambda^-(0)^{-1}K_{01}^\mathsf{T}\Lambda^+(0) & \Lambda^-(0)^{-1}K_{11}^\mathsf{T}\Lambda^-(L) \end{pmatrix} \begin{pmatrix} \varphi^+(t,0) \\ \varphi^-(t,L) \end{pmatrix}, \quad \forall t \in [0,T], \tag{A.55}$$

we get

$$\int_0^L \varphi(T,x)^\mathsf{T}\mathbf{R}(T,x)dx - \int_0^L \varphi(0,x)^\mathsf{T}\mathbf{R}_0(x) =$$

$$\int_0^T \int_0^L (\varphi_t^\mathsf{T} + \varphi_x^\mathsf{T}\Lambda + \varphi^\mathsf{T}(\Lambda_x - M))\mathbf{R}dxdt. \tag{A.56}$$

This leads to the following definition.

Definition A.3. A L^2-solution $\mathbf{R} : (0,+\infty) \times (0,L) \to \mathbb{R}^n$ of the Cauchy problem (A.7), (A.8), (A.9) is a map $\mathbf{R} \in C^0([0,+\infty); L^2((0,L); \mathbb{R}^n))$ satisfying (A.9) such that, for every $T > 0$ and every $\varphi \in C^1([0,T] \times [0,L]; \mathbb{R}^n)$ satisfying (A.55), we have (A.56).

With this definition, we have the following theorem.

Theorem A.4. For every $\mathbf{R}_0 \in L^2((0,L); \mathbb{R}^n)$, the Cauchy problem (A.7), (A.8), (A.9) has one and only one L^2-solution. Moreover, there exists $C_1 > 0$ such that, for every $\mathbf{R}_0 \in L^2((0,L); \mathbb{R}^n)$, the solution $\mathbf{R}$ to the Cauchy problem (A.7), (A.8), (A.9) satisfies

$$\|\mathbf{R}(t)\|_{L^2((0,L);\mathbb{R}^n)} \leq C_1 e^{C_1 t} \|\mathbf{R}_0\|_{L^2((0,L);\mathbb{R}^n)}, \quad \forall t \in [0,+\infty). \tag{A.57}$$

Proof. Let us first prove the existence part (together with the estimate (A.57)). Let $\mathbf{R}_0 \in L^2((0,L); \mathbb{R}^n)$. Since $D(A)$ is dense in $\mathcal{J}$ there exists a sequence $(\mathbf{R}_0^k)_{k\in\mathbb{N}}$ of elements in $D(A)$ such that

$$\mathbf{R}_0^k \to \mathbf{R}_0 \text{ in } L^2((0,L); \mathbb{R}^n) \text{ as } k \to +\infty. \tag{A.58}$$

By Theorem A.1, there exists a sequence $(\mathbf{R}^k)_{k\in\mathbb{N}}$ such that, for every $k \in \mathbb{N}$,

$$\mathbf{R}^k \in C^1([0, +\infty); L^2((0, L); \mathbb{R}^n)) \cap C^0([0, +\infty); H^1((0, L); \mathbb{R}^n)), \tag{A.59}$$

$$\mathbf{R}^k_t + \Lambda\mathbf{R}^k_x + M\mathbf{R}^k = 0, \ t \in [0, +\infty), \ x \in (0, L), \tag{A.60}$$

$$\begin{pmatrix} \mathbf{R}^{k+}(t, 0) \\ \mathbf{R}^{k-}(t, L) \end{pmatrix} = \mathbf{K} \begin{pmatrix} \mathbf{R}^{k+}(t, L) \\ \mathbf{R}^{k-}(t, 0) \end{pmatrix}, \ t \in [0, +\infty), \tag{A.61}$$

$$\mathbf{R}^k(0, x) = \mathbf{R}^k_0(x), \ \forall x \in (0, L), \tag{A.62}$$

$$\|\mathbf{R}^k(t, .)\|_{L^2((0,L);\mathbb{R}^n)} \leqslant C_0 e^{C_0 t}\|\mathbf{R}^k_0\|_{L^2((0,L);\mathbb{R}^n)}, \ \forall t \in [0, +\infty). \tag{A.63}$$

Still by Theorem A.1,

$$\|\mathbf{R}^k(t, .) - \mathbf{R}^l(t, .)\|_{L^2((0,L);\mathbb{R}^n)} \leqslant C_0 e^{C_0 t}\|\mathbf{R}^k_0 - \mathbf{R}^l_0\|_{L^2((0,L);\mathbb{R}^n)},$$

$$\forall t \in [0, +\infty), \ \forall k \in \mathbb{N}, \ \forall l \in \mathbb{N}.$$

which, together with (A.58) shows that $\mathbf{R}^k e^{-C_0 t}$ is a Cauchy sequence in $C^0([0, +\infty); L^2((0, L); \mathbb{R}^n)) \cap L^\infty((0, +\infty); L^2((0, L); \mathbb{R}^n))$. Hence, there exists $\mathbf{R} \in C^0([0, +\infty); L^2((0, L); \mathbb{R}^n))$ such that

$$\|\mathbf{R}(t, .)\|_{L^2((0,L);\mathbb{R}^n)} \leqslant C_0 e^{C_0 t}\|\mathbf{R}^0\|_{L^2((0,L);\mathbb{R}^n)}, \ \forall t \in [0, +\infty),$$

$$\mathbf{R}(0, x) = \mathbf{R}^0(x), \ x \in (0, L),$$

$$\sup\{e^{-C_0 t}\|\mathbf{R}^k(t, .) - \mathbf{R}(t, .)\|_{L^2((0,L);\mathbb{R}^n)}; \ t \in [0, +\infty)\} \to 0 \text{ as } k \to +\infty. \tag{A.64}$$

Let $T > 0$ and let $\varphi \in C^1([0, T] \times [0, L]; \mathbb{R}^n)$ satisfying (A.55). Using (A.59), (A.60), (A.61) and the computations we have done to justify Definition A.3, we get that

$$\int_0^L \varphi(T, x)^\top \mathbf{R}^k(T, x)dx - \int_0^L \varphi(0, x)^\top \mathbf{R}^k_0(x)$$

$$= -\int_0^T \int_0^L (\varphi_t^\top + \varphi_x^\top \Lambda + \varphi^\top \Lambda_x)\mathbf{R}^k dx dt. \tag{A.65}$$

Letting $k \to +\infty$ in (A.65) and using (A.64), we get (A.56). This shows the existence part of Theorem A.4 with $C_1 \overset{\Delta}{=} C_0$.

Let us now prove the uniqueness. By linearity it suffices to study the case where $\mathbf{R}^0 = \mathbf{0}$. Hence we assume that $\mathbf{R} \in C^0([0, +\infty); L^2((0, L); \mathbb{R}^n))$ is a solution of the Cauchy problem (A.7), (A.8), (A.9) with $\mathbf{R}_o = \mathbf{0}$ and we want to prove that $\mathbf{R} = \mathbf{0}$.

Let $T > 0$. Let us first point out that a simple density argument shows that (A.56) holds for every φ satisfying (A.55) and

$$\varphi \in C^1([0, T]; L^2((0, L); \mathbb{R}^n)) \cap C^0([0, T]; H^1((0, L); \mathbb{R}^n)). \tag{A.66}$$

Let $\varphi_T \in H^1((0, L); \mathbb{R}^n)$ be such that

$$\begin{pmatrix} \varphi_T^+(L) \\ \varphi_T^-(0) \end{pmatrix} = \begin{pmatrix} K_{00}^\mathsf{T} \Lambda^+(0) & -K_{10}^\mathsf{T} \Lambda^-(L) \\ -K_{01}^\mathsf{T} \Lambda^+(0) & K_{11}^\mathsf{T} \Lambda^-(L) \end{pmatrix} \begin{pmatrix} \varphi_T^+(0) \\ \varphi_T^-(L) \end{pmatrix}. \tag{A.67}$$

Let us assume for the moment the existence of $\varphi : [0, T] \times [0, L] \to \mathbb{R}$ satisfying (A.66) and

$$\varphi_t + \Lambda \varphi_x + (\Lambda_x - M)\varphi = 0, \tag{A.68}$$

$$\varphi(T, x) = \varphi_T(x), \quad \forall t \in [0, T] \tag{A.69}$$

and such that (A.56) holds. From (A.56), (A.66), (A.68), and (A.69), we get

$$\int_0^L \varphi_T(x)^\mathsf{T} \mathbf{R}(T, x) dx = 0. \tag{A.70}$$

Since the set of $\varphi_T \in H^1((0, L); \mathbb{R}^n)$ satisfying (A.67) is dense in $L^2((0, L); \mathbb{R}^n)$, it follows from (A.70) that $\mathbf{R}(T) = 0$.

It remains only to prove the existence of φ. This existence can be obtained by doing the change of variables $\varphi(t, x) = \widetilde{\varphi}(T - t, L - x)$ and then apply Theorem A.1 (with suitable m, Λ and $\mathbf{K}$). This concludes the proof of Theorem A.4. $\qquad \square$

In the last part of this appendix we briefly explain how the previous results can be extended to the case of differential boundary conditions of the form

$$\begin{pmatrix} \mathbf{R}^+(t, 0) \\ \mathbf{R}^-(t, L) \end{pmatrix} = \mathbf{K} \begin{pmatrix} \mathbf{R}^+(t, L) \\ \mathbf{R}^-(t, 0) \end{pmatrix} + \begin{pmatrix} N^+ \\ N^- \end{pmatrix} \mathbf{X} \tag{A.71}$$

with

$$\frac{d\mathbf{X}}{dt} = E^+ \mathbf{R}^+(t, L) + E^- \mathbf{R}^-(t, 0) + E_0 \mathbf{X}, \quad \mathbf{X} \in \mathbb{R}^p, \tag{A.72}$$

and

$$N^+ \in \mathbb{R}^{m \times p}, \ N^- \in \mathbb{R}^{(n-m) \times p}, \ E^+ \in \mathbb{R}^{p \times m}, \ E^- \in \mathbb{R}^{p \times (n-m)}, \ E_0 \in \mathbb{R}^{p \times p}.$$

The initial condition (A.9) is extended as follows:

$$\mathbf{R}(0, x) = \mathbf{R}_0(x), \quad \mathbf{X}(0) = \mathbf{X}_0. \tag{A.73}$$

Again to support the definition of the weak solutions of the Cauchy problem (A.7), (A.71), (A.72), (A.73), equation (A.7) is left-multiplied by $\varphi^\mathsf{T}(t, x)$ and integrated on $[0, T] \times [0, L]$ while equation (A.72) is left-multiplied by $\eta^\mathsf{T}(t)$ and integrated on $[0, T]$. The two resulting equalities are summed. Then using the new boundary conditions (A.71), equation (A.55) becomes

$$\begin{pmatrix} \varphi^+(t, L) \\ \varphi^-(t, 0) \end{pmatrix} = \begin{pmatrix} \Lambda^+(L)^{-1} K_{00}^\mathsf{T} \Lambda^+(0) & \Lambda^+(L)^{-1} K_{10}^\mathsf{T} \Lambda^-(L) \\ \Lambda^-(0)^{-1} K_{01}^\mathsf{T} \Lambda^+(0) & \Lambda^-(0)^{-1} K_{11}^\mathsf{T} \Lambda^-(L) \end{pmatrix} \begin{pmatrix} \varphi^+(t, 0) \\ \varphi^-(t, L) \end{pmatrix}$$

$$+ \begin{pmatrix} \Lambda^+(L) E^{+\mathsf{T}} \\ \Lambda^-(0) E^{-\mathsf{T}} \end{pmatrix} \eta \tag{A.74}$$

and (A.56) becomes

$$\int_0^L \varphi(T, x)^\mathsf{T} \mathbf{R}(T, x) dx - \int_0^L \varphi(0, x)^\mathsf{T} \mathbf{R}_0(x) + \eta^\mathsf{T}(T)\mathbf{X}(T) - \eta^\mathsf{T}(0)\mathbf{X}(0)$$

$$= \int_0^T \int_0^L (\varphi_t^\mathsf{T} + \varphi_x^\mathsf{T} \Lambda + \varphi^\mathsf{T}(\Lambda_x - M))\mathbf{R} dx dt$$

$$+ \int_0^T \left(\eta_t^\mathsf{T} + \eta^\mathsf{T} E_0 + \varphi^{-\mathsf{T}}(t, L)\Lambda^-(L)N^- + \varphi^{+\mathsf{T}}(t, 0)\Lambda^+(0)N^+ \right) \mathbf{X} dt. \tag{A.75}$$

This leads to the following definition.

Definition A.5. A solution $\mathbf{R} : (0, +\infty) \times (0, L) \to \mathbb{R}^n, \mathbf{X} : (0, \infty) \to \mathbb{R}^p$ of the Cauchy problem (A.7), (A.71), (A.72), (A.73) are maps $\mathbf{R} \in C^0([0, +\infty); L^2((0, L); \mathbb{R}^n))$, $\mathbf{X} \in C^0([0, +\infty); \mathbb{R}^p)$ satisfying (A.73) such that, for every $T > 0$, every $\varphi \in C^1([0, T] \times [0, L]; \mathbb{R}^n)$, and every $\eta \in C^1([0, T]; \mathbb{R}^p)$ satisfying (A.74), we have (A.75).

With this definition, proceeding as for the proof of Theorem A.4, we have the following theorem.

Theorem A.6. For every $\mathbf{R}_0 \in L^2((0, L); \mathbb{R}^n)$, for every $\mathbf{X}_0 \in \mathbb{R}^p$, the Cauchy problem (A.7), (A.71), (A.72), (A.73) has one and only one solution. Moreover, there exists $C_2 > 0$ such that, for every $\mathbf{R}_0 \in L^2((0, L); \mathbb{R}^n)$, for every $\mathbf{X}_0 \in \mathbb{R}^p$ the unique solution $\mathbf{R}, \mathbf{X}$ to the Cauchy problem (A.7), (A.71), (A.72), (A.73) satisfies

$$\|\mathbf{R}(t, .)\|_{L^2((0,L);\mathbb{R}^n)} + \|\mathbf{X}(t)\| \leq C_2 e^{C_2 t}(\|\mathbf{R}_0\|_{L^2((0,L);\mathbb{R}^n)} + \|\mathbf{X}_0\|), \ \forall t \in [0, +\infty).$$

$\square$

Appendix B
Well-Posedness of the Cauchy Problem
for Quasi-Linear Hyperbolic Systems

In this appendix we study a Cauchy problem associated with the quasi-linear hyperbolic system

$$\mathbf{Z}_t + A(\mathbf{Z}, x)\mathbf{Z}_x + B(\mathbf{Z}, x) = \mathbf{0}, \quad x \in [0, L], \quad t \in [0, +\infty),$$

under the boundary condition

$$\begin{pmatrix} \mathbf{Z}^+(t, 0) \\ \mathbf{Z}^-(t, L) \end{pmatrix} = \mathcal{H}\begin{pmatrix} \mathbf{Z}^+(t, L) \\ \mathbf{Z}^-(t, 0) \end{pmatrix}, \quad t \in [0, +\infty),$$

where

- $\mathbf{Z} : [0, +\infty) \times [0, L] \to \mathbb{R}^n$;
- $A : \mathbb{R}^n \times [0, L] \to \mathcal{M}_{n,n}(\mathbb{R})$ is of class C^2 and such that

$$A(\mathbf{0}, x) = \mathrm{diag}\{\lambda_1(x), \ldots, \lambda_m(x), -\lambda_{m+1}(x), \ldots, -\lambda_n(x)\}, \quad \lambda_i(x) > 0, \ \forall x \in [0, L];$$

- $B : \mathbb{R}^n \times [0, L] \to \mathbb{R}^n$ is of class C^2 and such that $B(\mathbf{0}, x) = \mathbf{0}$.
- $\mathcal{H} : \mathbb{R}^n \to \mathbb{R}^n$ is of class C^2 and such that $\mathcal{H}(\mathbf{0}) = \mathbf{0}$.

We consider the following Cauchy problem

$$\mathbf{Z}_t + A(\mathbf{Z}, x)\mathbf{Z}_x + B(\mathbf{Z}, x) = \mathbf{0}, \quad x \in [0, L], \quad t \in [0, +\infty), \tag{B.1}$$

$$\begin{pmatrix} \mathbf{Z}^+(t, 0) \\ \mathbf{Z}^-(t, L) \end{pmatrix} = \mathcal{H}\begin{pmatrix} \mathbf{Z}^+(t, L) \\ \mathbf{Z}^-(t, 0) \end{pmatrix}, \quad t \in [0, +\infty), \tag{B.2}$$

$$\mathbf{Z}(0, x) = \mathbf{Z}_\mathrm{o}(x), \quad x \in (0, L). \tag{B.3}$$

© Springer International Publishing Switzerland 2016
G. Bastin, J.-M. Coron, *Stability and Boundary Stabilization of 1-D Hyperbolic Systems*, Progress in Nonlinear Differential Equations and Their Applications 88, DOI 10.1007/978-3-319-32062-5

where $\mathbf{Z}_o \in L^2((0, L); \mathbb{R}^n)$ is given and satisfies the following compatibility conditions of order 1:

$$\begin{pmatrix} \mathbf{Z}_o^+(0) \\ \mathbf{Z}_o^-(L) \end{pmatrix} = \mathcal{H} \begin{pmatrix} \mathbf{Z}_o^+(L) \\ \mathbf{Z}_o^-(0) \end{pmatrix}, \tag{B.4}$$

$$A^+(\mathbf{Z}_o(0))\partial_x \mathbf{Z}_o(0) + B^+(\mathbf{Z}_o(0)) =$$
$$\left[\frac{\partial \mathcal{H}^+}{\partial \mathbf{Z}^+} \begin{pmatrix} \mathbf{Z}_o^+(L) \\ \mathbf{Z}_o^-(0) \end{pmatrix} \right] \left(A^+(\mathbf{Z}_o(L))\partial_x \mathbf{Z}_o(L) + B^+(\mathbf{Z}_o(L)) \right)$$
$$+ \left[\frac{\partial \mathcal{H}^+}{\partial \mathbf{Z}^-} \begin{pmatrix} \mathbf{Z}_o^+(L) \\ \mathbf{Z}_o^-(0) \end{pmatrix} \right] \left(A^-(\mathbf{Z}_o(0))\partial_x \mathbf{Z}_o(0) + B^-(\mathbf{Z}_o(0)) \right), \tag{B.5}$$

$$A^-(\mathbf{Z}_o(L))\partial_x \mathbf{Z}_o(L) + B^-(\mathbf{Z}_o(L)) =$$
$$\left[\frac{\partial \mathcal{H}^-}{\partial \mathbf{Z}^+} \begin{pmatrix} \mathbf{Z}_o^+(L) \\ \mathbf{Z}_o^-(0) \end{pmatrix} \right] \left(A^+(\mathbf{Z}_o(L))\partial_x \mathbf{Z}_o(L) + B^+(\mathbf{Z}_o(L)) \right)$$
$$+ \left[\frac{\partial \mathcal{H}^-}{\partial \mathbf{Z}^-} \begin{pmatrix} \mathbf{Z}_o^+(L) \\ \mathbf{Z}_o^-(0) \end{pmatrix} \right] \left(A^-(\mathbf{Z}_o(0))\partial_x \mathbf{Z}_o(0) + B^-(\mathbf{Z}_o(0)) \right) \tag{B.6}$$

where $A^+ \in \mathcal{M}_{m,n}(\mathbb{R})$, $A^- \in \mathcal{M}_{n-m,n}(\mathbb{R})$, $B^+ \in \mathbb{R}^m$, $B^- \in \mathbb{R}^{n-m}$, $\mathcal{H}^+ \in \mathbb{R}^m$, $\mathcal{H}^- \in \mathbb{R}^{n-m}$ are defined such that

$$A \triangleq \begin{pmatrix} A^+ \\ A^- \end{pmatrix}, \quad B \triangleq \begin{pmatrix} B^+ \\ B^- \end{pmatrix}, \quad \mathcal{H} \triangleq \begin{pmatrix} \mathcal{H}^+ \\ \mathcal{H}^- \end{pmatrix}.$$

Theorem B.1. There exists $\delta_0 > 0$ such that, for every $\mathbf{Z}_o \in H^2((0, L), \mathbb{R}^n)$ satisfying

$$\|\mathbf{Z}_o\|_{H^2((0,L);\mathbb{R}^n)} \leqslant \delta_0$$

and the compatibility conditions (B.4), (B.5), and (B.6), the Cauchy problem (B.1), (B.2), and (B.3) has a unique maximal classical solution

$$\mathbf{Z} \in C^0([0, T); H^2((0, L); \mathbb{R}^n))$$

with $T \in (0, +\infty]$. Moreover, if

$$\|\mathbf{Z}(t, \cdot)\|_{H^2((0,L);\mathbb{R}^n)} \leq \delta_0, \ \forall t \in [0, T),$$

then $T = +\infty$.   □

Proof. For a proof of this theorem, see for instance Kato (1975), Lax (1973), (Majda 1984, pp. 35–43), or (Serre 1999, Chapter 3). Actually these references deal with systems on the real line $\mathbb{R}$ instead of the finite interval of $[0, L]$, but the proofs given there can be adapted. Let us briefly explain how the adaptation can be done in order to get, for example, the existence of a solution $\mathbf{Z} \in C^0([0, T), H^2((0, L); \mathbb{R}^n))$ in the special case where $m = n$ (just to simplify the notations), for $T \in (0, +\infty)$ given, and for every $\mathbf{Z}_o \in H^2((0, L); \mathbb{R}^n)$ satisfying the compatibility conditions (B.4), (B.5), and (B.6) (when $m = n$, the third compatibility condition disappears) and such that $|\mathbf{Z}_o|_{H^2((0,L);\mathbb{R}^n)}$ is small enough (the smallness depending on T in general). We first deal with the case where

$$T \in (0, \min\{\lambda_1^{-1}, \ldots, \lambda_n^{-1}\}).$$

The basic ingredient is the following fixed point method, which is related to the one given in (Li and Yu 1985, page 97) (see also the pioneering works of Kato (1975) and Lax (1973), where the authors deal with $\mathbb{R}$ instead of $[0, L]$). For $\mathcal{N} > 0$ and for $\mathbf{Z}_o \in H^2((0, L); \mathbb{R}^n)$ satisfying the compatibility conditions (B.4), (B.5), and (B.6), let $C_{\mathcal{N}}(\mathbf{Z}_o)$ be the set of

$$\mathbf{Z} \in L^\infty((0, T); H^2((0, L); \mathbb{R}^n)) \cap W^{1,\infty}((0, T); H^1((0, L); \mathbb{R}^n))$$

$$\cap W^{2,\infty}((0, T); L^2((0, L); \mathbb{R}^n))$$

such that

$$|\mathbf{Z}|_{L^\infty((0,T);H^2((0,L);\mathbb{R}^n))} \leq \mathcal{N},$$

$$|\mathbf{Z}|_{W^{1,\infty}((0,T);H^1((0,L);\mathbb{R}^n))} \leq \mathcal{N},$$

$$|\mathbf{Z}|_{W^{2,\infty}((0,T);L^2((0,L);\mathbb{R}^n))} \leq \mathcal{N},$$

$$\mathbf{Z}(\cdot, L) \in H^2((0, T); \mathbb{R}^n) \text{ and } |\mathbf{Z}(\cdot, L)|_{H^2((0,T);\mathbb{R}^n)} \leq \mathcal{N}^2,$$

$$\mathbf{Z}(0, \cdot) = \mathbf{Z}_o,$$

$$\mathbf{Z}_t(0, \cdot) = -A(\mathbf{Z}_o, x)\partial_x\mathbf{Z}_o - B(\mathbf{Z}_o, x).$$

The set $C_N(\mathbf{Z}_0)$ is a closed subset of $L^\infty((0, T); L^2((0, L); \mathbb{R}^n))$ (at least if $|\mathbf{Z}_0|_{H^2((0,L);\mathbb{R}^n)}$ is small enough so that $|\mathbf{Z}_0|_{C^0([0,L];\mathbb{R}^n)} < \varepsilon_0$). Moreover, since $N > 0$, the set $C_N(\mathbf{Z}_0)$ is not empty if $|\mathbf{Z}_0|_{H^2((0,L);\mathbb{R}^n)}$ is small enough.

Let $\mathcal{F} : C_N(\mathbf{Z}_0) \to L^\infty((0, T); H^2((0, L); \mathbb{R}^n)) \cap W^{1,\infty}((0, T); H^1((0, L); \mathbb{R}^n)) \cap W^{2,\infty}((0, T); L^2((0, L); \mathbb{R}^n))$ be defined by $\mathcal{F}(\widetilde{\mathbf{Z}}) = \mathbf{Z}$, where $\mathbf{Z}$ is the solution of the linear hyperbolic Cauchy problem

$$\mathbf{Z}_t + A(\widetilde{\mathbf{Z}}, x)\mathbf{Z}_x + B(\widetilde{\mathbf{Z}}, x) = \mathbf{0}, \quad \mathbf{Z}(t, 0) = \mathcal{H}(\widetilde{\mathbf{Z}}(t, L)), \quad t \in [0, T], \qquad (B.7)$$

$$\mathbf{Z}(0, x) = \mathbf{Z}_0(x), \qquad x \in [0, L].$$

Using standard energy estimates and the finite speed of propagation inherent in (B.7), one gets the existence of $\mathcal{M} > 0$ and $N_0 > 0$ such that, for every $N \in (0, N_0]$, there exists $\delta > 0$ such that, for every $\mathbf{Z}_0 \in H^2((0, L); \mathbb{R}^n)$ such that $|\mathbf{Z}_0|_{H^2((0,L);\mathbb{R}^n)} \leq \delta$ and satisfying the compatibility conditions (B.4), (B.5), and (B.6),

$$\mathcal{F}(C_N(\mathbf{Z}_0)) \subset C_N(\mathbf{Z}_0) \qquad (B.8)$$

and

$$|\mathcal{F}(\widetilde{\mathbf{Z}}_2) - \mathcal{F}(\widetilde{\mathbf{Z}}_1)|_{L^\infty((0,T);L^2((0,L);\mathbb{R}^n))} + \mathcal{M}|\mathcal{F}(\widetilde{\mathbf{Z}}_2)(\cdot, L) - \mathcal{F}(\widetilde{\mathbf{Z}}_1)(\cdot, L)|_{L^2((0,L);\mathbb{R}^n)}$$

$$\leq \frac{1}{2}|\widetilde{\mathbf{Z}}_2 - \widetilde{\mathbf{Z}}_1|_{L^\infty((0,T);L^2((0,L);\mathbb{R}^n))} + \frac{\mathcal{M}}{2}|\widetilde{\mathbf{Z}}_2(\cdot, L) - \widetilde{\mathbf{Z}}_1(\cdot, L)|_{L^2((0,T);\mathbb{R}^n)}$$

$$\forall (\widetilde{\mathbf{Z}}_1, \widetilde{\mathbf{Z}}_2) \in C_N(\mathbf{Z}_0).$$

This allows us to prove that $\mathcal{F}$ has a fixed point $\mathbf{Z} \in C_N(\mathbf{Z}_0)$; i.e., there exists a solution $\mathbf{Z} \in C_N(\mathbf{Z}_0)$ to the Cauchy problem (B.1), (B.2), and (B.3). In order to get the extra regularity property $\mathbf{Z} \in C^0([0, T], H^2((0, L); \mathbb{R}^n))$, we can then adapt (Majda 1984, pp. 44–46) by observing that, when one uses usual energy estimates to get (B.8), one also gets, for $\mathbf{Z} \triangleq \mathcal{F}(\widetilde{\mathbf{Z}})$ with $\widetilde{\mathbf{Z}} \in C_N(\mathbf{Z}_0)$, the "hidden regularity" $\mathbf{Z}_{xx}(\cdot, L) \in L^2((0, T); \mathbb{R}^n)$ together with estimates on $|\mathbf{Z}_{xx}(\cdot, L)|_{L^2((0,T);\mathbb{R}^n)}$ which are sufficient to take care of the boundary terms which now appear when one does integrations by parts. The case of general $T \in (0+\infty)$ follows by applying the above result to $[0, T_1], [T_1, 2T_1], [2T_1, 3T_1], \ldots$, with T_1 given in $(0, \min\{\lambda_1^{-1}, \ldots, \lambda_n^{-1}\})$. This concludes our sketch of the proof of Proposition B.1. $\square$

Corollary B.2. The classical solution of the Cauchy problem (B.1), (B.2), and (B.3) $\mathbf{Z} \in C^1([0, T) \times [0, L]; \mathbb{R}^n))$.

Proof. From Theorem B.1 we have that $\mathbf{Z} \in C^0([0, T); H^2((0, L); \mathbb{R}^n))$ which implies that $\mathbf{Z}_x \in C^0([0, T); H^1((0, L); \mathbb{R}^n))$. Now

$$C^0([0, T); H^1((0, L); \mathbb{R}^n)) \subseteq C^0([0, T); C^0([0, L]; \mathbb{R}^n)) = C^0([0, T) \times [0, L]; \mathbb{R}^n).$$

Hence $\mathbf{Z}_x \in C^0([0, T) \times [0, L]; \mathbb{R}^n)$. Since $\mathbf{Z}_t = -A(\mathbf{Z}, x)\mathbf{Z}_x - B(\mathbf{Z}, x)$ it follows also that $\mathbf{Z}_t \in C^0([0, T) \times [0, L]; \mathbb{R}^n)$ and therefore that $\mathbf{Z} \in C^1([0, T) \times [0, L]; \mathbb{R}^n)$. □

Appendix C
Properties and Comparisons of the Functions $\bar{\rho}$, ρ_2 and ρ_∞

C.1 Properties of the Function ρ_2

In this appendix, taken from Coron et al. (2008), we first give some properties which are useful to estimate or compute $\rho_2(\mathbf{K})$. Some of these properties are used to prove Theorem 3.12 in the next section.

Proposition C.1. Let $l \in \{1, \ldots, n-1\}$. Let $K_1 \in \mathcal{M}_{l,l}(\mathbb{R})$, $K_2 \in \mathcal{M}_{l,n-l}(\mathbb{R})$, $K_3 \in \mathcal{M}_{n-l,l}(\mathbb{R})$, $K_4 \in \mathcal{M}_{n-l,n-l}(\mathbb{R})$ and let $\mathbf{K} \in \mathcal{M}_{n,n}(\mathbb{R})$ be defined by

$$\mathbf{K} \triangleq \begin{pmatrix} K_1 & K_2 \\ K_3 & K_4 \end{pmatrix}.$$

Then

$$\rho_2(\mathbf{K}) \geqslant \max\{\rho_2(K_1), \rho_2(K_4)\}. \tag{C.1}$$

Moreover, if $K_2 = 0$ or $K_3 = 0$, then

$$\rho_2(\mathbf{K}) = \max\{\rho_2(K_1), \rho_2(K_4)\}. \tag{C.2}$$

Proof. Let $D \in \mathcal{D}_n^+$. Let $D_1 \in \mathcal{D}_l^+$ and $D_2 \in \mathcal{D}_{n-l}^+$ be such that

$$D = \begin{pmatrix} D_1 & 0 \\ 0 & D_2 \end{pmatrix}.$$

Let

$$M \triangleq DKD^{-1}.$$

© Springer International Publishing Switzerland 2016
G. Bastin, J.-M. Coron, *Stability and Boundary Stabilization of 1-D Hyperbolic Systems*, Progress in Nonlinear Differential Equations and Their Applications 88, DOI 10.1007/978-3-319-32062-5

We have

$$M^\mathsf{T} M = \begin{pmatrix} M_{11} & M_{12} \\ M_{21} & M_{22} \end{pmatrix}$$

with

$$M_{11} \triangleq D_1^{-1} K_1^\mathsf{T} D_1^2 K_1 D_1^{-1} + D_1^{-1} K_3^\mathsf{T} D_2^2 K_3 D_1^{-1},$$

$$M_{12} \triangleq D_1^{-1} K_1^\mathsf{T} D_1^2 K_2 D_2^{-1} + D_1^{-1} K_3^\mathsf{T} D_2^2 K_4 D_2^{-1},$$

$$M_{21} \triangleq D_2^{-1} K_2^\mathsf{T} D_1^2 K_1 D_1^{-1} + D_2^{-1} K_4^\mathsf{T} D_2^2 K_3 D_1^{-1},$$

$$M_{22} \triangleq D_2^{-1} K_2^\mathsf{T} D_1^2 K_2 D_2^{-1} + D_2^{-1} K_4^\mathsf{T} D_2^2 K_4 D_2^{-1}.$$

For $X \in \mathbb{R}^l$, let $\widetilde{X} \in \mathbb{R}^n$ be defined by

$$\widetilde{X} \triangleq \begin{pmatrix} X \\ 0 \end{pmatrix}.$$

Note that $|\widetilde{X}| = |X|$ and that

$$\widetilde{X}^\mathsf{T} M^\mathsf{T} M \widetilde{X} = X^\mathsf{T} D_1^{-1} K_1^\mathsf{T} D_1^2 K_1 D_1^{-1} X + X^\mathsf{T} D_1^{-1} K_3^\mathsf{T} D_2^2 K_3 D_1^{-1} X$$

$$\geq X^\mathsf{T} D_1^{-1} K_1^\mathsf{T} D_1^2 K_1 D_1^{-1} X.$$

Hence

$$\max\{Z^\mathsf{T} M^\mathsf{T} M Z; Z \in \mathbb{R}^n, |Z| = 1\} \geq \max\{X^\mathsf{T} D_1^{-1} K_1^\mathsf{T} D_1^2 K_1 D_1^{-1} X; X \in \mathbb{R}^l, |X| = 1\}$$

$$\geq \rho_2(K_1)^2,$$

which implies that $\rho_2(K_1) \leq \rho_2(\mathbf{K})$. Similarly $\rho_2(K_4) \leq \rho_2(\mathbf{K})$. This proves (C.1).

Let us now prove (C.2). We deal only with the case $K_3 = 0$ (the case $K_2 = 0$ being similar). Let $\eta > 0$. Let $D_1 \in \mathcal{D}_l^+$ and $D_2 \in \mathcal{D}_{n-l}^+$ be such that

$$\|D_1 K_1 D_1^{-1}\| \leq \rho_2(K_1) + \eta, \ \|D_2 K_4 D_2^{-1}\| \leq \rho_2(K_4) + \eta. \tag{C.3}$$

Let $\varepsilon > 0$ and

$$D \triangleq \begin{pmatrix} \varepsilon D_1 & 0 \\ 0 & D_2 \end{pmatrix} \in \mathcal{D}_n^+, \ M \triangleq D K D^{-1} \in \mathcal{M}_{n,n}(\mathbb{R}).$$

Let $Z \in \mathbb{R}^n$ and let $X \in \mathbb{R}^l$ and $Y \in \mathbb{R}^{n-l}$ be such that

$$Z = \begin{pmatrix} X \\ Y \end{pmatrix}.$$

We have

$$Z^\mathsf{T} M^\mathsf{T} M Z = X^\mathsf{T} D_1^{-1} K_1^\mathsf{T} D_1^2 K_1 D_1^{-1} X + 2\varepsilon X^\mathsf{T} D_1^{-1} K_1^\mathsf{T} D_1^2 K_2 D_2^{-1} Y$$
$$+ \varepsilon^2 Y^\mathsf{T} D_2^{-1} K_2^\mathsf{T} D_1^2 K_2 D_2^{-1} Y + Y^\mathsf{T} D_2^{-1} K_4^\mathsf{T} D_2^2 K_4 D_2^{-1} Y.$$

Hence, there exists a constant $C > 0$ independent of Z and $\varepsilon > 0$ such that

$$Z^\mathsf{T} M^\mathsf{T} M Z \leqslant (\|D_1 K_1 D_1^{-1}\| \, |X|)^2 + (\|D_2 K_4 D_2^{-1}\| \, |Y|)^2 + C\varepsilon |Z|^2. \qquad (C.4)$$

From (C.3) and (C.4), we obtain that

$$\rho_2(\mathbf{K})^2 \leqslant \max\{(\rho_2(K_1) + \eta)^2, (\rho_2(K_4) + \eta)^2\} + C\varepsilon. \qquad (C.5)$$

Letting $\varepsilon \to 0$ and $\eta \to 0$ in (C.5), one gets that $\rho_2(\mathbf{K})^2 \leqslant \max\{\rho_2(K_1)^2, \rho_2(K_4)^2\}$. This completes the proof of Proposition C.1. $\square$

Proposition C.2. The map $\rho_2 : \mathcal{M}_{n,n}(\mathbb{R}) \to [0, +\infty)$ is continuous.

Proof. We proceed by induction on n. For $n = 1$ the function ρ_2 satisfies $\rho(k) = |k|$ for every $k \in \mathbb{R} = \mathcal{M}_{1,1}(\mathbb{R})$ and is therefore continuous. We now assume that ρ_2 is continuous on $\mathcal{M}_{p,p}(\mathbb{R})$ for every $p \in \{1, \ldots, n-1\}$ and prove that ρ_2 is continuous on $\mathcal{M}_{n,n}(\mathbb{R})$. Since, for every $D \in \mathcal{D}_n^+$ the function $\mathbf{K} \in \mathcal{M}_{n,n}(\mathbb{R}) \mapsto \|DKD^{-1}\| \in \mathbb{R}$ is continuous, it readily follows from its definition that ρ_2 is upper semi-continuous on $\mathcal{M}_{n,n}(\mathbb{R})$. It remains only to check that ρ_2 is lower semi-continuous. We argue by contradiction: let $\mathbf{K} \in \mathcal{D}_n(\mathbb{R})$ and let $(K_k)_{k \in \mathbb{N}}$ be a sequence of elements of $\mathcal{M}_{n,n}(\mathbb{R})$ such that

$$K_k \to \mathbf{K} \text{ as } k \to +\infty, \qquad (C.6)$$

$$\lim_{k \to +\infty} \rho_2(K_k) < \rho_2(\mathbf{K}). \qquad (C.7)$$

Let $(D_k)_{k \in \mathbb{N}}$ be a sequence of elements of $\mathcal{D}_n^+$ such that

$$\|D_k K_k D_k^{-1}\| \leqslant \rho_2(K_k) + k^{-1}, \ \forall k \in \mathbb{N} \setminus \{0\}. \qquad (C.8)$$

Note that, denoting by $(e_1, \ldots, e_n)$ the canonical basis of $\mathbb{R}^n$,

$$|A_{ij}| = |e_i^\mathsf{T} A e_j| \leqslant \|A\|, \forall A \in \mathcal{M}_{n,n}(\mathbb{R}), \ \forall i \in \{1, \ldots, n\}, \ \forall j \in \{1, \ldots, n\}.$$

Hence, if we denote by K_{ijk} the term on the ith line and jth column of the matrix K_k,

$$|K_{ijk}|\frac{d_{ik}}{d_{jk}} \leq \|D_k K_k D_k^{-1}\|, \ \forall (i,j) \in \{1,\ldots,n\}^2, \ \forall k \in \mathbb{N}, \tag{C.9}$$

where $(d_{ik})_{i\in\{1,\ldots,n\}}$ is defined by $D_k = \text{diag}\,(d_{1k},\ldots,d_{nk})$. After suitable reorderings (note that $\rho_2(\Sigma A \Sigma^{-1}) = \rho_2(A)$, for every $A \in \mathcal{M}_{n,n}(\mathbb{R})$ and for every permutation matrix Σ) and extracting subsequences if necessary, we may assume without loss of generality that

$$d_{1k} \leq d_{2k} \leq \ldots \leq d_{(n-1)k} \leq d_{nk}, \ \forall k \in \mathbb{N}. \tag{C.10}$$

A simple scaling argument also shows that, we may assume without loss of generality that

$$d_{1k} = 1, \ \forall k \in \mathbb{N}. \tag{C.11}$$

Extracting subsequences if necessary, there exist $l \in \{1,\ldots,n\}$, $(d_1,\ldots,d_l) \in [1,+\infty)^l$ such that

$$d_{ik} \to d_i \text{ as } k \to +\infty, \ \forall i \in \{1,\ldots,l\}, \tag{C.12}$$

$$d_{ik} \to +\infty \text{ as } k \to +\infty, \ \forall i \in \{l+1,\ldots,n\}. \tag{C.13}$$

We first treat the case where $l = n$. Let $D \triangleq \text{diag}\,(d_1,\ldots,d_n) \in \mathcal{D}_n^+$. From (C.12), we have

$$D_k \to D \text{ as } k \to +\infty. \tag{C.14}$$

From the definition (3.6) of $\rho_2(\mathbf{K})$, we have

$$\rho_2(\mathbf{K}) \leq \|DKD^{-1}\|, \tag{C.15}$$

which, together with (C.8) and (C.14), implies that

$$\liminf_{k\to+\infty} \rho_2(K_k) \geq \rho_2(\mathbf{K}),$$

in contradiction with (C.7).

It remains to deal with the case where $l < n$. Let us denote K_{ij} the term on the ith line and jth column of the matrix K. From (C.6), (C.7), (C.8), (C.9), (C.12), and (C.13), one gets that

$$K_{ij} = 0, \ \forall (i,j) \in \{l+1,\ldots,n\} \times \{1,\ldots,l\}. \tag{C.16}$$

Let $K^1 \in \mathcal{M}_{l,l}(\mathbb{R})$, $K^2 \in \mathcal{M}_{l,n-l}(\mathbb{R})$, $K^4 \in \mathcal{M}_{n-l,n-l}(\mathbb{R})$ be such that

$$\mathbf{K} = \begin{pmatrix} K^1 & K^2 \\ 0 & K^4 \end{pmatrix}.$$

From (C.2), we have

$$\rho_2(\mathbf{K}) = \max\{\rho_2(K^1), \rho_2(K^4)\}. \tag{C.17}$$

Similarly, for $k \in \mathbb{N}$, let $K_k^1 \in \mathcal{M}_{l,l}(\mathbb{R})$, $K_k^2 \in \mathcal{M}_{l,n-l}(\mathbb{R})$, $K_k^3 \in \mathcal{M}_{n-l,l}(\mathbb{R})$, $K_k^4 \in \mathcal{M}_{n-l,n-l}(\mathbb{R})$ be defined by

$$K_k \triangleq \begin{pmatrix} K_k^1 & K_k^2 \\ K_k^3 & K_k^4 \end{pmatrix}.$$

From (C.1), we have

$$\rho_2(K_k) \geqslant \max\{\rho_2(K_k^1), \rho_2(K_k^4)\}, \ \forall k \in \mathbb{N}. \tag{C.18}$$

From our induction hypothesis (the continuity of ρ_2 on $\mathcal{M}_{p,p}(\mathbb{R})$ for every $p \in \{1, \ldots, n-1\}$) and (C.6), we get that

$$\lim_{k \to +\infty} \rho_2(K_k^1) = \rho_2(K^1), \ \lim_{k \to +\infty} \rho_2(K_k^4) = \rho_2(K^4),$$

which, together with (C.17) and (C.18), leads again to a contradiction with (C.7). This concludes the proof of Proposition C.2.   □

Our next proposition shows a case where the value of $\rho_2(\mathbf{K})$ may be given directly. (For a converse of this proposition, see Proposition C.4.)

Proposition C.3. Let $l \in \{1, \ldots, n\}$. Let $(A_j)_{j \in \{1,\ldots,l\}}$ and $(B_j)_{j \in \{1,\ldots,l\}}$ be two sequences of vectors in $\mathbb{R}^n$ such that

$$A_j^{\mathsf{T}} A_k = B_j^{\mathsf{T}} B_k, \ \forall (j,k) \in \{1, \ldots, l\}^2, \tag{C.19}$$

$$\sum_{j=1}^{l} A_{ij}^2 = \sum_{j=1}^{l} B_{ij}^2, \ \forall i \in \{1, \ldots n\}, \tag{C.20}$$

where A_{ij} (resp. B_{ij}) is the element on the ith line of the vector A_j (resp. B_j). We assume that the l vectors $A_1, \ldots, A_l$ are linearly independent. Let $\sigma \geqslant 0$ and let $\mathbf{K} \in \mathcal{M}_{n,n}(\mathbb{R})$ be such that

$$\mathbf{K} A_j = \sigma B_j, \ \forall j \in \{1, \ldots, l\}, \tag{C.21}$$

$$|\mathbf{K}X| \leqslant \sigma |X|, \ \forall X \in \mathbb{R}^n \text{ such that } X^{\mathsf{T}} A_j = 0, \forall j \in \{1, \ldots, l\}. \tag{C.22}$$

Then $\rho_2(\mathbf{K}) = \sigma$.

Proof. It readily follows from the assumptions of this proposition that $\|\mathbf{K}\| = \sigma$. Hence it remains only to check that

$$\|DKD^{-1}\| \geq \sigma, \ \forall D \in \mathcal{D}_n^+. \tag{C.23}$$

Let $D \triangleq \mathrm{diag}\,(D_1, \ldots, D_n) \in \mathcal{D}_n^+$. For $j \in \{1, \ldots, l\}$, let us define

$$E_j \triangleq (E_{1j}, \ldots, E_{nj})^{\mathsf{T}} \in \mathbb{R}^n \setminus \{0\}, \ F_j \triangleq (F_{1j}, \ldots, F_{nj})^{\mathsf{T}} \in \mathbb{R}^n \setminus \{0\},$$

by

$$E_j \triangleq DA_j, \ F_j \triangleq DB_j.$$

We have, for every $j \in \{1, \ldots, l\}$,

$$DKD^{-1}E_j = \sigma F_j, \tag{C.24}$$

$$E_{ij} = D_i A_{ij}, \ F_{ij} = D_i B_{ij}, \ \forall i \in \{1, \ldots, n\}. \tag{C.25}$$

Using (C.20), (C.24), and (C.25), we get

$$\sum_{j=1}^{l} |DKD^{-1}E_j|^2 = \sigma^2 \sum_{j=1}^{l} \left(\sum_{i=1}^{n} F_{ij}^2 \right)$$

$$= \sigma^2 \sum_{i=1}^{n} D_i^2 \left(\sum_{j=1}^{l} B_{ij}^2 \right)$$

$$= \sigma^2 \sum_{i=1}^{n} D_i^2 \left(\sum_{j=1}^{l} A_{ij}^2 \right)$$

$$= \sigma^2 \sum_{j=1}^{l} |E_j|^2.$$

In particular, there exists $p \in \{1, \ldots l\}$, such that

$$|DKD^{-1}E_p|^2 \geq \sigma^2 |E_p|^2,$$

which, together with the fact that $E_p \neq 0$, implies that $\|DKD^{-1}\| \geq \sigma$. This concludes the proof of Proposition C.3. $\square$

C.2   Proof of Theorem 3.12

Statement of Theorem 3.12

(a) For every $n \in \{1, 2, 3, 4, 5\}$ and for every real $n \times n$ matrix $\mathbf{K}$, $\bar{\rho}(\mathbf{K}) = \rho_2(\mathbf{K})$.
(b) For every integer $n > 5$, there exist a real $n \times n$ matrix $\mathbf{K}$ such that $\bar{\rho}(\mathbf{K}) < \rho_2(\mathbf{K})$.

Proof. The proof relies on various independent propositions. The first proposition provides the converse (up to the D) to Proposition C.3 for generic $\mathbf{K} \in \mathcal{M}_{n,n}(\mathbb{R})$.

Proposition C.4. Let $\mathbf{K} \in \mathcal{M}_{n,n}(\mathbb{R})$ be such that, for every $M > 0$ there exists $\delta > 0$ such that

$$\left(D \triangleq (D_1, \ldots, D_n) \in \mathcal{D}_n^+, \ \sum_{i=1}^n D_i = 1, \ \min\{D_1, \ldots, D_n\} < \delta \right)$$

$$\Rightarrow (\|DKD^{-1}\| > M). \qquad (C.26)$$

(It is easily checked that this property holds, for example, if $K_{ij} \neq 0$, for every $(i,j) \in \{1, \ldots, n\}^2$ such that $i \neq j$, which is a generic property.). Then there exist $D \in \mathcal{D}_n^+$, an integer $l \in \{1, \ldots, n\}$, l vectors $A_j \in \mathbb{R}^n, j \in \{1, \ldots, l\}$ and l vectors $B_j \in \mathbb{R}^n, j \in \{1, \ldots, l\}$, such that (C.19) and (C.20) hold and

$$\text{the vectors } A_j \in \mathbb{R}^n, j \in \{1, \ldots, l\}, \text{ are linearly independent,} \qquad (C.27)$$

$$DKD^{-1}A_j = \rho_2(\mathbf{K})B_j, \ \forall j \in \{1, \ldots l\}, \qquad (C.28)$$

$$|DKD^{-1}X| \leq \rho_2(\mathbf{K})|X|, \ \forall X \in \mathbb{R}^n. \qquad (C.29)$$

Remark C.5. Proposition C.4 is false if assumption (C.26) is removed. Indeed, let us take $n = 2$ and

$$K = \begin{pmatrix} 0 & 1 \\ 0 & 0 \end{pmatrix}.$$

Then $\rho_2(\mathbf{K}) = 0$ and it is easily seen that the conclusion of Proposition C.4 does not hold.

Proof. From (C.26), one gets the existence of $\widetilde{D} \in \mathcal{D}_n^+$ such that

$$\|\widetilde{D}\mathbf{K}\widetilde{D}^{-1}\| = \rho_2(\mathbf{K}). \qquad (C.30)$$

Replacing $\mathbf{K}$ by $\widetilde{D}\mathbf{K}\widetilde{D}^{-1}$, we may assume without loss of generality that $\widetilde{D}$ is the identity map Id_n of $\mathbb{R}^n$. Then

$$\|\mathbf{K}\| = \rho_2(\mathbf{K}). \qquad (C.31)$$

Clearly (C.26) implies that $K \neq 0$ and therefore, by (C.31),

$$\rho_2(\mathbf{K}) \neq 0. \tag{C.32}$$

(In fact, if $\mathbf{K} = 0$, the conclusion of Proposition C.4 obviously holds.) Note that (C.31) implies (C.29) with $D \triangleq \mathrm{Id}_n$. Let $p \in \{1, \ldots, n\}$ be the dimension of the kernel of $\mathbf{K}^\mathsf{T}\mathbf{K} - \rho_2(\mathbf{K})^2 \mathrm{Id}_n$ and let $(X_1, \ldots, X_p)$ be an orthonormal basis of this kernel. For $j \in \{1, \ldots, p\}$, let $Y_j \triangleq KX_j$. One has

$$|Y_j|^2 = X_j^\mathsf{T} \mathbf{K}^\mathsf{T} \mathbf{K} X_j = \rho_2(\mathbf{K})^2 |X_j|^2, \ \forall j \in \{1, \ldots, p\}, \tag{C.33}$$

$$Y_k^\mathsf{T} Y_j = X_\mathbf{K}^\mathsf{T} \mathbf{K}_k^\mathsf{T} \mathbf{K} X_j = \rho_2(\mathbf{K})^2 X_k^\mathsf{T} X_j = 0, \ \forall (k,j) \in \{1, \ldots, p\}^2 \text{ such that } k \neq j. \tag{C.34}$$

For $i \in \{1, \ldots, n\}$ and $j \in \{1, \ldots, p\}$, let us denote by X_{ij} (resp. Y_{ij}) the ith component of X_j (resp. Y_j). For $j \in \{1, \ldots, p\}$, let us denote by E_j the element of $\mathbb{R}^n$ whose ith component is

$$E_{ij} \triangleq Y_{ij}^2 - X_{ij}^2. \tag{C.35}$$

Let us assume, for the moment, that

$$\forall \tau \in \mathbb{R}^n, \text{ there exists } j \in \{1, \ldots, p\} \text{ such that } \tau^\mathsf{T} E_j \geqslant 0. \tag{C.36}$$

Applying the separation principle for convex sets to $\{0\}$ and the convex hull of the vectors E_j, $j \in \{1, \ldots, p\}$ (see, e.g., Rudin (1973, Theorem 3.4 (b), page 58)), it follows from (C.36) that $\mathbf{0} \in \mathbb{R}^n$ is in the convex hull of the vectors $E_1, \ldots E_p$: there exists p nonnegative real numbers $t_1, \ldots, t_p$ such that

$$\sum_{j=1}^{p} t_i = 1, \ \sum_{j=1}^{p} t_i E_i = 0.$$

Let $l \in \{1, \ldots, p\}$ be the number of the t_i's which are not equal to 0. Reordering the X_i's if necessary, we may assume that

$$t_j > 0, \forall j \in \{1, \ldots, l\}, \ t_j = 0, \forall j \in \{l+1, \ldots p\}.$$

For $j \in \{1, \ldots, l\}$, we define $A_j \in \mathbb{R}^n$ and $B_j \in \mathbb{R}^n$ by

$$A_j \triangleq \sqrt{t_j} X_j, \ B_j \triangleq \sqrt{t_j} Y_j. \tag{C.37}$$

Then, it is easily checked that the vectors $A_1, \ldots, A_l$ are linearly independent, that (C.19) and (C.20) hold (one even has $A_k^\mathsf{T} A_j = B_k^\mathsf{T} B_j = 0$ for every $(k,j) \in \{1, \ldots, l\}^2$ such that $k \neq j$), that (C.28) holds with $D \triangleq \mathrm{Id}_n$.

It remains only to prove (C.36). Let $\tau \triangleq (\tau_1, \ldots, \tau_n)^\mathsf{T} \in \mathbb{R}^n$. For $s \in \mathbb{R}$, let

$$D(s) \triangleq \operatorname{diag} (1 + s\tau_1, \ldots, 1 + s\tau_n) \in \mathcal{D}_n.$$

For s small enough, $D(s) \in \mathcal{D}_n^+$ and therefore, by (C.31),

$$\|D(s)\mathbf{K}D(s)^{-1}\|^2 \geqslant \|\mathbf{K}\|^2 = \|D(0)\mathbf{K}D(0)^{-1}\|^2. \tag{C.38}$$

Let us estimate the left-hand side of (C.38). By a classical theorem due to Rellich (see, e.g., Reed and Simon (1978, Theorem XII.3, page 4)) on perturbations of the spectrum of self-adjoint operators, there exist $\varepsilon > 0$, p real functions $\lambda_1, \ldots, \lambda_p$ of class C^1 from $(-\varepsilon, \varepsilon)$ into $\mathbb{R}$, p maps $x_1, \ldots, x_p$ of class C^1 from $(-\varepsilon, \varepsilon)$ into $\mathbb{R}^n$ such that

$$\lambda_j(0) = \rho_2(\mathbf{K})^2, \; x_j(0) = X_j, \; \forall j \in \{1, \ldots, p\}, \tag{C.39}$$

$$D(s)^{-1}\mathbf{K}^\mathsf{T}D(s)^2 K D(s)^{-1} x_j(s) = \lambda_j(s) x_j(s), \; \forall s \in (-\varepsilon, \varepsilon), \; \forall j \in \{1, \ldots, p\}, \tag{C.40}$$

$$x_j(s)^\mathsf{T} x_j(s) = 1, \; \forall s \in (-\varepsilon, \varepsilon), \; \forall j \in \{1, \ldots, p\}, \tag{C.41}$$

$$x_j(s)^\mathsf{T} x_k(s) = 0, \; \forall s \in (-\varepsilon, \varepsilon), \; \forall (j, k) \in \{1, \ldots, p\}^p \text{ such that } k \neq j, \tag{C.42}$$

$$\|D(s)\mathbf{K}D(s)^{-1}\|^2 = \max\{\lambda_1(s), \ldots, \lambda_p(s)\}, \; \forall s \in (-\varepsilon, \varepsilon). \tag{C.43}$$

Differentiating (C.40) with respect to s and using (C.35), (C.39), (C.41), and (C.42), one gets

$$\lambda_j'(0) = 2\rho_2(\mathbf{K})^2 \tau^\mathsf{T} E_j, \; \forall j \in \{1, \ldots, p\}. \tag{C.44}$$

Property (C.36) follows from (C.32), (C.38), (C.43), and (C.44). This concludes the proof of Proposition C.4. □

The number l appearing in Proposition C.4 turns out to be important to compare $\bar{\rho}$ and ρ_2: we have the following proposition.

Proposition C.6. Let $K \in \mathcal{M}_{n,n}(\mathbb{R})$, $D \in \mathcal{D}_n^+$, $l \in \{1, \ldots, n\}$, l vectors $A_j \in \mathbb{R}^n$, $j \in \{1, \ldots, l\}$, and l vectors $B_j \in \mathbb{R}^n$, $j \in \{1, \ldots, l\}$ be such that (C.20), (C.27), (C.28), and (C.29) hold. If $l = 1$, there exist $X \in \mathbb{R}^n$ and $\Upsilon \triangleq \operatorname{diag} (\Upsilon_1, \ldots, \Upsilon_n) \in \mathcal{D}_n$ such that

$$|X| \neq 0, \tag{C.45}$$

$$\Upsilon_i \in \{1, -1\}, \; \forall i \in \{1, \ldots, n\}, \tag{C.46}$$

$$KX = \rho_2(\mathbf{K})\Upsilon X. \tag{C.47}$$

If $l = 2$, there exist $X \in \mathbb{C}^n$ and $(\Upsilon_1, \ldots, \Upsilon_n) \in \mathbb{C}^n$ such that

$$|X| \neq 0, \tag{C.48}$$

$$|\Upsilon_i| = 1, \ \forall i \in \{1, \ldots, n\}, \tag{C.49}$$

$$KX = \rho_2(K)\mathrm{diag}\,(\Upsilon_1, \ldots, \Upsilon_n)X. \tag{C.50}$$

In both cases ($l = 1$ or $l = 2$), we have

$$\bar{\rho}(K) = \rho_2(K). \tag{C.51}$$

Proof. Let us first consider the case $l = 1$. Let $i \in \{1, \ldots, n\}$. From (C.20), one has $|A_{i1}| = B_{i1}$ and, therefore, there exists $\Upsilon_i \in \{-1, 1\}$ such that $B_{i1} = \varepsilon_i A_{i1}$. From (C.28), one gets (C.47) if one defines X by $X \triangleq D^{-1}A_1$. Let us check equality (C.51). Let $(\theta_1, \ldots, \theta_n)^\mathsf{T} \in \mathbb{R}^n$ be defined by

$$\theta_i = 0 \text{ if } \Upsilon_i = 1, \ \theta_i = -\pi \text{ if } \Upsilon_i = -1.$$

Then (C.47) implies that

$$\mathrm{diag}\,(e^{i\theta_1}, \ldots, e^{i\theta_n})KX = \rho_2(K)X. \tag{C.52}$$

From the definition of $\bar{\rho}$ (see (3.17)), (C.45), and (C.52), we get that

$$\bar{\rho}(K) \geq \rho_2(K), \tag{C.53}$$

which, together with Theorem 3.11 gives (C.51).

Let us now turn to the case $l = 2$. Let $i \in \{1, \ldots, n\}$. From (C.20), one has

$$|A_{i1} + \iota A_{i2}| = |B_{i1} + \iota B_{i2}|$$

and, therefore, there exists $\Upsilon_i \in \mathbb{C}$ such that $|\Upsilon_i| = 1$ and $B_{i1} + \iota B_{i2} = \Upsilon_i(A_{i1} + \iota A_{i2})$. From (C.28), one gets (C.47) if one defines X by $X \triangleq D^{-1}(A_1 + \iota A_2)$. Finally, the proof that $\bar{\rho}(K) = \rho_2(K)$ is the same as in the case $l = 1$. This concludes the proof of Proposition C.6. $\qquad\square$

The next proposition deals with the case $n = l$.

Proposition C.7. Let $K \in \mathcal{M}_{n,n}(\mathbb{R})$, $D \in \mathcal{D}_n^+$, $l \in \{1, \ldots, n\}$, l vectors $A_j \in \mathbb{R}^n$, $j \in \{1, \ldots, l\}$, and l vectors $B_j \in \mathbb{R}^n$, $j \in \{1, \ldots, l\}$ be such that (C.20), (C.27), (C.28), and (C.29) hold. If $l = n$, there exist $X \in \mathbb{C}^n$ satisfying (C.48) and $\theta \in \mathbb{R}$ such that

$$KX = e^{-\iota\theta}\rho_2(K)X \tag{C.54}$$

and (C.51).

Proof. If $\rho_2(\mathbf{K}) = 0$, then $\mathbf{K} = 0$ and the conclusion of Proposition C.7 holds. If $\rho_2(\mathbf{K}) > 0$, it follows from (C.19), (C.27), and (C.28) and the assumption $l = n$ that $\rho_2(\mathbf{K})^{-1} KDKD^{-1}$ is an isometry. Hence, there exists $Y \in \mathbb{C}^n \setminus \{0\}$ and $\theta \in \mathbb{R}$ such that $\rho_2(\mathbf{K})^{-1} DKD^{-1} Y = e^{-i\theta} Y$, which implies (C.54) if $X \triangleq D^{-1} Y$. Finally (C.51) follows again from (C.48) and (C.54). This concludes the proof of Proposition C.7.
□

Note that $\bar{\rho}$ is continuous. Hence, from Proposition C.2, Proposition C.4, Proposition C.6, and Proposition C.7, in order to get $\bar{\rho}(\mathbf{K}) = \rho_2(\mathbf{K})$ for every $n \in \{1, \ldots, 5\}$ as stated in Theorem 3.12, it remains to address, with the notations of the conclusion of Proposition C.4, the cases $(l, n) = (3, 4)$, $(l, n) = (3, 5)$, and $(l, n) = (4, 5)$. This is done in the following proposition.
□

Proposition C.8. Let $K \in \mathcal{M}_{n,n}(\mathbb{R})$, $D \in \mathcal{D}_n^+$, $l \in \{1, \ldots, n\}$, l vectors $A_j \in \mathbb{R}^n$, $j \in \{1, \ldots, l\}$, and l vectors $B_j \in \mathbb{R}^n$, $j \in \{1, \ldots, l\}$, be such that (C.20), (C.27), (C.28), and (C.29) hold. If $(l, n) \in \{(3, 4), (3, 5), (4, 5)\}$, there exist $X \in \mathbb{C}^n$ and $(\Upsilon_1, \ldots, \Upsilon_n) \in \mathbb{C}^n$ such that (C.48), (C.49), and (C.50) hold. In particular $\bar{\rho}(\mathbf{K}) = \rho_2(\mathbf{K})$.

Proof. The fact that $\bar{\rho}(\mathbf{K}) = \rho_2(\mathbf{K})$ is implied by the assumptions of Proposition C.8, (C.48), (C.49), and (C.50) has already pointed out in the proof of Proposition C.6. The case $(l, n) = (3, 4)$ follows from the case $(l, n) = (3, 5)$ by replacing $\mathbf{K} \in \mathcal{M}_{4,4}(\mathbb{R})$ by the matrix

$$\widetilde{\mathbf{K}} \triangleq \begin{pmatrix} \mathbf{K} & 0 \\ 0 & 0 \end{pmatrix}.$$

Hence we may assume that $n = 5$. Taking $X \triangleq D^{-1}(Y_1 A_1 + Y_2 A_2 + \ldots + Y_l A_l)$ it suffices to prove the existence of $Y \triangleq (Y_1, Y_2, \ldots, Y_l)^\top \in \mathbb{C}^l \setminus \{0\}$ such that

$$|Y_1 B_{i1} + Y_2 B_{i2} + \ldots Y_l B_{il}|^2 - |Y_1 A_{i1} + Y_2 A_{i2} + \ldots Y_l A_{il}|^2 = 0, \; \forall$$

$$i \in \{1, 2, 3, 4, 5\}. \tag{C.55}$$

For $p \in \mathbb{N} \setminus \{0\}$, let us denote by $\mathcal{S}_p$ the set of elements $Q \in \mathcal{M}_{p,p}$ such that $Q^\top = Q$. For $i \in \{1, 2, 3, 4, 5\}$, there exists a unique $Q_i \in \mathcal{S}_l$ such that, for every $Y \triangleq (Y_1, Y_2, \ldots, Y_l)^\top \in \mathbb{C}^l$,

$$Y^\top Q_i \bar{Y} = |Y_1 B_{i1} + Y_2 B_{i2} + \ldots Y_l B_{il}|^2 - |Y_1 A_{i1} + Y_2 A_{i2} + \ldots Y_l A_{il}|^2,$$

with $\bar{Y} \triangleq (\bar{Y}_1, \bar{Y}_2, \ldots, \bar{Y}_l)$ ($\bar{z}$ denoting the complex conjugate of $z \in \mathbb{C}$). Then (C.2) is equivalent to

$$Y^\top Q_i \bar{Y} = 0, \; \forall i \in \{1, 2, 3, 4, 5\}. \tag{C.56}$$

For a matrix $M \in \mathcal{M}_{p,p}(\mathbb{C})$, let us denote by $\mathrm{tr}(M)$ its trace. Using (C.20) we have that

$$\mathrm{tr}(Q_i) = 0, \forall i \in \{1, 2, 3, 4, 5\}.$$

Using (C.19), one gets that

$$Y^\mathsf{T} Q_1 \bar{Y} + Y^\mathsf{T} Q_2 \bar{Y} + Y^\mathsf{T} Q_3 \bar{Y} + Y^\mathsf{T} Q_4 \bar{Y} + Y^\mathsf{T} Q_5 \bar{Y} = 0, \ \forall Y \in \mathbb{C}^l.$$

Hence (C.56) is equivalent to

$$Y^\mathsf{T} Q_i \bar{Y} = 0, \ \forall i \in \{1, 2, 3, 4\}.$$

Therefore Proposition C.8 is a consequence of the following proposition due to Claire Voisin Voisin (2007). □

Proposition C.9. Let $l \in \{3, 4\}$. Let Q_1, Q_2, Q_3, and Q_4 be four elements of $\mathcal{S}_l$ such that

$$\mathrm{tr}(Q_i) = 0, \ \forall i \in \{1, 2, 3, 4\}. \tag{C.57}$$

Then there exists $Y \in \mathbb{C}^l \setminus \{0\}$ such that

$$Y^\mathsf{T} Q_i \bar{Y} = 0, \ \forall i \in \{1, 2, 3, 4\}. \tag{C.58}$$

Proof. We reproduce the proof of Voisin (2007). For $l \in \mathbb{N}$, let $\overline{\mathcal{S}_{l,+}}$ be the set of semi definite positive $S \in \mathcal{S}_l$. The first step is the following lemma.

Lemma C.10. Let l, p, and n be three positive integers. Let $Q_i, i \in \{1, \ldots, n\}$ be n elements of $\mathcal{S}_l$. Assume that

$$\mathrm{tr}(Q_i) = 0, \ \forall i \in \{1, \ldots, n\}, \tag{C.59}$$

$$n < \frac{(p+1)(p+2)}{2} - 1. \tag{C.60}$$

Then there exists $S \in \overline{\mathcal{S}_{l,+}} \setminus \{0\}$ such that

$$\text{the rank of S is less than or equal to } p, \tag{C.61}$$

$$\mathrm{tr}(SQ_i) = 0, \ \forall i \in \{1, \ldots, n\}. \tag{C.62}$$

Proof. Let

$$C \triangleq \{S \in \overline{\mathcal{S}_{l,+}} \ ; \ \mathrm{tr}(S) = l, \ \mathrm{tr}(SQ_i) = 0, \ \forall i \in \{1, \ldots, n\}\}.$$

The set C is a closed convex bounded subset of $\mathcal{M}_{l,l}(\mathbb{R})$. By (C.59), $\mathrm{Id}_l \in C$ and therefore C is not empty. Hence, by the Krein-Milman theorem (see, e.g., Rudin (1973, Theorem 3.21, page 70)), the convex set C has at least an extreme point. Let S be an extreme point of C. Then $S \in \overline{\mathcal{S}_{l,+}} \setminus \{0\}$ and satisfies (C.62). It remains only to check that (C.61) holds. Let k be the rank of S. There exists an orthonormal matrix $O \in \mathcal{M}_{l,l}(\mathbb{R})$ and a definite positive matrix $S_0 \in \mathcal{S}_k$ such that

$$S = O^{\mathsf{T}} \begin{pmatrix} S_0 & 0 \\ 0 & 0 \end{pmatrix} O. \tag{C.63}$$

Let

$$\Pi \triangleq \left\{ O^{\mathsf{T}} \begin{pmatrix} S' & 0 \\ 0 & 0 \end{pmatrix} O; \ S' \in \mathcal{S}_k, \ \mathrm{tr}(S') = 0 \right\} \subset \mathcal{S}_l. \tag{C.64}$$

Let us assume that

$$n < \frac{k(k+1)}{2} - 1. \tag{C.65}$$

Since Π is a vector subspace of $\mathcal{S}_l$ of dimension $(k(k+1)/2) - 1$, (C.65) implies that there exists $S_0 \in \Pi \setminus \{0\}$ such that

$$\mathrm{tr}(S_0 Q_i) = 0, \ \forall i \in \{1, \ldots, n\}. \tag{C.66}$$

Then, for $\tau \in \mathbb{R}$ with $|\tau|$ small enough, $S + \tau S_0$ is in C, which contradicts the fact that S is an extreme point of C. Hence (C.65) does not hold, which, together with (C.60), implies that $k \leq p$. This concludes the proof of Lemma C.10. □

Let us go back to the proof of Proposition C.9. We apply Lemma C.10 with $n = 4$ and $p = 2$ (then (C.60) holds). We get the existence of $S \in \overline{\mathcal{S}_{l,+}} \setminus \{0\}$ satisfying

$$\text{the rank of } S \text{ is less than or equal to } 2, \tag{C.67}$$

$$\mathrm{tr}(S Q_i) = 0, \ \forall i \in \{1, \ldots, 4\}. \tag{C.68}$$

Let $\lambda_1 > 0$, $\lambda_2 \geq 0$ and 0 be the eigenvalues of S. Let

$$S_0 = \begin{pmatrix} \lambda_1 & 0 & 0 \\ 0 & \lambda_2 & 0 \\ 0 & 0 & 0 \end{pmatrix} \in \overline{\mathcal{S}_{l,+}}. \tag{C.69}$$

There exists an orthonormal matrix O such that

$$S = O^{\mathsf{T}} S_0 O. \tag{C.70}$$

Let $Z \triangleq (\sqrt{\lambda_1}, \iota\sqrt{\lambda_2}, 0) \in \mathbb{C}^l \setminus \{0\}$ and $Y \triangleq O^\mathsf{T} Z \in \mathbb{C}^l \setminus \{0\}$. Then, using (C.68) and (C.70), one gets that, for every $i \in \{1, \ldots, 4\}$,

$$2Y^\mathsf{T} Q_i \bar{Y} = \mathrm{tr}((Y\bar{Y}^\mathsf{T} + \bar{Y}Y^\mathsf{T})Q_i) = \mathrm{tr}(O^\mathsf{T}(Z\bar{Z}^\mathsf{T} + \bar{Z}Z^\mathsf{T})OQ_i)$$
$$= 2\mathrm{tr}(O^\mathsf{T} S_0 O Q_i) = \mathrm{tr}(SQ_i) = 0,$$

which concludes the proof of Proposition C.9 and therefore the proof of Proposition C.8.   □

Finally, in order to complete the proof of Theorem 3.12, it remains only to check that, for $n = 6$ and therefore for every $n \geqslant 6$, there exists $\mathbf{K} \in \mathcal{M}_{n,n}(\mathbb{R})$ such that $l = 3$ and $\bar{\rho}(\mathbf{K}) < \rho_2(\mathbf{K})$. This is done in the following example.

Example C.11. Let $(u_1, v_1, w_1)^\mathsf{T} \in \mathbb{R}^3$, $(u_2, v_2, w_2)^\mathsf{T} \in \mathbb{R}^3$, $(x_1, y_1, z_1)^\mathsf{T} \in \mathbb{R}^3$ and $(x_2, y_2, z_2)^\mathsf{T} \in \mathbb{R}^3$. We define $A_1 \in \mathbb{R}^6$, $A_2 \in \mathbb{R}^6$, $A_3 \in \mathbb{R}^6$, $B_1 \in \mathbb{R}^6$, $B_2 \in \mathbb{R}^6$ and $B_3 \in \mathbb{R}^6$ by

$$A_1 \triangleq \begin{pmatrix} 1 \\ 0 \\ 0 \\ 1 \\ u_1 \\ u_2 \end{pmatrix}, A_2 = \begin{pmatrix} 0 \\ 1 \\ 0 \\ 0 \\ v_1 \\ v_2 \end{pmatrix}, A_3 \triangleq \begin{pmatrix} 0 \\ 0 \\ 1 \\ 0 \\ w_1 \\ w_2 \end{pmatrix},$$

$$B_1 \triangleq \begin{pmatrix} 0 \\ 0 \\ 0 \\ 1/\sqrt{2} \\ x_1 \\ x_2 \end{pmatrix}, B_2 \triangleq \begin{pmatrix} 1 \\ 0 \\ 1/\sqrt{2} \\ -1/\sqrt{2} \\ y_1 \\ y_2 \end{pmatrix}, B_3 \triangleq \begin{pmatrix} 0 \\ 1 \\ 1/\sqrt{2} \\ 0 \\ z_1 \\ z_2 \end{pmatrix}.$$

One easily checks that (C.20) holds if (and only if)

$$u_1^2 + v_1^2 + w_1^2 = x_1^2 + y_1^2 + z_1^2, \tag{C.71}$$
$$u_2^2 + v_2^2 + w_2^2 = x_2^2 + y_2^2 + z_2^2. \tag{C.72}$$

Similarly (C.19) holds if (and only if)

$$\frac{3}{2} + u_1^2 + u_2^2 - x_1^2 - x_2^2 = 0, \tag{C.73}$$
$$-1 + v_1^2 + v_2^2 - y_1^2 - y_2^2 = 0, \tag{C.74}$$

$$-\frac{1}{2} + w_1^2 + w_2^2 - z_1^2 - z_2^2 = 0, \tag{C.75}$$

$$\frac{1}{2} + u_1 v_1 + u_2 v_2 - x_1 y_1 - x_2 y_2 = 0, \tag{C.76}$$

$$u_1 w_1 + u_2 w_2 - x_1 z_1 - x_2 z_2 = 0, \tag{C.77}$$

$$-\frac{1}{2} + v_1 w_1 + v_2 w_2 - y_1 z_1 - y_2 z_2 = 0, \tag{C.78}$$

Note that (C.71), (C.73), (C.74), and (C.75) imply (C.72).

We take $l \triangleq 3$ and $R \triangleq 1$. We define $K \in \mathcal{M}_{6,6}(\mathbb{R})$ by requiring (C.21) and

$$\mathbf{K}X = 0, \quad \forall X \in \mathbb{R}^6 \text{ such that } X^\mathsf{T} A_1 = X^\mathsf{T} A_2 = X^\mathsf{T} A_3 = 0.$$

From Proposition C.3 we get that, if (C.71) and (C.73) to (C.78) hold, then

$$\rho_2(\mathbf{K}) = 1.$$

Let us assume, for the moment, that (C.73) to (C.78) hold. If $\bar{\rho}(\mathbf{K}) \not< \rho_2(\mathbf{K})$, we have $\bar{\rho}(\mathbf{K}) = \rho_2(\mathbf{K}) = 1$ and therefore there exist $X \in \mathbb{C}^6$ and $(\Upsilon_1, \ldots, \Upsilon_6)\mathsf{T} \in \mathbb{C}^n$ such that (C.48), (C.49) and (C.50) hold. Clearly

$$|\mathbf{K}X| = |X|. \tag{C.79}$$

Since

$$|\mathbf{K}(Y + Z)| = |Y|, \quad \forall Y \in \mathbb{C}A_1 + \mathbb{C}A_2 + \mathbb{C}A_3,$$

$$\forall Z \in \mathbb{C}^n \text{ such that } Z^\mathsf{T} A_1 = Z^\mathsf{T} A_2 = Z^\mathsf{T} A_3 = 0,$$

it follows from (C.79) that $X \in \mathbb{C}A_1 + \mathbb{C}A_2 + \mathbb{C}A_3$. Hence, there exist $\xi_1 \in \mathbb{C}, \xi_2 \in \mathbb{C}$ and $\xi_3 \in \mathbb{C}$ such that

$$X = \xi_1 A_1 + \xi_2 A_2 + \xi_3 A_3. \tag{C.80}$$

Using (C.50), one gets $(KX)_1 = \Upsilon_1 X_1$ and $(KX)_2 = \Upsilon_2 X_2$, which together with (C.21) and (C.49) implies that

$$|\xi_1| = |\xi_2| = |\xi_3|. \tag{C.81}$$

Using (C.48) and (C.81), without loss of generality, we may assume that

$$\xi_1 = 1, \ |\xi_2| = |\xi_3| = 1.$$

Hence there exist $\theta_2 \in \mathbb{R}$ and $\theta_3 \in \mathbb{R}$ such that

$$\xi_2 = e^{\iota\theta_2}, \ \xi_3 = e^{\iota\theta_3}. \tag{C.82}$$

Using now $|(\mathbf{K}X)_3| = |X_3|$, one gets

$$|\xi_2 + \xi_3| = \sqrt{2},$$

which, together with (C.82), is equivalent to

$$\cos(\theta_3 - \theta_2) = 0,$$

i.e., there exists $\varepsilon_3 \in \{1, -1\}$ such that

$$\xi_3 = \varepsilon_3 \iota \xi_2 \tag{C.83}$$

Proceeding similarly with the 4th of KX, one gets the existence of $\varepsilon_2 \in \{1, -1\}$ such that

$$\xi_2 = \varepsilon_2 \iota, \tag{C.84}$$

Then, $|(KX)_5| = |X_5|$ and $|(KX)_6| = |X_6|$ are equivalent to

$$(u_1 + \varepsilon_1 w_1)^2 + v_1^2 = (x_1 + \varepsilon_1 z_1)^2 + y_1^2 \tag{C.85}$$

$$(u_2 + \varepsilon_1 w_2)^2 + v_2^2 = (x_2 + \varepsilon_1 z_2)^2 + y_2^2 \tag{C.86}$$

with

$$\varepsilon_1 \triangleq -\varepsilon_2 \varepsilon_3 \in \{1, -1\}.$$

Let

$$F : \qquad\qquad \mathbb{R}^{12} \qquad\qquad\qquad \to \ \mathbb{R}^7$$
$$P \triangleq (u_1, v_1, w_1, x_1, y_1, z_1, u_2, v_2, w_2, x_2, y_2, z_2)^{\mathsf{T}} \mapsto F(P) \tag{C.87}$$

be defined by

$$
F(P) \triangleq
\begin{pmatrix}
\dfrac{3}{2} + u_1^2 + u_2^2 - x_1^2 - x_2^2 \\[4pt]
-1 + v_1^2 + v_2^2 - y_1^2 - y_2^2 \\[4pt]
-\dfrac{1}{2} + w_1^2 + w_2^2 - z_1^2 - z_2^2 \\[4pt]
\dfrac{1}{2} + u_1 v_1 + u_2 v_2 - x_1 y_1 - x_2 y_2 \\[4pt]
u_1 w_1 + u_2 w_2 - x_1 z_1 - x_2 z_2 \\[4pt]
-\dfrac{1}{2} + v_1 w_1 + v_2 w_2 - y_1 z_1 - y_2 z_2 \\[4pt]
u_1^2 + v_1^2 + w_1^2 - x_1^2 - y_1^2 - z_1^2
\end{pmatrix} .
$$

Let Σ be the subset of $\mathbb{R}^{12}$ defined by

$$
\Sigma := \{ P \in \mathbb{R}^{12};\ F(P) = 0 \text{ and the rank of } F'(P) \text{ is } 7 \}.
$$

Let

$$
\widetilde{P} \triangleq \left(0, 1, 0, 1, 0, 0, -\frac{1}{4}, \frac{1}{2}, \frac{3}{4}, \frac{3}{4}, \frac{1}{2}, -\frac{1}{4} \right)^{\mathsf{T}} \in \mathbb{R}^{12}.
$$

One easily checks that $F(\widetilde{P}) = 0$. Straightforward computations give

$$
F'(\widetilde{P}) =
\begin{pmatrix}
0 & 0 & 0 & -2 & 0 & 0 & -\dfrac{1}{2} & 0 & 0 & -\dfrac{3}{2} & 0 & 0 \\[4pt]
0 & 2 & 0 & 0 & 0 & 0 & 0 & 1 & 0 & 0 & -1 & 0 \\[4pt]
0 & 0 & 0 & 0 & 0 & 0 & 0 & 0 & \dfrac{3}{2} & 0 & 0 & \dfrac{1}{2} \\[4pt]
1 & 0 & 0 & 0 & -1 & 0 & \dfrac{1}{2} & -\dfrac{1}{4} & 0 & -\dfrac{1}{2} & \dfrac{3}{4} & 0 \\[4pt]
0 & 0 & 0 & 0 & 0 & -1 & \dfrac{3}{4} & 0 & -\dfrac{1}{4} & \dfrac{1}{4} & 0 & -\dfrac{3}{4} \\[4pt]
0 & 0 & 1 & 0 & 0 & 0 & \dfrac{3}{4} & \dfrac{1}{2} & 0 & \dfrac{1}{4} & -\dfrac{1}{2} \\[4pt]
0 & 2 & 0 & -2 & 0 & 0 & 0 & 0 & 0 & 0 & 0
\end{pmatrix} .
\tag{C.88}
$$

In particular, the rank of $F'(\widetilde{P})$ is 7. Hence $\widetilde{P}$ is in Σ and the set Σ is not empty and is a submanifold of $\mathbb{R}^{12}$ of dimension $12 - 7 = 5$. The tangent space to this manifold at $\widetilde{P}$ is $\operatorname{Ker} F'(\widetilde{P})$. Let G_+ be the map

$$
G_+ : \qquad \mathbb{R}^{12} \qquad\qquad\qquad \to \quad \mathbb{R}^2
$$
$$
P \triangleq (u_1, v_1, w_1, x_1, y_1, z_1, u_2, v_2, w_2, x_2, y_2, z_2)^{\mathsf{T}} \mapsto G_+(P)
$$

defined by

$$G_+(P) \triangleq \begin{pmatrix} (u_1 + w_1)^2 + v_1^2 - (x_1 + z_1)^2 - y_1^2 \\ (u_2 + w_2)^2 + v_2^2 - (x_2 + z_2)^2 - y_2^2 \end{pmatrix}.$$

Similarly, let G_- be the map

$$G_- : \qquad\qquad\qquad \mathbb{R}^{12} \qquad\qquad\qquad \to \quad \mathbb{R}^2$$
$$P \triangleq (u_1, v_1, w_1, x_1, y_1, z_1, u_2, v_2, w_2, x_2, y_2, z_2)^\mathsf{T} \mapsto G_-(P)$$

defined by

$$G_-(P) \triangleq \begin{pmatrix} (u_1 - w_1)^2 + v_1^2 - (x_1 - z_1)^2 - y_1^2 \\ (u_2 - w_2)^2 + v_2^2 - (x_2 - z_2)^2 - y_2^2 \end{pmatrix}.$$

Let $S_+ \subset \mathbb{R}^{12}$ and $S_- \subset \mathbb{R}^{12}$ be defined by

$$S_+ \triangleq \{P \in \mathbb{R}^{12}; \ G_+(P) = 0\},$$
$$S_- \triangleq \{P \in \mathbb{R}^{12}; \ G_-(P) = 0\}.$$

It suffices to check that

$$\Sigma \text{ is not a subset of } S_- \cup S_+. \tag{C.89}$$

Indeed (C.89) implies the existence of $P \in \Sigma$ arbitrary close to $\widetilde{P}$ such that $G_+(\widetilde{P}) \neq 0$ and $G_-(\widetilde{P}) \neq 0$. For such a P, it follows from the above analysis (see in particular (C.85) and (C.86)) that $\bar{\rho}(\mathbf{K}) < \rho_2(\mathbf{K})$.

Note that $\widetilde{P} \in S_- \cap S_+$ and

$$G'_-(\widetilde{P}) = \begin{pmatrix} 0 & 2 & 0 & -2 & 0 & 2 & 0 & 0 & 0 & 0 & 0 & 0 \\ 0 & 0 & 0 & 0 & 0 & 0 & -2 & 1 & 2 & -2 & -1 & 2 \end{pmatrix}, \tag{C.90}$$

$$G'_+(\widetilde{P}) = \begin{pmatrix} 0 & 2 & 0 & -2 & 0 & -2 & 0 & 0 & 0 & 0 & 0 & 0 \\ 0 & 0 & 0 & 0 & 0 & 0 & 1 & 1 & 1 & -1 & -1 & -1 \end{pmatrix}. \tag{C.91}$$

In particular the rank of $G'_-(\widetilde{P})$ and the rank of $G'_+(\widetilde{P})$ are both equal to 2. Hence, if $r > 0$ is small enough, the set $\{P \in S_-; \ |P - \widetilde{P}| < r\}$ and the set $\{P \in S_+; \ |P - \widetilde{P}| < r\}$ are submanifolds of $\mathbb{R}^{12}$ of dimension $12 - 2 = 10$ whose tangent spaces at $\widetilde{P}$ are $\operatorname{Ker} G'_-(\widetilde{P})$ and $\operatorname{Ker} G'_+(\widetilde{P})$ respectively. Therefore (C.89) holds if

$$\operatorname{Ker} F'(\widetilde{P}) \text{ is not a subset of } \operatorname{Ker} G'_-(\widetilde{P}) \cup \operatorname{Ker} G'_+(\widetilde{P}). \tag{C.92}$$

Property (C.92) follows from (C.88), (C.90), and (C.91).

This concludes the example C.11 and therefore the proof of Theorem 3.12.   □

C.3   Proof of Proposition 4.7

Statement of Proposition 4.7
  For every $\mathbf{K} \in \mathcal{M}_{n,n}(\mathbb{R})$,

$$\rho_2(\mathbf{K}) \leq \rho_\infty(\mathbf{K}).$$

Let us first prove the following lemma.

Lemma C.12. For every $\mathbf{K} \in \mathcal{M}_{n,n}(\mathbb{R})$, for every $D \in \mathcal{D}_n^+$, for every $\Delta \in \mathcal{D}_n^+$, for every $X \in \mathbb{R}^n$, and for every $Y \in \mathbb{R}^n$,

$$Y^\mathsf{T} \Delta \mathbf{K} \Delta^{-1} X \leq \frac{1}{2} \mathcal{R}_\infty(D\Delta^{-1}\mathbf{K}^\mathsf{T}\Delta D^{-1})|X|^2 + \frac{1}{2}\mathcal{R}_\infty(D\Delta\mathbf{K}\Delta^{-1}D^{-1})|Y|^2. \tag{C.93}$$

Proof. Replacing, if necessary, $\mathbf{K}$ by $\Delta\mathbf{K}\Delta^{-1}$, we may assume without loss of generality that Δ is the identity map of $\mathbb{R}^n$. We write $X \triangleq (X_1, \ldots, X_n)^\mathsf{T} \in \mathbb{R}^n$, $Y \triangleq (Y_1, \ldots, Y_n)^\mathsf{T} \in \mathbb{R}^n$, $D \triangleq \mathrm{diag}(D_1, \ldots, D_n)$. We have

$$Y^\mathsf{T}\mathbf{K}X = \sum_{i=1}^n Y_i \left(\sum_{j=1}^n K_{ij}X_j \right) = \sum_{i=1}^n \sum_{j=1}^n \frac{K_{ij}}{D_i D_j} D_i Y_i D_j X_j$$

$$\leq \frac{1}{2}Q_1 + \frac{1}{2}Q_2, \tag{C.94}$$

with

$$Q_1 \triangleq \sum_{i=1}^n \sum_{j=1}^n \frac{|K_{ij}|}{D_i D_j}D_j^2 X_j^2 \quad \text{and} \quad Q_2 \triangleq \sum_{i=1}^n \sum_{j=1}^n \frac{|K_{ij}|}{D_i D_j}D_i^2 Y_i^2.$$

Note that

$$Q_1 = \sum_{j=1}^n \left(\sum_{i=1}^n |K_{ij}|D_i^{-1}D_j \right) X_j^2 = \sum_{j=1}^n \left(\sum_{i=1}^n |(D^{-1}\mathbf{K}D)^\mathsf{T})_{ji}| \right) X_j^2$$

$$\leq \sum_{j=1}^n \mathcal{R}_\infty((D^{-1}\mathbf{K}D)^\mathsf{T})X_j^2 = \mathcal{R}_\infty(D\mathbf{K}^\mathsf{T}D^{-1})\|X\|^2, \tag{C.95}$$

where $\mathcal{R}_\infty$ is defined in (4.15). Similarly,

$$Q_2 = \sum_{i=1}^{n} \left(\sum_{j=1}^{n} D_i |K_{ij}| D_j^{-1} \right) Y_i^2 = \sum_{i=1}^{n} \left(\sum_{j=1}^{n} |(DKD^{-1})_{ij}| \right) Y_i^2$$

$$\leq \sum_{i=1}^{n} \mathcal{R}_\infty (DKD^{-1}) Y_i^2 = \mathcal{R}_\infty (DKD^{-1}) |Y|^2. \tag{C.96}$$

Inequality (C.93) follows from (C.94), (C.95), and (C.96). This concludes the proof of Lemma C.12.   □

Let us now go back to the proof of Proposition 4.7. It is easily seen that

$$\{(D\Delta^{-1}, D\Delta); \ D \in \mathcal{D}_n^+, \ \Delta \in \mathcal{D}_n^+\} = \mathcal{D}_n^+ \times \mathcal{D}_n^+. \tag{C.97}$$

Equality (C.97) implies that

$$\rho_2(\mathbf{K}^{\mathsf{T}}) + \rho_2(\mathbf{K}) =$$

$$\text{Inf } \{\mathcal{R}_\infty (D\Delta^{-1} \mathbf{K}^{\mathsf{T}} \Delta D^{-1}) + \mathcal{R}_\infty (D\Delta K\Delta^{-1} D^{-1});$$

$$notag \ D \in \mathcal{D}_n^+, \ \Delta \in \mathcal{D}_n^+\}. \tag{C.98}$$

Using (4.16), we have

$$\rho_\infty(\mathbf{K}^{\mathsf{T}}) = \rho(|\mathbf{K}^{\mathsf{T}}|) = \rho(|\mathbf{K}|^{\mathsf{T}}) = \rho(|\mathbf{K}|) = \rho_\infty(\mathbf{K}), \tag{C.99}$$

which, combined with (C.98), gives

$$\text{Inf } \{\mathcal{R}_\infty (D\Delta^{-1} |\mathbf{K}|^{\mathsf{T}} \Delta D^{-1}) + \mathcal{R}_\infty (D\Delta |\mathbf{K}| \Delta^{-1} D^{-1}); \ D \in \mathcal{D}_n^+, \ \Delta \in \mathcal{D}_{n,+}\}$$

$$= 2\rho_\infty(\mathbf{K}). \tag{C.100}$$

Finally, let us note that, for every Δ in $\mathcal{D}_n^+$,

$$\text{Sup } \{Y^{\mathsf{T}} \Delta K \Delta^{-1} X; \ X \in \mathbb{R}^n, \ Y \in \mathbb{R}^n, \ |X| = |Y| = 1\} = \|\Delta K \Delta^{-1}\| \tag{C.101}$$

$$\geq \rho_2(K).$$

Proposition 4.7 then follows from (C.93), (C.100), and (C.101).

Appendix D
Proof of Lemma 4.12 (b) and (c)

Lemma 4.12 (b).
There exists $\mu_2 > 0$ such that, $\forall \mu \in (0, \mu_2)$, there exist positive real constants $\alpha_2, \beta_2, \delta_2$ such that, if $|\mathbf{R}|_0 < \delta_2$,

$$\frac{1}{\beta_2} \int_0^L \|\mathbf{R}_t\|^2 dx \leqslant \mathbf{V}_2 \leqslant \beta_2 \int_0^L \|\mathbf{R}_t\|^2 dx, \tag{D.1}$$

$$\frac{d\mathbf{V}_2}{dt} \leqslant -\alpha_2 \mathbf{V}_2 + \beta_2 \int_0^L \|\mathbf{R}_t\|^3 dx. \tag{D.2}$$

Proof. From (4.71), we know that

$$\frac{d\mathbf{V}_2}{dt} = -\int_0^L 2\mathbf{R}_t^\mathsf{T} P(\mu x)\Big(\Lambda(\mathbf{R})\mathbf{R}_{tx} - [\Lambda'(\mathbf{R})\mathbf{R}_t]\Lambda^{-1}(\mathbf{R})\mathbf{R}_t\Big) dx. \tag{D.3}$$

Using integration by parts, we can decompose $d\mathbf{V}_2/dt$ in the following way:

$$\frac{d\mathbf{V}_2}{dt} = \mathcal{T}_{21} + \mathcal{T}_{22} + \mathcal{T}_{23}, \tag{D.4}$$

with

$$\mathcal{T}_{21} \triangleq -\int_0^L \left(\mathbf{R}_t^\mathsf{T} P(\mu x)\Lambda(\mathbf{R})\mathbf{R}_t\right)_x dx, \tag{D.5}$$

$$\mathcal{T}_{22} \triangleq \int_0^L \mathbf{R}_t^\mathsf{T} P(\mu x)\Big([\Lambda'(\mathbf{R})\mathbf{R}_t]\Lambda^{-1}(\mathbf{R}) - 2[\Lambda'(\mathbf{R})\Lambda^{-1}(\mathbf{R})\mathbf{R}_t]\Big)\mathbf{R}_t dx, \tag{D.6}$$

$$\mathcal{T}_{23} \triangleq -\mu \int_0^L \left(\mathbf{R}_t^\mathsf{T} P(\mu x)|\Lambda(\mathbf{R})|\mathbf{R}_t\right) dx. \tag{D.7}$$

© Springer International Publishing Switzerland 2016
G. Bastin, J.-M. Coron, *Stability and Boundary Stabilization of 1-D Hyperbolic Systems*, Progress in Nonlinear Differential Equations and Their Applications 88, DOI 10.1007/978-3-319-32062-5

Analysis of the First Term $\mathcal{T}_{21}$. We have

$$\mathcal{T}_{21} = -\left[\mathbf{R}_t^\mathsf{T} P(\mu x)\Lambda(\mathbf{R})\mathbf{R}_t\right]_0^L = -\left[\mathbf{R}_t^\mathsf{T} P(\mu x)\Lambda(0)\mathbf{R}_t\right]_0^L + \text{ h.o.t.}$$

We observe that this expression is analog to expression (4.82) for $\mathcal{T}_1$ in the proof of Lemma 4.12(a). Therefore, following the same argumentation as in the proof of Lemma 4.12(a), we deduce that there exist $\mu_2 > 0$ and $\delta_{21} > 0$ such that $\mathcal{T}_{21} \leqslant 0$ for all $\mu \in (0, \mu_2)$ if $|\mathbf{R}^{+\mathsf{T}}(t, L), \mathbf{R}^{-\mathsf{T}}(t, 0)|_0 \leqslant \delta_{21}$.

Analysis of the Second Term $\mathcal{T}_{22}$. The integrand of $\mathcal{T}_{22}$ is, at least, linear with respect to $\mathbf{R}$ and cubic with respect to $\mathbf{R}_t$. It follows that β_2 can be selected sufficiently large such that

- inequality (D.1) holds for every $\mu \in (0, \mu_2)$,
- there exists $\delta_{22} > 0$ such that

$$\mathcal{T}_{22} \leqslant \beta_2 \int_0^L \|\mathbf{R}_t\|^3 dx$$

if $|\mathbf{R}|_0 \leqslant \delta_{23}$.

Analysis of the Third Term $\mathcal{T}_{23}$. By (D.1) there exist $\alpha_2 > 0$ and $0 < \delta_{23}$ such that

$$\mathcal{T}_{23} \leqslant -\alpha_2 V_2$$

for every $\mu \in (0, \mu_2)$ if $|\mathbf{R}|_0 \leqslant \delta_{23}$.

   Then we conclude that

$$\frac{dV_2}{dt} = \mathcal{T}_{21} + \mathcal{T}_{22} + \mathcal{T}_{23} \leqslant -\alpha_2 V_2 + \beta_2 \int_0^L \|\mathbf{R}_t\|^3 dx$$

if $|\mathbf{R}|_0 \leqslant \delta_2 \triangleq \min(\delta_{21}, \delta_{22}, \delta_{23})$. $\square$

Lemma 4.12 (c). There exists $\mu_3 > 0$ such that, $\forall \mu \in (0, \mu_3)$, there exist positive real constants $\alpha_3, \beta_3, \delta_3$ such that, if $|\mathbf{R}|_0 + |\mathbf{R}_t|_0 < \delta_3$,

$$\frac{1}{\beta_3} \int_0^L \|\mathbf{R}_{tt}\|^2 dx \leqslant V_3 \leqslant \beta_3 \int_0^L \|\mathbf{R}_{tt}\|^2 dx, \tag{D.8}$$

$$\frac{dV_3}{dt} \leqslant -\alpha_3 V_3 + \beta_3 \int_0^L (\|\mathbf{R}_t\|^2 \|\mathbf{R}_{tt}\| + \|\mathbf{R}_t\| \|\mathbf{R}_{tt}\|^2) dx. \tag{D.9}$$

Proof. From (4.72), we know that

$$\frac{d\mathbf{V}_3}{dt} = -\int_0^L 2\mathbf{R}_{tt}^\mathsf{T} P(\mu x)\Big(\Lambda(\mathbf{R})\mathbf{R}_{ttx}$$

$$+ 2[\Lambda'(\mathbf{R})\mathbf{R}_t]\mathbf{R}_{tx} - [\Lambda'(\mathbf{R})\mathbf{R}_t]_t\Lambda^{-1}(\mathbf{R})\mathbf{R}_t\Big)dx. \qquad (D.10)$$

Using integration by parts, we can decompose $d\mathbf{V}_3/dt$ in the following way:

$$\frac{d\mathbf{V}_3}{dt} = \mathcal{T}_{31} + \mathcal{T}_{32} + \mathcal{T}_{33}, \qquad (D.11)$$

with

$$\mathcal{T}_{31} \triangleq -\int_0^L \left(\mathbf{R}_{tt}^\mathsf{T} P(\mu x)\Lambda(\mathbf{R})\mathbf{R}_{tt}\right)_x dx, \qquad (D.12)$$

$$\mathcal{T}_{32} \triangleq \int_0^L \Big[\mathbf{R}_{tt}^\mathsf{T} P(\mu x)\Big([\Lambda'(\mathbf{R})\Lambda^{-1}(\mathbf{R})\mathbf{R}_t] - 4[\Lambda'(\mathbf{R})\mathbf{R}_t]\Lambda^{-1}(\mathbf{R})\Big)\mathbf{R}_{tt}$$

$$- 4\mathbf{R}_{tt}^\mathsf{T} P(\mu x)[\Lambda'(\mathbf{R})\mathbf{R}_{tt}]\Lambda^{-1}(\mathbf{R})\mathbf{R}_t$$

$$+ 4\mathbf{R}_{tt}^\mathsf{T} P(\mu x)\Big([\Lambda'(\mathbf{R})\mathbf{R}_t]^2\Lambda^{-2}(\mathbf{R}) - [\Lambda''(\mathbf{R})\mathbf{R}_t]\Big)\mathbf{R}_t\Big]dx, \qquad (D.13)$$

$$\mathcal{T}_{33} \triangleq -\mu \int_0^L \left(\mathbf{R}_{tt}^\mathsf{T} P(\mu x)|\Lambda(\mathbf{R})|\mathbf{R}_{tt}\right)dx. \qquad (D.14)$$

Analysis of the First Term $\mathcal{T}_{31}$. We have

$$\mathcal{T}_{31} = -\Big[\mathbf{R}_{tt}^\mathsf{T} P(\mu x)\Lambda(\mathbf{R})\mathbf{R}_{tt}\Big]_0^L = -\Big[\mathbf{R}_{tt}^\mathsf{T} P(\mu x)\Lambda(0)\mathbf{R}_{tt}\Big]_0^L + \text{h.o.t.}$$

We observe that this expression is analog to expression (4.82) for $\mathcal{T}_1$ in the proof of Lemma 4.12(a). Therefore, following the same argumentation as in the proof of Lemma 4.12(a), we deduce that there exist $\mu_3 > 0$ and $\delta_{31} > 0$ such that $\mathcal{T}_{31} \leqslant 0$ for all $\mu \in (0, \mu_3)$ if $|\mathbf{R}^{+\mathsf{T}}(t, L), \mathbf{R}^{-\mathsf{T}}(t, 0)|_0 + |\mathbf{R}_t^{+\mathsf{T}}(t, L), \mathbf{R}_t^{-\mathsf{T}}(t, 0)|_0 \leqslant \delta_{31}$.

Analysis of the Second Term $\mathcal{T}_{32}$. The first two terms of the integrand of $\mathcal{T}_{32}$ are, at least, linear with respect to $\mathbf{R}_t$ and quadratic with respect to $\mathbf{R}_{tt}$. The last term of the integrand of $\mathcal{T}_{32}$ is, at least, quadratic with respect to $\mathbf{R}_t$ and linear with respect to $\mathbf{R}_{tt}$. It follows that β_3 can be selected sufficiently large such that

- inequality (D.8) holds for every $\mu \in (0, \mu_3)$,
- there exists $\delta_{32} > 0$ such that

$$\mathcal{T}_{32} \leq \beta_2 \int_0^L (\|\mathbf{R}_t\|^2 \|\mathbf{R}_{tt}\| + \|\mathbf{R}_t\| \|\mathbf{R}_{tt}\|^2) dx$$

if $|\mathbf{R}|_0 + |\mathbf{R}_t|_0 \leq \delta_{32}$.

Analysis of the Third Term $\mathcal{T}_{33}$. By (D.8) there exist $\alpha_3 > 0$ and $0 < \delta_{33}$ such that

$$\mathcal{T}_{33} \leq -\alpha_3 \mathbf{V}_3$$

for every $\mu \in (0, \mu_3)$ if $|\mathbf{R}|_0 + |\mathbf{R}_t|_0 \leq \delta_{33}$.

Then we conclude that

$$\frac{d\mathbf{V}_3}{dt} = \mathcal{T}_{31} + \mathcal{T}_{32} + \mathcal{T}_{33} \leq -\alpha_3 \mathbf{V}_3 + \beta_3 \int_0^L (\|\mathbf{R}_t\|^2 \|\mathbf{R}_{tt}\| + \|\mathbf{R}_t\| \|\mathbf{R}_{tt}\|^2) dx$$

if $|\mathbf{R}|_0 \leq \delta_2 \triangleq \min(\delta_{21}, \delta_{22}, \delta_{23})$.   □

Appendix E
Proof of Theorem 5.11

In Theorem 5.11, the concerned dynamics are represented by the following system of $2n$ linear balance laws with constant coefficients:

$$\partial_t R_i(t,x) + \lambda_i \partial_x R_i(t,x) + \gamma_i R_i(t,x) + \delta_i R_{n+i}(t,x) = 0$$

$$i = 1,\ldots,n,$$

$$\partial_t R_{n+i}(t,x) - \lambda_{n+i} \partial_x R_{n+i}(t,x) + \gamma_i R_i(t,x) + \delta_i R_{n+i}(t,x) = 0$$

$$\text{(E.1)}$$

where $R_i : [0,+\infty) \times [0,L] \to \mathbb{R}$ for $i = 1,\ldots,2n$, and

$$0 < \lambda_{n+i} < \lambda_i, \quad 0 < \gamma_i < \delta_i \text{ and } \gamma_i \lambda_i > \delta_i \lambda_{n+i} \quad \text{for } i = 1,\ldots,n. \tag{E.2}$$

The boundary conditions are

$$R_1(t,0) = -\frac{\lambda_{n+1}}{\lambda_1} R_{n+1}(t,0), \tag{E.3a}$$

$$R_{i+1}(t,0) = \frac{(\lambda_i + \lambda_{n+i})(\lambda_i + k_i \lambda_{n+i})}{\lambda_{i+1}(\lambda_{i+1} + \lambda_{n+i+1})} R_i(t,L) - \frac{\lambda_{n+i+1}}{\lambda_{i+1}} R_{n+i+1}(t,0)$$

$$+ \frac{\lambda_{n+i}(\lambda_i + \lambda_{n+i})k_{n+i}}{\lambda_{i+1}(\lambda_{i+1} + \lambda_{n+i+1})} X_i(t), \quad i = 1,\ldots,n-1, \tag{E.3b}$$

$$R_{n+i}(t,L) = k_i R_i(t,L) + k_{n+i} X_i(t), \quad i = 1,\ldots,n, \tag{E.3c}$$

with

$$X_i(t) \triangleq \int_0^t (R_i(\tau,L) - R_{n+i}(\tau,L))\, d\tau, \quad i = 1,\ldots,n.$$

© Springer International Publishing Switzerland 2016
G. Bastin, J.-M. Coron, *Stability and Boundary Stabilization of 1-D Hyperbolic Systems*, Progress in Nonlinear Differential Equations and Their Applications 88, DOI 10.1007/978-3-319-32062-5

Theorem 5.11. If the dissipativity conditions

$$|k_i| < \sqrt{\frac{\gamma_i}{\delta_i} \frac{\lambda_i}{\lambda_{n+i}}}, \qquad k_{n+i} > 0, \qquad i = 1, \ldots, n, \tag{E.4}$$

are satisfied, then the system (E.1), (E.3) is exponentially stable for the L^2-norm.

Proof. We use the following candidate Lyapunov function:

$$\mathbf{V} = \int_0^L \left[\sum_{i=1}^n \frac{p_i}{\lambda_i} R_i^2(t, x) e^{-\mu x/\lambda_i} + \sum_{i=n+1}^{2n} \frac{p_{n+i}}{\lambda_i} R_i^2(t, x) e^{\mu x/\lambda_{n+i}} \right] dx$$

$$+ \frac{1}{2} \sum_{i=1}^n q_i X_i^2(t). \tag{E.5}$$

with $\mu > 0$, $p_i > 0$, $p_{n+i} > 0$, and $q_i > 0$ for all $i = 1, \ldots, n$.

Using integration by part, the time derivative of $\mathbf{V}$ along the C^1-solutions of (E.1) is

$$\frac{d\mathbf{V}}{dt} = \mathcal{T}(t) + \mathcal{I}(t) \tag{E.6}$$

with

$$\mathcal{T}(t) \triangleq - \left[\sum_{i=1}^n p_i e^{-\mu x/\lambda_i} R_i^2(t, x) \right]_0^L + \left[\sum_{i=1}^n p_{n+i} e^{\mu x/\lambda_{n+i}} R_{n+i}^2(t, x) \right]_0^L$$

$$+ \sum_{i=1}^n q_i X_i(t) (R_i(t, L) - R_{n+i}(t, L)), \tag{E.7}$$

and

$$\mathcal{I}(t) \triangleq - \sum_{i=1}^n \int_0^L (R_i(t, x) \ R_{n+i}(t, x)) \ \Omega_i \begin{pmatrix} R_i(t, x) \\ R_{n+i}(t, x) \end{pmatrix} dx, \tag{E.8}$$

with

$$\Omega_i \triangleq \begin{pmatrix} \frac{p_i}{\lambda_i}(\mu + 2\gamma_i)e^{-\mu x/\lambda_i} & \frac{p_i}{\lambda_i}\delta_i e^{-\mu x/\lambda_i} + \frac{p_{n+i}}{\lambda_{n+i}}\gamma_i e^{\mu x/\lambda_{n+i}} \\ \frac{p_i}{\lambda_i}\delta_i e^{-\mu x/\lambda_i} + \frac{p_{n+i}}{\lambda_{n+i}}\gamma_i e^{\mu x/\lambda_{n+i}} & \frac{p_{n+i}}{\lambda_{n+i}}(\mu + 2\delta_i)e^{\mu x/\lambda_{n+i}} \end{pmatrix}. \tag{E.9}$$

Analysis of $\mathcal{I}(t)$.

Clearly we have trace$(\Omega_i) > 0$. Let us consider the determinant of Ω_i as a function of μ denoted

$$[\det \Omega_i](\mu) = \frac{p_i p_{n+i}}{\lambda_i \lambda_{n+i}}(\mu^2 + 2(\gamma_i + \delta_i)\mu) \exp\left(-\mu x(\frac{1}{\lambda_i} - \frac{1}{\lambda_{n+i}})\right)$$
$$- \left(\frac{p_i}{\lambda_i}\delta_i e^{-\mu x/\lambda_i} - \frac{p_{n+i}}{\lambda_{n+i}}\gamma_i e^{\mu x/\lambda_{n+i}}\right)^2. \qquad \text{(E.10)}$$

The parameters $p_i > 0$ and $p_{n+i} > 0$ are selected such that

$$\frac{p_i}{\lambda_i}\delta_i = \frac{p_{n+i}}{\lambda_{n+i}}\gamma_i. \qquad \text{(E.11)}$$

Then $[\det \Omega_i](0) = 0$ and $[\det \Omega_i]'(0) > 0$. Hence there exists $\mu > 0$ sufficiently small such that $[\det \Omega_i](\mu) > 0$ for all $x \in [0, L]$ and all $i = 1, \ldots, n$. Consequently $\mathcal{I}(t)$ is a negative definite quadratic form for all t if μ is sufficiently small.

The next step of the proof is to show that, in addition to satisfying equality (E.11), the parameters p_i, p_{n+i}, and q_i can also be selected in such a way that $\mathcal{T}(t)$ is negative definite for all t.

Analysis of $\mathcal{T}(t)$.
The function $\mathcal{T}(t)$ is expressed as follows when $\mu = 0$:

$$\mathcal{T}(t) = \mathcal{W}_0 R_{n+1}^2(t, 0)$$
$$+ \sum_{i=1}^{n-1} \left(R_i(t, L)\, R_{n+i+1}(t, 0)\, X_i(t)\right) \mathcal{W}_i \begin{pmatrix} R_i(t, L) \\ R_{n+i+1}(t, 0) \\ X_i(t) \end{pmatrix}$$
$$+ \left(R_n(t, L)\, X_n(t)\right) \mathcal{W}_n \begin{pmatrix} R_n(t, L) \\ X_n(t) \end{pmatrix}, \qquad \text{(E.12)}$$

where,

$$\mathcal{W}_0 \triangleq (-p_{n+1} + \kappa_1^2 p_1) \qquad \text{(E.13)}$$

and for $i = 1, \ldots, n - 1$,

$$\mathcal{W}_i \triangleq$$
$$\begin{pmatrix} -p_i + \eta_i^2 p_{i+1} + k_i^2 p_{n+i} & -\eta_i \kappa_{i+1} p_{i+1} & \eta_i \zeta_i p_{i+1} + k_i k_{n+i} p_{n+i} + \frac{1}{2}(1 - k_i)q_i \\ -p_{i+1}\eta_i \kappa_{i+1} & -p_{n+i+1} + \kappa_{i+1}^2 p_{i+1} & -\kappa_{i+1}\zeta_i p_{i+1} \\ \eta_i \zeta_i p_{i+1} + k_i k_{n+i} p_{n+i} + \frac{1}{2}(1 - k_i)q_i & -\kappa_{i+1}\zeta_i p_{i+1} & \zeta_i^2 p_{i+1} + k_{n+i}^2 p_{n+i} - k_{n+i}q_i \end{pmatrix},$$
$$\text{(E.14)}$$

and, finally,

$$
W_n \triangleq \begin{pmatrix} -p_n + k_n^2 p_{2n} & k_n k_{2n} p_{2n} + \frac{1}{2}(1 - k_n) q_n \\ k_n k_{2n} p_{2n} + \frac{1}{2}(1 - k_n) q_n & k_{2n}^2 p_{2n} - k_{2n} q_n \end{pmatrix}, \tag{E.15}
$$

with

$$
\eta_i \triangleq \frac{(\lambda_i + \lambda_{n+i})(\lambda_i + k_i \lambda_{n+i})}{\lambda_{i+1}(\lambda_{i+1} + \lambda_{n+i+1})}, \quad \zeta_i \triangleq \frac{\lambda_{n+i}(\lambda_i + \lambda_{n+i})}{\lambda_{i+1}(\lambda_{i+1} + \lambda_{n+i+1})} k_{n+i}, \quad \kappa_i \triangleq \frac{\lambda_{n+i}}{\lambda_i}. \tag{E.16}
$$

From (E.2) and (E.11) we know that the parameters p_i, $i = 1, \ldots, n$, satisfy

$$
\frac{p_i}{p_{n+i}} = \frac{\gamma_i \lambda_i}{\delta_i \lambda_{n+i}} \triangleq \sigma_i > 1. \tag{E.17}
$$

We have $W_0 < 0$ because

$$
W_0 = -p_{n+1} + \kappa_1^2 p_1 < 0 \iff \frac{\gamma_1}{\delta_1} < \frac{\lambda_1}{\lambda_{n+1}} \tag{E.18}
$$

and this latter inequality is satisfied by assumption (E.2).

Each matrix W_i ($i = 1, \ldots, n - 1$) is negative definite if and only if

(i) $-p_i + \eta_i^2 p_{i+1} + k_i^2 p_{n+i} < 0$,

(ii) $M \triangleq \det \begin{pmatrix} -p_i + \eta_i^2 p_{i+1} + k_i^2 p_{n+i} & -\eta_i \kappa_{i+1} p_{i+1} \\ -\eta_i \kappa_{i+1} p_{i+1} & -p_{n+i+1} + \kappa_{i+1}^2 p_{i+1} \end{pmatrix} > 0$,

(iii) $\det(W_i) < 0$.

Condition (i): From (E.2) and (E.11) we know that the parameters k_i and p_i, $i = 1, \ldots, n$, satisfy

$$
\frac{p_i}{p_{n+i}} = \frac{\gamma_i \lambda_i}{\delta_i \lambda_{n+i}} \triangleq \sigma_i > 1.
$$

Moreover, from Assumption (E.4), we have

$$
k_i^2 < \sigma_i. \tag{E.19}
$$

Let us now assume that the parameters are selected such that

$$
\frac{p_{i+1}}{p_i} = \varepsilon > 0. \tag{E.20}
$$

Then, condition (i) is satisfied if ε is sufficiently small since

$$-p_i + \eta_i^2 p_{i+1} + k_i^2 p_{n+i} = -p_i(1 - \varepsilon\eta_i^2 - \frac{k_i^2}{\sigma_i}) < 0 \text{ if } \varepsilon < \eta_i^{-2}\left(1 - \frac{k_i^2}{\sigma_i}\right).$$

$$(E.21)$$

Condition (ii): Using (E.17) and (E.20), we have

$$M = \varepsilon p_i^2 \left[\left(\frac{1 - \sigma_{i+1}\kappa_{i+1}^2}{\sigma_{i+1}}\right)\left(1 - \frac{k_i^2}{\sigma_i}\right) - \varepsilon\frac{\eta_i^2}{\sigma_{i+1}}\right].$$

From (E.2), (E.16) and (E.17), we know that $\sigma_{i+1}\kappa_{i+1}^2 < 1$. Hence, using (E.19), it follows that $M > 0$ if ε is taken sufficiently small.
Condition (iii): Using (E.17) and (E.20), we have

$$W_i = p_i^3$$

$$\begin{pmatrix} -1 + \varepsilon\eta_i^2 + \dfrac{k_i^2}{\sigma_i} & -\varepsilon\kappa_{i+1}\eta_i & \varepsilon\eta_i\zeta_i + \dfrac{k_ik_{n+i}}{\sigma_i} + \frac{1}{2}(1 - k_i)\dfrac{q_i}{p_i} \\ -\varepsilon\kappa_{i+1}\eta_i & -\dfrac{\varepsilon}{\sigma_{i+1}} + \varepsilon\kappa_{i+1}^2 & -\varepsilon\kappa_{i+1}\zeta_i \\ \varepsilon\eta_i\zeta_i + \dfrac{k_ik_{n+i}}{\sigma_i} + \frac{1}{2}(1 - k_i)\dfrac{q_i}{p_i} & -\varepsilon\kappa_{i+1}\zeta_i & \varepsilon\zeta_i^2 + \dfrac{k_{n+i}^2}{\sigma_i} - k_{n+i}\dfrac{q_i}{p_i} \end{pmatrix}.$$

$$(E.22)$$

The determinant of W_i is

$$\det(W_i) = -\varepsilon p_i \left(\frac{1 - \sigma_{i+1}\kappa_{i+1}^2}{\sigma_{i+1}}\right) \mathcal{P}_i + \mathcal{O}(\varepsilon^2),$$

$$\text{with } \mathcal{P}_i \triangleq -\frac{1}{4}(1 - k_i)^2\left(\frac{q_i}{p_i}\right)^2 + k_{n+i}\left(1 - \frac{k_i}{\sigma_i}\right)\left(\frac{q_i}{p_i}\right) - \frac{k_{n+i}^2}{\sigma_i}.$$

From (E.2), (E.16) and (E.17), we know that $\sigma_{i+1}\kappa_{i+1}^2 < 1$. Hence, $\det(W_i) < 0$ for sufficiently small ε if $\mathcal{P}_i > 0$. If $k_i \neq 1$, $\mathcal{P}_i$ is a degree-2 polynomial in q_i/p_i. It is positive if the discriminant is positive and q_i/p_i is located between the two real roots. The discriminant is

$$\Delta = k_{n+i}^2\left(1 - \frac{1}{\sigma_i}\right)\left(1 - \frac{k_i^2}{\sigma_i}\right).$$

By (E.17) and (E.19), we then have $\Delta > 0$.

The two polynomial real roots are given by

$$\pi_{\pm} = \frac{2k_{n+i}}{(1-k_i)^2}\left[\left(1-\frac{k_i}{\sigma_i}\right) \pm \sqrt{\left(1-\frac{1}{\sigma_i}\right)\left(1-\frac{k_i^2}{\sigma_i}\right)}\right]. \qquad (E.23)$$

Let us now observe that, using again inequalities (E.17) and (E.19), we have

$$(1-\sigma_i) + (k_i^2 - \sigma_i) < 0 < 2\sqrt{(\sigma_i - 1)(\sigma_i - k_i^2)}.$$

These inequalities are equivalent to

$$(1-\sigma_i) + (k_i^2 - \sigma_i) < 0 < 2\sqrt{(\sigma_i - 1)(\sigma_i - k_i^2)}$$

i.e.

$$1 + k_i^2 < 2\sigma_i + 2\sqrt{(\sigma_i - 1)(\sigma_i - k_i^2)}$$

or

$$\tfrac{1}{2}(1-k_i)^2 < (\sigma_i - k_i) + \sqrt{(\sigma_i - 1)(\sigma_i - k_i^2)}.$$

Using (E.23), this latter inequality is equivalent to

$$0 < \frac{k_{n+i}}{\sigma_i} < \pi_+.$$

Hence, it follows that $q_i > 0$ can be selected such that

$$0 < \max\left\{\pi_-, \frac{k_{n+i}}{\sigma_i}\right\} < \frac{q_i}{p_i} < \pi_+$$

and therefore such that $\mathcal{P}_i > 0$.

If $k_i = 1$, it is readily seen that $\mathcal{P}_i > 0$ if $q_i > 0$ is selected such that

$$\frac{q_i}{p_i} > \frac{k_{n+i}}{\sigma_i - 1}. \qquad (E.24)$$

Note that by (E.17) $\sigma_i > 1$.

Let us now consider the matrix $\mathcal{W}_n$ which, using (E.17), is written

$$\mathcal{W}_n = p_n \begin{pmatrix} -1 + \dfrac{k_n^2}{\sigma_n} & \dfrac{1}{2}(1 - k_n)\dfrac{q_n}{p_n} + \dfrac{k_n k_{2n}}{\sigma_n} \\[3mm] \dfrac{1}{2}(1 - k_n)\dfrac{q_n}{p_n} + \dfrac{k_n k_{2n}}{\sigma_n} & \dfrac{k_{2n}^2}{\sigma_n} - k_{2n}\dfrac{q_n}{p_n} \end{pmatrix}. \tag{E.25}$$

From (E.19), we know that $(-1 + k_n^2/\sigma_n) < 0$. Hence, $\mathcal{W}_n$ is negative definite if $\det(\mathcal{W}_n) > 0$. We have

$$\det(\mathcal{W}_n) = p_n^2 \left(-\frac{1}{4}(1 - k_n)^2 \left(\frac{q_n}{p_n}\right)^2 + k_{2n}\left(1 - \frac{k_n}{\sigma_n}\right)\left(\frac{q_n}{p_n}\right) - \frac{k_{2n}^2}{\sigma_n} \right). \tag{E.26}$$

We note that $\det(\mathcal{W}_n)/p_n^2$ has exactly the same form as $\mathcal{P}_i$ above. Consequently, using the same argumentation, $q_n > 0$ can be selected such that $\det(\mathcal{W}_n) > 0$.

So we have shown that the parameters $\varepsilon > 0$, $p_i > 0$, $p_{n+i} > 0$ and $q_i > 0$, $i = 1, \ldots, n$, can be selected such that $\mathcal{T}(t)$ is a negative definite quadratic form in the special case where $\mu = 0$. It follows that there exists $\mu > 0$ sufficiently small such that $\mathcal{T}(t)$ remains negative definite with the same parameters ε, p_i, p_{n+i}, and q_i.

□

Appendix F
Notations

In this appendix, we recall some of the notations that appear most often in the book.

Sets

$\mathcal{B}_\sigma$	open ball of radius σ in $\mathbb{R}^n$
$\mathcal{M}_{m,n}(\mathbb{R})$	set of $m \times n$ real matrices
$\mathcal{D}_n$	set of $n \times n$ diagonal real matrices
$\mathcal{D}_n^+$	set of $n \times n$ positive diagonal real matrices

Models of Hyperbolic Systems
General physical quasi-linear models

$$\mathbf{Y}_t + F(\mathbf{Y})\mathbf{Y}_x + G(\mathbf{Y}) = \mathbf{0}, \qquad F(\mathbf{Y}^*)\mathbf{Y}_x^* + G(\mathbf{Y}^*) = \mathbf{0}. \tag{F.1}$$

Linear models derived from (F.1) by linearization around the steady state

$$\mathbf{Y}_t + A(x)\mathbf{Y}_x + B(x)\mathbf{Y} = \mathbf{0},$$

$$\mathbf{Y}_t + A\mathbf{Y}_x + B\mathbf{Y} = \mathbf{0}.$$

Quasi-linear models in Riemann coordinates derived from (F.1) by diagonalization of $F(\mathbf{Y})$

$$\mathbf{R}_t + \Lambda(\mathbf{R})\mathbf{R}_x + C(\mathbf{R}) = \mathbf{0},$$

$$\mathbf{R}_t + \Lambda(\mathbf{R}, x)\mathbf{R}_x + C(\mathbf{R}, x) = \mathbf{0}, \quad C(\mathbf{0}, x) = \mathbf{0}.$$

Linear models in Riemann coordinates derived from (F.1) by diagonalization of $F(\mathbf{Y})$ and linearization around the steady state

© Springer International Publishing Switzerland 2016
G. Bastin, J.-M. Coron, *Stability and Boundary Stabilization of 1-D Hyperbolic Systems*, Progress in Nonlinear Differential Equations and Their Applications 88, DOI 10.1007/978-3-319-32062-5

$$\mathbf{R}_t + \Lambda(x)\mathbf{R}_x + M(x)\mathbf{R} = \mathbf{0},$$

$$\mathbf{R}_t + \Lambda\mathbf{R}_x + M\mathbf{R} = \mathbf{0}.$$

General quasi-linear models with zero steady state derived from (F.1) by diagonalization of $F(\mathbf{Y}^*)$

$$\mathbf{Z}_t + A(\mathbf{Z}, x)\mathbf{Z}_x + B(\mathbf{Z}, x) = \mathbf{0}, \quad A(\mathbf{0}, x) \in \mathcal{D}_n^+, \quad B(\mathbf{0}, x) = \mathbf{0},$$

$$\mathbf{Z}_t + A(\mathbf{Z}, x)\mathbf{Z}_x + M(\mathbf{Z}, x)\mathbf{Z} = \mathbf{0}, \quad A(\mathbf{0}, x) \in \mathcal{D}_n^+.$$

Functions for Dissipative Boundary Conditions

$\bar{\rho}(M)$	:	$\max\{\rho(\mathrm{diag}\{e^{-i\theta_1}, \dots, e^{-i\theta_n}\}M); (\theta_1, \dots, \theta_n)^{\mathsf{T}} \in \mathbb{R}^n\}$
$\rho_p(M)$	:	$\inf\{\|\Delta M\Delta^{-1}\|_p, \ \Delta \in \mathcal{D}_n^+\}, \ 1 \leqslant p \leqslant \infty$
$\rho_\infty(M)$	:	$\inf\{\mathcal{R}_\infty(\Delta M\Delta^{-1}); \Delta \in \mathcal{D}_n^+\}$
$\mathcal{R}_\infty(M)$	:	$\max\{\sum_{j=1}^n |M_{ij}|; i \in \{1, \dots, n\}\}$

References

Aamo, O.-M. (2013). Disturbance rejection in 2×2 linear hyperbolic systems. *IEEE Transactions on Automatic Control, 58*(5), 1095–1106.

Allievi, L. (1903). Teoria generale del moto perturbato dell'acqua nei tubi in pressione (colpo d'ariete). *Annali della Società degli Ingegneri ed Architetti Italiani, 17*(5), 285–325.

Amin, S., Hante, F., & Bayen, A. (2012). Exponential stability of switched linear hyperbolic initial-boundary value problems. *IEEE Transactions on Automatic Control, 57*(2), 291–301.

Armbruster, D., Goettlich, S., & Herty, M. (2011). A continuous model for supply chains with finite buffers. *SIAM Journal on Applied Mathematics, 71*(4), 1070–1087.

Armbruster, D., Marthaler, D., & Ringhofer, C. (2003). Kinetic and fluid model hierarchies for supply chains. *Multiscale Modeling and Simulation, 2*(1), 43–61

Armbruster, D., Marthaler, D., Ringhofer, C., Kempf, K., & Jo, T.-C. (2006). A continuum model for a re-entrant factory. *Operations Research, 54*(5), 933–950.

Åström, K., & Murray, R. (2009). *Feedback systems*. Princeton: Princeton University Press.

Audusse, E., Briteau, M.-O., Perthame, B., & Sainte-Marie, J. (2011). A multilayer Saint-Venant system with mass exchanges for shallow-water flows. Derivation and numerical validation. *ESAIM Mathematical Modelling and Numerical Analysis, 45*, 169–200.

Aw, A., & Rascle, M. (2000). Resurrection of "second-order" models for traffic flow. *SIAM Journal on Applied Mathematics, 60*, 916–938.

Barker, G. P., Berman, A., & Plemmons, R. J. (1978). Positive diagonal solutions to the Lyapunov equations. *Linear and Multilinear Algebra, 5*(3), 249–256.

Barnard, A., Hunt, W., Timlake, W., & Varley, E. (1966). Theory of fluid flow in compliant tubes. *Biophysical Journal, 6*, 717–724.

Barré de Saint-Venant, A.-C. (1871). Théorie du mouvement non permanent des eaux, avec application aux crues des rivières et à l'introduction des marées dans leur lit. *Comptes rendus de l'Académie des Sciences de Paris, Série 1, Mathématiques, 53*, 147–154.

Bastin, G., & Coron, J.-M. (2011). On boundary feedback stabilization of non-uniform 2×2 hyperbolic systems over a bounded interval. *Systems and Control Letters, 60*(11), 900–906.

Bastin, G., & Coron, J.-M. (2016). Stability of semi-linear systems for the H^1-norm. In *Proceedings of Praly's Fest*.

Bastin, G., Coron, J.-M., & d'Andréa-Novel, B. (2009). On Lyapunov stability of linearised Saint-Venant equations for a sloping channel. *Networks and Heterogeneous Media, 4*(2), 177–187.

© Springer International Publishing Switzerland 2016

G. Bastin, J.-M. Coron, *Stability and Boundary Stabilization of 1-D Hyperbolic Systems*, Progress in Nonlinear Differential Equations and Their Applications 88, DOI 10.1007/978-3-319-32062-5

Bastin, G., Coron, J.-M., d'Andréa-Novel, B., & Moens, L. (2005). Boundary control for exact cancellation of boundary disturbances in hyperbolic systems of conservation laws. In *Proceedings 44th IEEE Conference on Decision and Control and the European Control Conference 2005* (pp. 1086–1089), Seville, Spain.

Bastin, G., Coron, J.-M., & Tamasoiu, S. (2015). Stability of linear density-flow hyperbolic systems under PI boundary control. *Automatica, 53*, 37–42.

Bastin, G., & Guffens, V. (2006). Congestion control in compartmental network systems. *Systems and Control Letters, 55*(18), 689–696.

Bedjaoui, N., & Weyer, E. (2011). Algorithms for leak detection, estimation, isolation and localization in open water channels. *Control Engineering Practice, 19*, 564–573.

Bedjaoui, N., Weyer, E., & Bastin, G. (2009). Methods for the localization of a leak in an open water channel. *Networks and Heterogeneous Media, 4*(2), 189–210.

Bellman, R., & Cooke, K. (1963). *Differential-difference equations*. Number R-374-PR. Ran Corporation. www.rand.org/pubs/reports/R374.html (availability web-only).

Bernard, O., Boulanger, A.-C., Bristeau, M.-O., & Sainte-Marie, J. (2013). A 2d model for hydrodynamics and biology coupling applied to algae growth simulations. *ESAIM Mathematical Modelling and Numerical Analysis, 47*(5), 1387–1412.

Bernard, P., & Krstic, M. (2014). Adaptive output-feedback stabilization of non-local hyperbolic PDEs. In *Preprints of 19th World IFAC Congress*, Cape Town.

Bernot, G., Comet, J.-P., Richard, A., Chaves, M., Gouzé, J.-L., & Dayan, F. (2013). Modeling and analysis of gene regulatory networks. In F. Cazals & P. Kornprobst (Eds.), *Modeling in computational biology and biomedicine: A multidisciplinary endeavor* (Chapter 2). New York: Springer.

Branicky, M.-S. (1998). Multiple Lyapunov functions and other analysis tools for switched and hybrid systems. *IEEE Transactions on Automatic Control, 43*(4), 475–482.

Brezis, H. (1983). *Analyse fonctionnelle, théorie et applications. Collection Mathématiques appliquées pour la maîtrise*. Paris: Masson.

Bridges, D., & Schuster, P. (2006). A simple constructive proof of Kronecker's density theorem. *Elemente der Mathematik, 61*(4), 152–154.

Buckley, S. & Leverett, M. (1942). Mechanism of fluid displacements in sands. *Transactions of the AIME, 146*, 107–116.

Burgers, J. (1939). Mathematical examples illustrating relations occurring in the theory of turbulent fluid motion. *Transactions of the Royal Netherlands Academy of Science, 17*, 1–53.

Calvez, V., Doumic, M., & Gabriel, P. (2012). Self-similarity in a general aggregation-fragmentation problem. Application to fitness analysis. *Journal de mathématiques pures et appliquées (9), 98*(1), 1–27.

Calvez, V., Lenuzza, N., Doumic, M., Deslys, J.-P., Mouthon, F., & Perthame, B. (2010). Prion dynamics with size dependency–strain phenomena. *Journal of Biological Dynamics, 4*(1),28–42.

Cantoni, M., Weyer, E., Li, Y., Mareels, I., & Ryan, M. (2007). Control of large-scale irrigation networks. *Proceedings of the IEEE, 95*(1), 75–91.

Castillo, F., Witrant, E., Prieur, C., & Dugard, L. (2012). Dynamic boundary stabilization of linear and quasi-linear hyperbolic systems. In *Proceedings of IEEE 51st Annual Conference on Decision and Control* (pp. 2952–2957).

Castillo, F., Witrant, E., Prieur, C., & Dugard, L. (2013). Boundary observers for linear and quasi-linear hyperbolic systems with application to flow control. *Automatica, 49*(11), 3180–3188.

Castro Diaz, M., Fernandez-Nieto, E., & Ferreiro, A. (2008). Sediment transport models in shallow water equations and numerical approach by high order finite volume methods. *Computers and Fluids, 37*, 299–316.

Chalons, C., Goatin, P., & Seguin, N. (2013). General constrained conservation laws. Application to pedestrian flow modeling. *Networks and Heterogeneous Media, 8*(2), 433–463.

Chen, G.-Q., & Li, Y. (2004). Stability of Riemann solutions with large oscillation for the relativistic Euler equations. *Journal of Differential Equations, 202*, 332–353.

Colombo, R., Corli, A., & Rosini, M. (2007). Non local balance laws in traffic models and crystal growth. *Journal of Applied Mathematics and Mechanics, 87*(6), 449–461.

Coron, J.-M. (1999). On the null asymptotic stabilization of the two-dimensional incompressible Euler equations in a simply connected domain. *SIAM Journal of Control and Optimization, 37*(6), 1874–1896.

Coron, J.-M. (2007). *Control and Nonlinearity. Mathematical surveys and monographs* (vol. 136). Providence: American Mathematical Society.

Coron, J.-M., & Bastin, G. (2015). Dissipative boundary conditions for one-dimensional quasilinear hyperbolic systems: Lyapunov stability for the C^1-norm. *SIAM Journal of Control and Optimization, 53*(3), 1464–1483.

Coron, J.-M., Bastin, G., & d'Andréa-Novel, B. (2008). Dissipative boundary conditions for one dimensional nonlinear hyperbolic systems. *SIAM Journal of Control and Optimization, 47*(3), 1460–1498.

Coron, J.-M., & d'Andréa-Novel, B. (1998). Stabilization of a rotating body beam without damping. *IEEE Transactions on Automatic Control, 43*(5), 608–618.

Coron, J.-M., d'Andréa-Novel, B., & Bastin, G. (1999). A Lyapunov approach to control irrigation canals modeled by Saint-Venant equations. In *Proceedings European Control Conference*, Karlsruhe.

Coron, J.-M., d'Andréa-Novel, B., & Bastin, G. (2007). A strict Lyapunov function for boundary control of hyperbolic systems of conservation laws. *IEEE Transactions on Automatic Control, 52*(1), 2–11.

Coron, J.-M., Ervedoza, S., Ghoshal, S. S., Glass, O., & Perrollaz, V. (2015). Dissipative boundary conditions for 2 x 2 hyperbolic systems of conservation laws for entropy solutions in BV. Preprint, https://hal.archives-ouvertes.fr/hal-01244280/file/Feedback-13-11-2015.pdf.

Coron, J.-M., Glass, O., & Wang, Z. (2009). Exact boundary controllability for 1-D quasilinear hyperbolic systems with a vanishing characteristic speed. *SIAM Journal on Control and Optimization, 48*(5), 3105–3122.

Coron, J.-M., Kawski, M., & Wang, Z. (2010). Analysis of a conservation law modeling a highly re-entrant manufacturing system. *Discrete and Continuous Dynamical Systems. Series B, 14*(4), 1337–1359.

Coron, J.-M., & Lü, Q. (2014). Local rapid stabilization for Korteweg-de Vries equation with a Neumann boundary control on the right. *Journal de Mathématiques Pures et Appliquées, 102*, 1080–1120.

Coron, J.-M., & Nguyen, H.-M. (2015). Dissipative boundary conditions for nonlinear 1-D hyperbolic systems: Sharp conditions through an approach via time-delay systems. *SIAM Journal on Mathematical Analysis, 47*(3):2220–2240.

Coron, J.-M., & Tamasoiu, S. (2015). Feedback stabilization for a scalar conservation law with PID boundary control. *Chinese Annals of Mathematics. Series B, 36*(5):763–776.

Coron, J.-M., Vazquez, R., Krstic, M., & Bastin, G. (2013). Local exponential H^2 stabilization of a 2×2 quasilinear hyperbolic system using backstepping. *SIAM Journal of Control and Optimization, 51*(3), 2005–2035.

Coron, J.-M., & Wang, Z. (2013). Output feedback stabilization for a scalar conservation law with a nonlocal velocity. *SIAM Journal on Mathematical Analysis, 45*(5), 2646–2665.

Daafouz, J., Tucsnak, M., & Valein, J. (2014). Nonlinear control of a coupled PDE/ODE system modeling a switched power converter with a transmission line. *Systems and Control Letters, 70*, 92–99.

Dafermos, C. (2000). *Hyperbolic conservation laws in continuum physics. A series on comprehensive studies in mathematics* (vol. 325). New York: Springer.

Dafermos, C., & Pan, R. (2009). Global BV solutions for the p-system with frictional damping. *SIAM Journal on Mathematical Analysis, 41*(3), 1190–1205.

Dal Cin, C., Moens, L., Dierickx, P., Bastin, G., & Zech, Y. (2005). An integrated approach for real-time flood-map forecasting on the Belgian Meuse river. *Natural Hazards, 36*, 237–256.

d'Andréa-Novel, B., Fabre, B., & Coron, J.-M. (2010). An acoustic model for the automatic control of a slide flute. *Acta Acustica, 96*, 713–721.

D'Apice, C., Manzo, R., & Piccoli, B. (2006). Packets flow on telecommunication networks. *SIAM Journal on Mathematical Analysis, 38*(3), 717–740.

de Halleux, J., Prieur, C., Coron, J.-M., d'Andréa-Novel, B., & Bastin, G. (2003). Boundary feedback control in networks of open-channels. *Automatica, 39*(8), 1365–1376.

De Schutter, B., & Heemels, W. (2011). *Modeling and control of hybrid systems*. Delft Technical University.

Di Meglio, F., Rabbani, T., Litrico, X., & Bayen, A. (2008). Feed-forward river flow control using differential flatness. In *Proceedings 47th IEEE Conference on Decision and Control* (pp. 3895–3902).

Di Meglio, F., Vazquez, R., & Krstic, M. (2013). Stabilization of a system of $n + 1$ coupled first-order hyperbolic linear PDEs with a single boundary input. *IEEE Transactions on Automatic Control, 58*(12), 3097–3111.

Diagne, A., Bastin, G., & Coron, J.-M. (2012). Lyapunov exponential stability of linear hyperbolic systems of balance laws. *Automatica, 48*(1), 109–114.

Diagne, A., & Drici, A. (2012). Personal communication. Unpublished.

Diagne, A., & Sène, A. (2013). Control of shallow water and sediment continuity coupled system. *Mathematics of Control, Signal and Systems (MCSS), 25*(3), 387–406.

Diagne, M., Shang, P., & Wang, Z. (2016a). Feedback stabilization for the mass balance equations of a food extrusion process. *IEEE Transactions on Automatic Control, 61*(3):760–765.

Diagne, M., Shang, P., & Wang, Z. (2016b). Well-posedness and exact controllability for the mass balance equations of an extrusion process. *Mathematical Methods in Applied Sciences, 39*(10):2659–2670.

Dick, M., Gugat, M., & Leugering, G. (2010). Classical solutions and feedback stabilisation for the gas flow in a sequence of pipes. *Networks and Heterogeneous Media, 5*(4), 691–709.

Djordjevic, S., Bosgra, O., & van den Hof, P. (2011). Boundary control of two-phase fluid flow using the Laplace-space domain. In *Proceedings American Control Conference* (pp. 3283–3288).

Djordjevic, S., Bosgra, O., van den Hof, P., & Jeltsema, D. (2010). Boundary actuation structure of linearized two-phase flow. In *Proceedings American Control Conference* (pp. 3759–3764).

Dos Santos, V., Bastin, G., Coron, J.-M., & d'Andréa-Novel, B. (2008). Boundary control with integral action for hyperbolic systems of conservation laws: Stability and experiments. *Automatica, 44*(5), 1310–1318.

Dos Santos Martins, V. (2013). Introduction of a non constant viscosity on an extrusion process: Improvements. In Le Gorrec, Y. (Ed.), *Proceedings 1st IFAC Workshop on Control of Systems Governed by Partial Differential Equations* (pp. 215–220), Paris.

Dos Santos Martins, V., & Prieur, C. (2008). Boundary control of open channels with numerical and experimental validations. *IEEE Transactions on Control Systems Technology, 16*(6), 1252–1264.

Dos Santos Martins, V., & Rodrigues, M. (2011). A proportional integral feedback for open channels control trough LMI design. In *Preprints of the 18th IFAC World Congress* (pp. 4107–4112).

Dower, P., & Farrel, P. (2006). On linear control of backward pumped Raman amplifiers. In *Proceedings IFAC Symposium on System Identification* (pp. 547–552), Newcastle.

Dower, P., Farrel, P., & Nesic, D. (2008). Extremum seeking control of cascaded Raman optical amplifiers. *IEEE Transactions on Control Systems Technology, 16*(3), 396–407.

Drici, A. (2011). *Boundary control with integral action for 2×2 hyperbolic systems of conservation laws* (Technical report). UPMC.

Engel, K.-J., Fijavž, M., Nagel, R., & Sikolya, E. (2008). Vertex control of flows in networks. *Networks and Heterogeneous Media, 3*(4), 709–722.

Euler, L. (1755). Principes généraux du mouvement des fluides. *Mémoires de l'Académie des Sciences de Berlin, 11*, 274–315.

Evans, G. (1910). Volterra's integral equation of the second kind with discontinuous kernel. *Transactions of the American Mathematical Society, 11*(4), 393–413.

Exner, F. (1920). Zur physik der dünen. *Akademie der Wissenschaften in Wien Mathematisch-Naturwissenschaftliche Klasse, 129*(2a), 929–952.

Exner, F. (1925). Über die wechselwirkung zwischen wasser und geschiebe in flüssen. *Akademie der Wissenschaften in Wien Mathematisch-Naturwissenschaftliche Klasse, 134*(2a), 165–204.

Fliess, M., Lévine, J., Martin, P., & Rouchon, P. (1995). Flatness and defect of non-linear systems: Introductory theory and examples. *International Journal of Control, 61*(6), 1327–1361.

Fovet, O., Litrico, X., & Belaud, G. (2010). Modeling and control of algae detachment in regulated canal networks. In *Proceedings IEEE Conference on Control Applications* (pp. 1881–1886).

Garavello, M., & Piccoli, B. (2006). *Traffic flow on networks. Applied mathematics* (vol. 1). Springfield: American Institute of Mathematical Sciences.

Ghidaoui, M., Zhao, M., McInnis, D., & Axworthy, D. (2005). A review of water hammer - Theory and practice. *Applied Mechanics Reviews, 58*, 49–76.

Glass, O. (2007). On the controllability of the 1-D isentropic Euler equation. *Journal of the European Mathematical Society (JEMS), 9*(3), 427–486.

Glass, O. (2014). On the controllability of the non-isentropic 1-D Euler equation. *Journal of Difference Equations, 257*(3), 638–719.

Goatin, P. (2006). The Aw-Rascle vehicular traffic model with phase transitions. *Mathematical and Computer Modelling, 44*, 287–303.

Godlewski, E., & Raviart, P.-A. (1996). *Numerical approximation of hyperbolic systems of conservation laws. Applied mathematical sciences* (vol. 118). New York: Springer

Goldstein, S. (1951). On diffusion by discontinuous movements, and the telegraph equation. *The Quarterly Journal of Mechanics and Applied Mathematics, 4*, 129–156.

Greenberg, J., & Li, T. (1984). The effect of boundary damping for the quasilinear wave equations. *Journal of Differential Equations, 52*, 66–75.

Greenshields, B. (1935). A study of traffic capacity. *Highway Research Board Proceedings, 14*, 448–477.

Gugat, M., Hante, F., Hirsch-Dick, M., & Leugering, G. (2015). Stationary state in gas networks. *Networks and Heterogeneous Media, 10*(2), 298–320.

Gugat, M., & Herty, M. (2011). Existence of classical solutions and feedback stabilisation for the flow in gas networks. *ESAIM Control Optimisation and Calculus of Variations, 17*(1), 28–51.

Hale, J., & Verduyn-Lunel, S. (1993). *Introduction to functional-differential equations. Applied mathematical sciences* (vol. 99). New York: Springer.

Hale, J., & Verduyn-Lunel, S. (2002). Strong stabilization of neutral functional differential equations. *IMA Journal of Mathematical Control and Information, 19*(1–2), 5–23.

Hasan, A., & Imsland, L. (2014). Moving horizon estimation in managed pressure drilling using distributed models. In *Proceedings IEEE Conference on Control Applications* (pp. 605–610).

Haut, B. (2007). *Modelling and control of road traffic networks* (PhD thesis). Ecole Polytechnique de Louvain, from http://hdl.handle.net/2078.1/5172

Haut, B., Bastin, G., & Van Dooren, P. (2009). Maximal stability region of a perturbed nonnegative matrix. *International Journal of Robust and Nonlinear Control, 19*(3), 364–376.

Heaviside, O. (1892). Electromagnetic induction and its propagation. In *Electrical Papers* (vol. II, 2nd ed.). London: Macmillan and Co.

Hethcote, H. (2000). The mathematics of infectious diseases. *SIAM Review, 42*(4), 599–653.

Hsiao, L., & Marcati, P. (1988). Nonhomogeneous quasilinear hyperbolic system arising in chemical engineering. *Annali della Scuola Normale Superiore di Pisa, Classe di Scienze 4e série, 15*(1), 65–97.

Hu, L., & Di Meglio, F. (2015). Finite-time backstepping boundary stabilization of 3×3 hyperbolic systems. In *Proceedings European Control Conference 2015.*

Hu, L., Di Meglio, F., Vazquez, R., & Krstic, M. (2016). Control of homodirectional and general heterodirectional linear coupled hyperbolic PDEs. *IEEE Transactions on Automatic Control* (in press), (99).

Hu, L., Vazquez, R., Di Meglio, F., & Krstic, M. (2015b). Boundary exponential stabilization of 1-D inhomogeneous quasilinear hyperbolic systems. Preprint.

Hu, L., & Wang, Z. (2015). On boundary control of a hyperbolic system with a vanishing characteristic speed. *ESAIM: Control, Optimisation and Calculus of Variations, 22*(1), 134–147.

Hudson, J., & Sweby, P. (2003). Formulations for numerically approximating hyperbolic systems governing sediment transport. *Journal of Scientific Computing, 19*(1–3), 225–252.

Kac, M. (1956). A stochastic model related to the telegrapher's equation. *Rocky Mountain Journal of Mathematics, 4*, 497–509.

Karafyllis, I., Daoutidis, P. (2002). Control of hot spots in plug flow reactors. *Computers and Chemical Engineering, 26*, 10B7–1094.

Kato, T. (1975). The Cauchy problem for quasi-linear symmetric hyperbolic systems. *Archive for Rational Mechanics and Analysis, 58*(3), 181–205.

Kermack, W., & McKendrick, A. (1927). A contribution to the mathematical theory of epidemics. *Proceedings Royal Society. Series A, Mathematical and Physical Sciences, 115*(772), 700–721.

Khalil, H. (1996). *Nonlinear systems*. Upper Saddle River: Prentice Hall.

Krstic, M., Guo, B.-Z., Balogh, A., & Smyshlyaev, A. (2008). Output-feedback stabilization of an unstable wave equation. *Automatica, 44*, 63–74.

Krstic, M., Kanellakopoulos, I., & Kokotovic, P. (1995). *Nonlinear and adaptive control design. Adaptive and learning systems for signal processing, communications, and control.* New York: Wiley.

Krstic, M., & Smyshlyaev, A. (2008a). Backstepping boundary control for first-order hyperbolic PDEs and application to systems with actuator and sensor delays. *Systems and Control Letters, 57*(9), 750–758.

Krstic, M., & Smyshlyaev, A. (2008b). *Boundary control of PDEs: A course on backstepping designs. Advances in design and control* (vol. 16). Philadelphia: Society for Industrial and Applied Mathematics.

Lamare, P.-O., Bekiaris-Liberis, N., & Bayen, A. (2015a). Control of 2 × 2 linear hyperbolic systems: Backstepping-based trajectory generation and PI-based tracking. In *Proceedings European Control Conference 2015*, Linz, Austria.

Lamare, P.-O., Girard, A., & Prieur, C. (2013). Lyapunov techniques for stabilization of switched linear systems of conservation laws. In *Proceedings 52nd IEEE Conference on Decision and Control* (pp. 448–453), Firenze.

Lamare, P.-O., Girard, A., & Prieur, C. (2015b). Switching rules for stabilization of linear systems of conservation laws. *SIAM Journal of Control and Optimization, 53*(3):1599–1624.

Langmuir, I. (1916). The constitution and fundamental properties of solids and liquids - Part 1 - Solids. *Journal of American Chemical Society, 38*(11), 2221–2295.

Lax, P. (1973). Hyperbolic systems of conservation laws and the mathematical theory of shock waves. In *Conference Board of the Mathematical Sciences Regional Conference Series in Applied Mathematics, N° 11*. Society for Industrial and Applied Mathematics, Philadelphia.

Lee, Y., & Liu, H. (2015). Thresholds for shock formation in traffic flow models with Arrhenius look-ahead dynamics. *Discrete and Continuous Dynamical Systems. Series A, 35*(1), 323–339.

LeFloch, P., & Yamazaki, M. (2007). Entropy solutions of the Euler equations for isothermal relativistic fluids. *International Journal of Dynamical Systems and Differential Equations, 1*(1), 20–37.

Leugering, G., & Schmidt, J.-P. (2002). On the modelling and stabilisation of flows in networks of open canals. *SIAM Journal of Control and Optimization, 41*(1), 164–180.

LeVeque, R. J. (1992). *Numerical methods for conservation laws. Lectures in mathematics ETH Zürich* (2nd ed.). Basel: Birkhäuser.

Li, T.-T. (1994). *Global classical solutions for quasi-linear hyperbolic systems. Research in applied mathematics.* Paris: Masson.

Li, T.-T. (2010). *Controllability and observability for quasilinear hyperbolic systems.* Springfield: American Institute for Mathematic Sciences.

Li, T.-T., & Canic, S. (2009). Critical thresholds in a quasilinear hyperbolic model of blood flow. *Networks and Heterogeneous Media, 4*, 527–536.

Li, T.-T., Rao, B., & Wang, Z. (2010). Exact boundary controllability and observability for first order quasilinear hyperbolic systems with a kind of nonlocal boundary conditions. *Discrete and Continuous Dynamical Systems. Series A, 28*(1), 243–257.

Li, T.-T., & Yu, W.-C. (1985). *Boundary value problems for quasilinear hyperbolic systems. Duke University mathematics series* (vol. V). Durham: Duke University Mathematics Department.

Li, Y., & De Schutter, B. (2010). Control of a string of identical pools using non-identical feedback controllers. In *Proceedings 49th IEEE Conference on Decision and Control* (pp. 120–125), Atlanta.

Liberzon, D. (2003). *Switching in systems and control. Systems and control: Foundations and applications.* Boston: Birkhäuser.

Lichtner, M. (2008). Spectral mapping theorem for linear hyperbolic systems. *Proceedings of the American Mathematical Society, 136*(6), 2091–2101.

Lighthill, M., & Whitham, G. (1955). On kinematic waves. I: Flood movement in long rivers. II: A theory of traffic flow on long crowded roads. *Proceedings Royal Society. Series A, Mathematical and Physical Sciences, 229*(1178), 281–345.

Litrico, X., & Fromion, V. (2006). Boundary control of linearized Saint-Venant equations oscillating modes. *Automatica, 42,* 967–972.

Litrico, X., & Fromion, V. (2009). *Modeling and control of hydrosystems. A frequency domain approach.* New York: Springer.

Litrico, X., Fromion, V., Baume, J.-P., Arranja, C., & Rijo, M. (2005). Experimental validation of a methodology to control irrigation canals based on Saint-Venant equations. *Control Engineering Practice, 13,* 1425–1437.

Liu, W. (2003). Boundary feedback stabilization of an unstable heat equation. *SIAM Journal of Control and Optimization, 42,* 1033–1043.

Liu, W., & Krstic, M. (2000). Backstepping boundary control of Burger's equation with actuator dynamics. *Systems and Control Letters, 41*(4), 291–303.

Lorenz, H., Sheehan, P., & Seidel-Morgenstern, A. (2001). Coupling of simulated moving bed chromatography and fractional crystallisation for efficient enantioseparation. *Journal of Chromatography, 908,* 201–214.

Luskin, M., & Temple, B. (1982). The existence of a global weak solution to the non-linear waterhammer problem. *Communications in Pure and Applied Mathematics, 35,* 697–735.

Lutscher, F. (2002). Modeling alignment and movement of animals and cells. *Journal of Mathematical Biology, 45,* 234–260.

Lutscher, F., & Stevens, A. (2002). Emerging patterns in a hyperbolic model for locally interacting cell systems. *Journal of Nonlinear Science, 12,* 619–640.

Majda, A. (1984). *Compressible fluid flow and systems of conservation laws in several space variables. Applied mathematical sciences* (vol. 53). New York: Springer.

Manning, R. (1891). On the flow of water in open channels and pipes. *Transactions of the Institution of Civil Engineers of Ireland, 20,* 161–207.

Marigo, A. (2007). Equilibria for data networks. *Networks and Heterogeneous Media, 2*(3), 497–528.

Michiels, W., Engelborghs, K., Roose, D., & Dochain, D. (2001). Sensitivity to infinitesimal delays in neutral equations. *SIAM Journal of Control and Optimization, 40*(4), 1134–1158.

Michiels, W., & Niculescu, S.-I. (2007). *Stability and stabilization of time-delay systems. Advances in design and control* (vol. 12). Philadelphia: Society for Industrial and Applied Mathematics.

Michiels, W., & Vyhlidal, T. (2005). An eigenvalue based approach for the stabilization of linear time-delay systems of neutral type. *Automatica, 41*(6), 991–998.

Neves, A., de Souza Ribeiro, H., & Lopes, O. (1986). On the spectrum of evolution operators generated by hyperbolic systems. *Journal of Functional Analysis, 67*(3), 320–344.

Nicolet, C. (2007). *Hydroacoustic modelling and numerical simulation of unsteady operation of hydroelectric systems* (PhD thesis). Ecole Polytechnique Federale de Lausanne, Switzerland.

Nobre, C., Teixeira, J., Rodrigues, L., Severino, A., Retamal, C., de Weireld, G., et al. (2013). *Operating conditions of a simulated moving bed chromatography unit for the purification of fructo-oligosaccharides* (Technical report). University of Minho - University of Mons.

Ooi, S., Kutzen, M., & Weyer, E. (2005). On physical and data driven modelling of irrigation channels. *Control Engineering Practice, 13*, 461–471.

Pavel, L. (2013). Classical solutions in Sobolev spaces for a class of hyperbolic Lotka–Volterra systems. *SIAM Journal of Control and Optimization, 51*(3), 2132–2151.

Pavel, L., & Chang, L. (2012). Lyapunov-based boundary control for a class of hyperbolic Lotka-Volterra systems. *IEEE Transactions on Automatic Control, 57*(3), 701–714.

Pazy, A. (1983). *Semigroups of linear operators and application to partial differential equations. Applied mathematical sciences* (vol. 44). New York: Springer.

Perollaz, V. (2013). Asymptotic stabilization of entropy solutions to scalar conservation laws through a stationary feedback law. *Annales de l'Institut Henri Poincaré, AN 30*, 879–915.

Perollaz, V., & Rosier, L. (2013). Finite-time stabilization of 2 × 2 hyperbolic systems on tree-shaped networks. *SIAM Journal of Control and Optimization, 52*(1), 143–163.

Perthame, B. (2007). *Transport equations in biology. Frontiers in mathematics*. Basel: Birkhäuser.

Prieur, C., Girard, A., & Witrant, E. (2014). Stability of switched linear hyperbolic systems by Lyapunov techniques. *IEEE Transactions on Automatic Control, 59*(8), 2196–2202.

Prieur, C., & Mazenc, F. (2012). ISS-Lyapunov functions for time-varying hyperbolic systems of balance laws. *Mathematics of Control, Signal and Systems (MCSS), 24*(1–2), 111–134.

Prieur, C., Winkin, J., & Bastin, G. (2008). Robust boundary control of systems of conservation laws. *Mathematics of Control, Signal and Systems (MCSS), 20*, 173–197.

Qin, T.-H. (1985). Global smooth solutions of dissipative boundary value problems for first order quasilinear hyperbolic systems. *Chinese Annals Of Mathematics Series B, 6*(3), 289–298.

Rabbani, T. S., Di Meglio, F., Litrico, X., & Bayen, A. (2010). Feed-forward control of open channel flow using differential flatness. *IEEE Transactions on Control Systems Technology, 18*(1), 213–221.

Raman, C., & Krishnan, K. (1928). A new type of secondary radiation. *Nature, 121*, 501–502.

Rauch, J., & Taylor, M. (1974). Exponential decay of solutions to hyperbolic equations in bounded domains. *Indiana University Mathematics Journal, 24*(1), 79–86.

Reed, M., & Simon, B. (1978). *Methods of modern mathematical physics. IV. Analysis of operators*. New York: Academic Press.

Reilly, J., Krichene, W., Delle Monache, M.-L., Samaranayake, S., Goatin, P., and Bayen, A. (2015). Adjoint-based optimization on a network of discretized scalar conservation law PDEs with applications to coordinated ramp metering. *Journal of Optimization Theory and Applications, 167*(2):733–760.

Renardy, M. (1993). On the type of certain C_0-semigroups. *Communications on Partial Differential Equations, 18*(7–8), 1299–1307.

Rudin, W. (1973). *Functional analysis*. McGraw-Hill Series in Higher Mathematics. New York: McGraw-Hill.

Russell, D. (1978). Controllability and stabilizability theory for linear partial differential equations: Recent progress and open questions. *SIAM Review, 20*(4), 639–739.

Serre, D. (1999). *Systems of conservation laws. 1. Hyperbolicity, entropies, shock waves*. Cambridge: Cambridge University Press.

Shang, P., & Wang, Z. (2011). Analysis and control of a scalar conservation law modeling a highly re-entrant manufacturing system. *Journal of Differential Equations, 250*, 949–982.

Shorten, R., Mason, O., & King, C. (2009). An alternative proof of the Barker, Berman, Plemmons (BBP) result on diagonal stability and extensions. *Linear Algebra and Its Applications, 430*, 34–40.

Shorten, R., Wirth, F., Mason, O., Wulff, K., & King, C. (2007). Stability criteria for switched and hybrid systems. *SIAM Review, 49*(4), 545–592.

Silkowski, R. (1976). *Star shaped regions of stability in hereditary systems* (PhD thesis). Brown University, Providence.

Silva, G., Datta, A., & Bhattacharyya, S. (2005). *PID controllers for time delay systems*. Boston: Birkhäuser.

Slemrod, M. (1983). Boundary feedback stabilization for a quasilinear wave equation. In *Control theory for distributed parameter systems. Lecture notes in control and information sciences* (vol. 54, pp. 221–237). New York: Springer.

Smits, W., Veening, J.-W., & Kuipers, O. (2008). Phenotypic variation and bistable switching in bacteria. In El-Sharoud, W. (Ed.), *Bacterial physiology: A molecular approach* (Chapter 12). Berlin/Heidelberg: Springer.

Smolen, P., Baxter, D., & Byrne, J. (2000). Modeling transcriptional control in gene network - methods, recent results and future directions. *Bulletin of Mathematical Biology, 62*, 247–292.

Smoller, J., & Temple, B. (1993). Global solutions of the relativistic Euler equations. *Communications in Mathematical Physics, 156*, 67–99.

Smyshlyaev, A., Cerpa, E., & Krstic, M. (2010). Boundary stabilisation of a 1-D wave equation with in-domain antidamping. *SIAM Journal of Control and Optimization, 48*(6), 4014–4031.

Smyshlyaev, A., Guo, B.-Z., & Krstic, M. (2009). Arbitrary decay rate for Euler-Bernoulli beam equation. *IEEE Transactions on Automatic Control, 54*(5), 1134–1140.

Smyshlyaev, A., & Krstic, M. (2004). Closed-form boundary state feedbacks for a class of 1-D partial integro-differential equations. *IEEE Transactions on Automatic Control, 49*(12), 2185–2202.

Smyshlyaev, A., & Krstic, M. (2009). Boundary control of an anti-stable wave equation with anti-damping on the uncontrolled boundary. *Systems and Control Letters, 58*(8), 617–623.

Suvarov, P., Vande Wouwer, A., & Kienle, A. (2012). A simple robust control for simulated moving bed chromatographic separation. In *Proceedings IFAC Symposium on Advanced Control of Chemical processes* (pp. 137–142), Singapore.

Tang, Y., Prieur, C., & Girard, A. (2015a). *Singular perturbation approximation of linear hyperbolic systems of balance laws (full version)* (Technical Report hal-01175825). Gipsa-Lab.

Tang, Y., Prieur, C., & Girard, A. (2015b). Tikhonov theorem for linear hyperbolic systems. *Automatica, 57*, 1–10.

Tchousso, A., Besson, T., & Xu, C.-Z. (2009). Exponential stability of distributed parameter systems governed by symmetric hyperbolic partial differential equations using Lyapunov's second method. *ESAIM Control Optimisation and Calculus of Variations, 15*(2), 403–425.

Thieme, H. (2003). *Mathematics in population biology. Princeton series in theoretical and computational biology*. Princeton/Oxford: Princeton University Press.

van Overloop, P., & Bombois, X. (2012). Identification of properties of open water channels for controller design. In *Proceedings IFAC Symposium on System Identification* (pp. 1019–1024), Brussels.

Van Pham, T., Georges, D., & Besançon, G. (2010). Infinite-dimensional receding horizon optimal control for an open-channel system. In *Proceedings IFAC Symposium on Nonlinear Control Systems* (pp. 391–396). University of Bologna.

Van Pham, T., Georges, D., & Besançon, G. (2014). Predictive control with guaranteed stability for water hammer equations. *IEEE Transactions on Automatic Control, 59*(2), 465–470.

Vazquez, R., & Krstic, M. (2013). Marcum Q-functions and explicit kernels for stabilization of 2×2 linear hyperbolic systems with constant coefficients. In *Proceedings 52nd IEEE Conference on Decision and Control* (pp. 466–471), Florence.

Vazquez, R., Krstic, M., & Coron, J.-M. (2011). Backstepping boundary stabilization and state estimation of a 2×2 linear hyperbolic system. In *Proceedings 50th IEEE Conference on Decision and Control and European Control Conference* (pp. 4937–4942), Orlando, FL.

Vazquez, R., Trelat, E., & Coron, J.-M. (2008). Control for fast and stable laminar-to- high-Reynolds-numbers transfer in a 2D Navier-Stokes channel flow. *Discrete and Continuous Dynamical Systems. Series B, 10*, 925–956.

Voisin, C. (2007). Private communication. Unpublished.

Volterra, V. (1896). Sulla inversione degli integrali definiti. *Rendiconti della Reale Accademia dei Lincei, 5*(5), 177–185.

Walton, K., & Marshall, J. (1987). Direct method for TDS stability analysis. *IEE Proceedings – Control Theory and Applications, 134*(2), 101–107.

Wang, Z. (2006). Exact controllability for nonautonomous first order quasilinear hyperbolic systems. *Chinese Annals of Mathematics. Series B, 27*(6), 643–656.

Xu, C.-Z., & Sallet, G. (1999). Proportional and integral regulation of irrigation canal systems governed by the Saint-Venant equation. In *Proceedings 14-th IFAC World congress*, Beijing.

Xu, C.-Z., & Sallet, G. (2002). Exponential stability and transfer functions of processes governed by symmetric hyperbolic systems. *ESAIM Control Optimisation and Calculus of Variations, 7,* 421–442.

Xu, C.-Z., & Sallet, G. (2014). Multivariable boundary PI control and regulation of a fluid flow system. *Mathematical Control and Related Fields, 4*(4), 501–520.

Zhao, Y.-C. (1986). *Classical solutions for quasilinear hyperbolic systems* (Thesis). Fudan University (in Chinese).

Index

A
Activation energy, 25, 163
Aw-Rascle equations, 31

B
Backstepping, 219
Balance law, 1
    linear, 159
    with nonlocal source term, 53
Blood flow, 30
Boltzmann constant, 163
Boundary condition, 6
    differential, 97
    dissipative, 88, 119, 136
    linear, 85
    moving, 30
    switching, 110
Boundary control, 6, 184
    backstepping, 219
    disturbance rejection, 6
    implementation, 233
    local vs. nonlocal, 234
    of a density-flow system, 67
    of a string of pools, 194
    of an open channel, 181
    of networks, 130
    output tracking, 6
    proportional-integral, 70, 190
    static, 7
Buckley-Leverett equation, 44
Burgers equation, 45

C
Cauchy problem, 7, 8
    well-posedness, 119, 137, 243, 255
Change of coordinates, 3, 9, 96, 144, 176, 206
Characteristic curve, 6, 57
Characteristic equation, 65, 186
Characteristic form, 3
Characteristic polynomial, 26
Characteristic velocities, 2
Chemotaxis, 33
Chromatography, 38
    simulated moving bed (SMB), 39, 111
Closed loop, 8, 68, 71, 238
Compartmental system, 132
Compatibility condition, 8, 59, 153, 204
Conservation law, 5
    nonlinear, 117
    linear, 85
    scalar, 43
Continuity equation, 31
Control error propagation, 196
Control Lyapunov function, 177
Convergence rate, 63, 65, 96, 227
Cooling fluid, 24

D
Darcy's law, 53
Dead-beat control, 69
Decay rate, 63
Density-flow system, 67, 184
Diagonally stable matrix, 166

© Springer International Publishing Switzerland 2016
G. Bastin, J.-M. Coron, *Stability and Boundary Stabilization of 1-D Hyperbolic
Systems*, Progress in Nonlinear Differential Equations and Their Applications 88,
DOI 10.1007/978-3-319-32062-5

Differential boundary condition, 97
Disease transmission rate, 36
Dissipative boundary conditions, 88, 119, 136
Disturbance rejection, 6

E
Elastic tube, 30
Electrical line, 10, 46
    lossless, 65, 99
Endemic equilibrium, 37
Energy balance, 26
Entropic solutions, 218
Epidemiology, 35
Equilibrium, 4
Euler equations, 26
    heat capacity ratio, 26
    isentropic, 27, 53
    linearization, 28
    steady state, 28
Exner equation, 18
Exothermic chemical reaction, 24
Exponential stability, 86, 89, 120, 137, 160
Extrusion process, 53

F
Feedback control, 6
    full state, 219
    static, 7
Feedforward control, 69, 240
Fick's law, 241

G
Gas pipe, 47
Gas pipeline, 27
Genetic regulatory network, 50, 106

H
Hayami model, 240
Heat capacity ratio, 26
Heat exchanger, 23
Hybrid system, 110
Hydraulic gate, 15, 172, 193
Hyperbolic system, 2
    quasi-linear, 2, 118, 203
    semi-linear, 2, 25, 33, 216
    strict, 8, 204, 206

I
Initial condition, 7
    compatibility, 8, 59, 153, 204

Input-to-state stability, 201
Isentropic Euler equations, 27, 53

K
Kac-Goldstein equations, 33
Kermack-McKendrick model, 35

L
Lagrangian derivative, 31
Langmuir isotherm, 38
LaSalle invariance principle, 157
Limitation of stabilizability, 197
Linearization, 4
    of Euler equations, 28
    of Saint-Venant equations, 16
    of Saint-Venant-Exner model, 168
Load disturbance, 70, 238
Local control, 68, 234
Lossless electrical line, 65, 99
Lotka-Volterra interactions, 53
LWR model, 44, 134
Lyapunov stability, 160

M
Mass balance, 26, 30
Matrix inequality, 161, 162, 208, 217
Method of characteristics, 57, 157
Meuse river, 229
Momentum balance, 26, 30
Moving boundary condition, 30
Musical wind instrument, 29

N
Navigable river, 15, 48, 193, 229
Network
    electrical lines, 46
    genetic regulatory, 50, 106
    scalar conservation laws, 130
Nonlocal control, 234
Nonuniform steady state, 4

O
Observer, 223
Observer-controller form, 219
Oil well drilling process, 53
Open channel, 13, 181
    subcritical flow, 14
Open loop, 238
Output injection, 223
Output tracking, 6

P

Pedestrian flow model, 44
Piezometric head, 20
Plug flow chemical reactor, 24, 163
    fluidized bed, 54
Pole, 65, 68
Porous medium (Flow in), 44, 53
Positive system, 36
Preissman scheme, 230
Proportional-Integral control, 70, 190

Q

Quasi-linear hyperbolic system, 2, 118, 203

R

Raceway process, 53
Raman amplifier, 218
Raman amplifiers, 12
Ramp metering, 33, 132–134
Reference model, 220
Riemann coordinate
    around a steady state, 4
    definition, 3
Riemann invariant, 6
Rigid pipe, 26
Road traffic, 31
    LWR model, 44, 134
    ramp metering, 33, 132, 133
    traffic pressure, 31
Robust stability, 94

S

Saint-Venant equations, 13, 44, 48, 230
    linearization, 16
    multilayer, 53
    steady state, 16
Saint-Venant-Exner model, 18, 167, 201
Scalar conservation law, 43
    control, 130
Semi-linear hyperbolic system, 2, 25, 33
Set-point regulation, 238
Shower control problem, 21
Simulated moving bed chromatography, 39, 111
SIR epidemiologic model, 35
    endemic equilibrium, 37
Slide flute, 29
Small gain, 89

Sobolev inequality, 142
Solution
    L^2-solution, 251
Sound velocity, 20, 26, 27
Source term, 2, 159, 203
Stability, 6
    condition, 55, 95
    exponential, 86, 89, 120, 137, 160
    for the C^0-norm, 59
    for the C^1-norm, 145
    for the C^p-norm, 153
    for the H^2-norm, 136, 205
    for the H^p-norm, 156
    for the L^2-norm, 59, 64
    for the L^∞-norm, 57
    Lyapunov, 160
    nonnegative matrix, 135
    robust, 94
Stabilizability
    limitation, 197
Stabilization, 6, 55
Steady state, 4
    nonuniform, 4, 181
    uniform, 4, 203
Subcritical flow, 14
Supply chain, 53
Switching boundary conditions, 110, 116, 201

T

Target system, 220
Telecommunication networks (packet flow), 53
Telegrapher equations, 10
Toricelli formula, 44
Traffic pressure, 31
Transfer function, 185, 196

U

Uniform steady state, 4, 203

V

Volterra transformation, 221

W

Water hammer, 22, 67
Well-posedness of the Cauchy problem, 8, 10, 34, 119, 137, 243, 255

obtained

s/38/P